CO_2 Sequestration and Valorization

CO_2 Sequestration and Valorization

Editor

Hemming Lindberg

scitus academics

CO_2 Sequestration and Valorization
Edited by **Hemming Lindberg**

ISBN: 978-1-68117-283-5
Library of Congress Control Number: 2016937439

www.scitusacademics.com
Box No. 4766, 616 Corporate Way,
Suite 2, Valley Cottage,
NY 10989

Printed in United States of America on Acid Free Paper

Preface

The reconciliation of economic development, social justice and the reduction of greenhouse gas emissions is currently one of the major political challenges. Carbon dioxide (CO2) capture and sequestration is a set of technologies capable of significantly reducing CO2 emissions from new and existing coal and gas power plants and major industrial sources. Carbon sequestration describes long-term carbon dioxide storage or other carbon forms to mitigate or delay global warming and prevent climate change. It was proposed as a way to slow the accumulation of greenhouse gasses in the atmosphere and at sea released by fossil fuels. Carbon dioxide is naturally obtained from the atmosphere through biological, chemical or physical processes. Artificial processes were designed to produce similar effects, including large-scale industrial CO2 capture and sequestration using sub-surface saline aquifers, reservoirs, ocean water, aging oil fields or other carbon sinks. Carbon dioxide (CO2) capture and sequestration could play a major role in reducing greenhouse gas emissions while allowing low-carbon power generation from power plants. Carbon dioxide capture and sequestration (CCS) technologies are currently available and can significantly reduce CO2 emissions from fossil-fuel-consuming power stations (80-90 percent).

This book offers strategies to mitigate CO2 emissions on a large scale using sequestration, storage and carbon technologies. This book is a comprehensive source of information for researchers, policymakers and anyone with an inquiring mind on this subject.

Table of Contents

Chapter 1

A Cost-Optimal Scenario of CO_2 Sequestration in a Carbon-Constrained World Through to 2050

Takayuki Takeshita

Transdisciplinary Initiative for Global Sustainability, The University of Tokyo, Tokyo, Japan

ABSTRACT

In this paper, a regionally disaggregated global energy system model with a detailed treatment of the whole chain of CO_2 capture and storage (CCS) is used to derive the cost-optimal global pattern of CO_2 sequestration in regional detail over the period 2010-2050 under the target of halving global energy-related CO_2 emissions in 2050 compared to the 2005 level. The major conclusions are the following. First, enhanced coalbed methane recovery will become a key early opportunity for CO_2 sequestration, so coalrich regions such as the US, China, and India will play a leading role in global CO_2 sequestration. Enhanced oil recovery will also have a participation in global CO_2 sequestration from the initial stage of CCS deployment, which may be applied mainly in China, southeastern Asia, and West Africa in 2030 and mainly in the Middle East in 2050. Second, CO_2 sequestration will be carried out in an increasing number of world regions over time. In particular, CCS will be deployed extensively in today's developing countries. Third, an increasing amount of the captured CO_2 will be stored in aquifers in many parts of the world due to their abundant and widespread availability and their low cost. It is shown that the share of aquifers in global CO_2 sequestration reaches 82.0% in 2050.

Keywords: CO_2 Sequestration; Simulation; Optimization; Global Energy System Model

1. INTRODUCTION

Avoiding dangerous climate change is an increasingly formidable challenge. CO_2 capture and storage (CCS) is now recognized as an important option for mitigating climate change. Research, development, and demonstration for CCS are ongoing not only in developed countries, but also in at least 19 developing countries. Although it has been indicated that CCS has advantages over other CO_2 mitigation options in terms of CO_2 emissions reduction potential and cost-effectiveness (e.g., [1,2]), there are still major hurdles to widespread deployment of CCS. First of all, it must be proven that CO_2 can be permanently and safely stored underground. Second, public acceptance of storing CO_2 underground must be gained.

Under these circumstances, identifying in advance the future likely CO_2 storage sites is considered to be highly useful in overcoming the above two hurdles. This is because much time and effort can be spent on better understanding the geological properties of potential CO_2 storage sites, on giving potential host communities for CO_2 storage sites a detailed explanation on the necessity, scope, and safety of the CO_2 storage project, and on building a reliable relationship with them. Also, taking into account that an important feature of CCS is that it is capital intensive, this would help all stakeholders (including governments, utilities, and CCS industries) make rational decisions on CCS infrastructure design.

Thus, the purpose of this paper is to derive the costoptimal global pattern of CO_2 sequestration in regional detail over the period to 2050 under a stringent CO_2 emissions reduction constraint (i.e., a halving of global energy-related CO_2 emissions in 2050 compared to the 2005 level). The cost-optimal global pattern of interregional CO_2 transportation to CO_2 storage sites is also drawn and analyzed under this constraint. These analyses are done by using the global energy system model REDGEM70 (an acronym for a REgionally Disaggregated Global Energy Model with 70 regions) [3,4], which treats the whole chain of CCS in detail.

2. METHODOLOGY

2.1. Overview of the REDGEM70 Model

REDGEM70 is a technology-rich, bottom-up global energy systems optimization model formulated as an intertemporal linear programming problem. Figure 1 schematically illustrates the structure of the model. With a 5% discount rate, the model is designed to determine the cost-optimal energy strategy (e.g., the cost-optimal

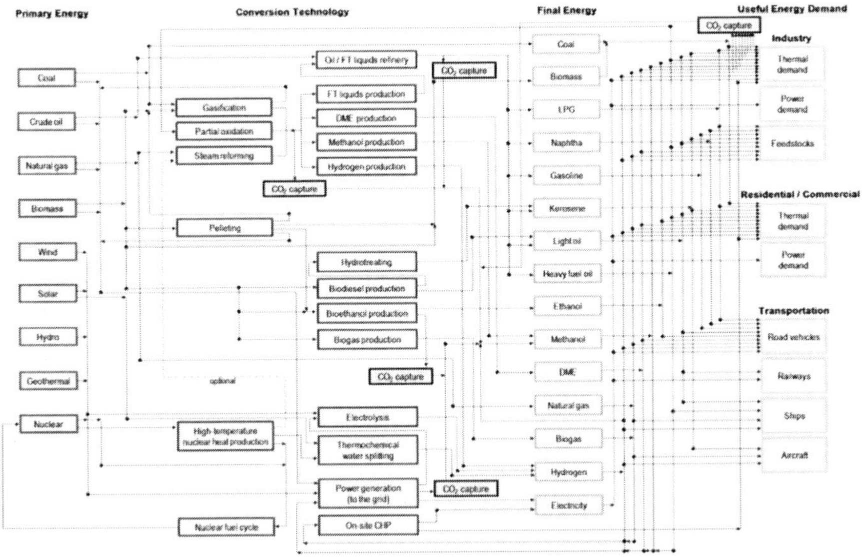

Figure 1. Schematic representation of the structure of REDGEM70[a]. [a]FT: Fischer-Tropsch; DME: dimethyl ether; LPG: liquefied petroleum gas; CHP: combined heat and power.

choice of technology options) from 2010 to 2050 at 10-year intervals for each of 70 world regions so that total discounted energy system costs are minimized under constraints on the satisfaction of exogenously given energy end-use demands, the availability of primary energy resources, material and energy balances, the maximum growth rates of new technologies, etc. In the model, price-induced energy demand reductions and energy efficiency improvements, fuel switching to less carbonintensive fuels, and CCS in geologic formations are the three options for CO_2 emissions reduction.

Furthermore, in the current version of the model used in this study, there is also a constraint that global energyrelated CO_2 emissions in 2050 are to be halved compared to the 2005 level. This constraint is the same as that given in the International Energy Agency's BLUE Map scenario [1,5]. The model has a full flexibility in where CO_2 emissions reduction is achieved to meet this constraint.

Figure 2 shows how the 70 world regions are defined in REDGEM70. These 70 regions are categorized into "energy production and consumption regions" and "energy production regions". The whole world was first divided into the 48 energy production and consumption regions to which future energy end-use demands are allocated. The 22 energy production regions, which are defined as geographical points, were then distinguished from the energy production and consumption regions to represent the geographical characteristics of the areas endowed with large amounts of fossil energy resources. While the 48 energy production and consumption regions cover the global final energy consumption, all the energy-related activities except final energy consumption are conducted in each of the two region types in the model. Such a detailed regional disaggregation enables the explicit consideration of regional characteristics in terms of energy resource supply, energy demands, CO_2 storage capacity, geography, and climate.

Future trajectories for energy end-use demands were estimated as a function of those for socio-economic driving forces such as population and income in the intermediate B2 scenario developed by [6]. Allocation of the energy end-use demand estimates to the 48 energy production and consumption regions was done by using countryand state-level statistics/estimates (and projections if available) on population, income, geography, energy use by type, and transport activity by mode, and by taking into account the underlying storyline of the B2 scenario that regional diversity might be somewhat preserved throughout the 21st century.

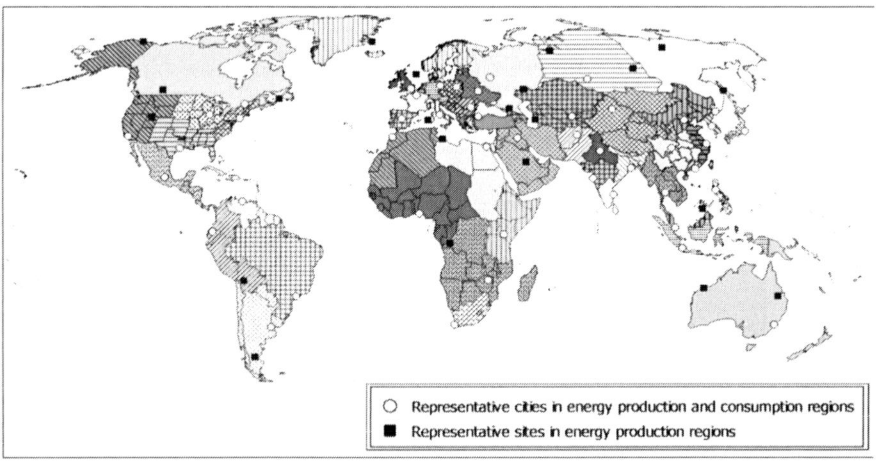

Figure 2. Regional disaggregation of REDGEM70.

Assumptions on the availability and extraction cost of fossil energy resources are taken from [7]. For biomass resources, REDGEM70 considers not only terrestrial biomass (such as energy crops and modern fuelwood), but also waste biomass. The availability of these biomass resources and excess cropland that can be used for energy purposes without conflicting with other biomass uses such as food production was estimated for each region and each time point. They were estimated assuming that biomass is produced in a sustainable way so that biomass-derived energy carriers can be regarded as carbon neutral. Data for these biomass resources (e.g., resource availability, yields per hectare of land, and supply costs) are provided in [8]. These resource availability estimates were then allocated to the 70 model regions by using country-, state-, and site-level statistics/estimates.

2.2. CCS Sector

In REDGEM70 as shown in Figure 1, the CO_2 generated from power plants (excluding those used for on-site CHP production and biomass-fired steam-cycle power generation), synthetic fuels production plants (excluding those used for converting stranded gas and decentralized small-scale hydrogen production), ethanol production plants, oil/FT refinery plants, and industrial processes can be captured for subsequent sequestration in geologic formations. It is assumed that the captured CO_2 is transported intraregionally (and interregionally if needed) to a storage site and then stored in geologic formations. Data for CO_2 capture technologies are provided in [8,9]. The costs of all types of CO_2 capture technologies, which represent the bulk of the overall CCS costs, are assumed to be reduced by 15.6% from 2010 to 2050 [10].

The cost of intraregionally transporting the captured CO_2 from fossil-fueled plants to a storage site was estimated to be 24.1 US $\$_{2007}$ per tonne of carbon, assuming that it is transported intraregionally through 250 km of pipeline [8,10]. On the other hand, due to the dispersed nature of biomass feedstocks, the intraregional transportation of the captured CO_2 from biomass-fueled plants to a storage site is assumed to suffer from diseconomies of small scale. Based on [8], the cost of intraregionally transporting the captured CO_2 from biomass-fueled plants to a storage site was estimated to be in the range of 59.0 - 95.4 US $\$_{2007}$ per tonne of carbon depending on secondary energy carriers produced from biomass.

REDGEM70 treats the interregional transportation of CO_2 between representative cities/sites in the 70 model regions and is able to determine its cost-optimal evolution path. Figure 3 shows the interregional CO_2 transportation costs as a function of transportation distance for each mode. In advance of model simulations, possible interregional transportation routes were given between representative cities/sites in the 70 model regions and their distances were calculated using geographic information system (GIS) land cover data, GIS land use data, and land elevation data. Thus, the cost of transporting a tonne of carbon between every couple of representative cities/sites in the 70 model regions is an input to the model.

The geologic CO_2 sequestration options included are enhanced oil recovery (EOR), enhanced coalbed methane recovery (ECBMR), depleted gas-field disposal, and aquifer disposal. Table 1 shows the input data for CO_2

sequestration options. In addition to CO_2 sequestration costs listed in Table 1, a monitoring cost of 0.864 US $_{2007}$ per tonne of carbon stored is assumed to be required [13].

Figure 4 shows the regional distribution of CO_2 storage capacity. The regional distribution of CO_2 storage capacity in aquifers was derived from data reported in

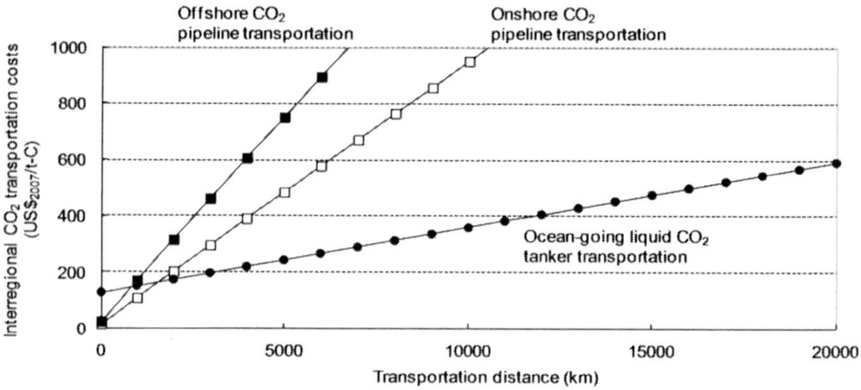

Figure 3. Interregional CO_2 transportation costs as a function of transportation distance.

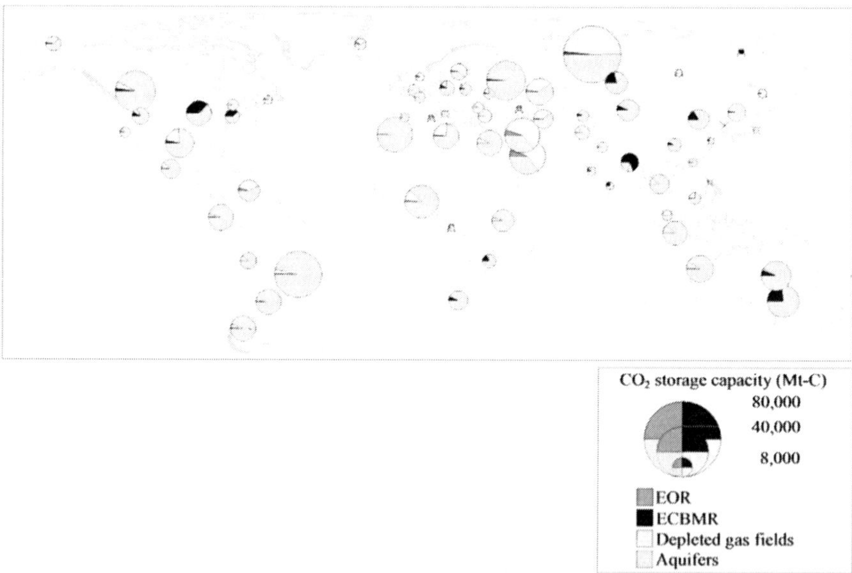

Figure 4. Estimated regional distribution of CO_2 storage capacity[a]. [a]Towers indicate representative sites in energy production regions, while crosses indicate representative cities in energy production and consumption regions.

Table 1. Data for geologic CO_2 sequestration options[a].

Option type	Sequestration cost (US $\$_{2007}$/t-C)	CO_2 storage capacity (Gt-C)
EOR	128 - 184	9
ECBMR	41 - 953	40
Depleted gas field	56 - 80	172
Aquifer	15 - 221	706

[a]Sources: [2,11-13].

[11]. Following the approach taken by [12], the regional distribution of CO_2 storage capacity in EOR sinks, ECBMR sinks, and depleted gas fields was estimated from that of conventional oil resource base, anthracite and bituminous coal resource base, and conventional natural gas resource base, respectively.

3. RESULTS AND DISCUSSION

3.1. Globally Aggregated Results for CO_2 Sequestration

Before examining the cost-optimal pattern of CO_2 sequestration on a regional basis, the focus is placed on the globally aggregated results. Figure 5" target="_self"> Figure 5 shows the evolution of the global amount of CO_2 sequestration by type of storage option. This result implies that large-scale commercial deployment of CCS is likely to begin around 2030 under the stringent CO_2 emissions reduction constraint. As technological progress in the field of CCS technologies advances over time, they make an increasing contribution to CO_2 emissions reduction. The share of CCS in global energy-related CO_2 emissions reduction compared to the business-as-usual case without any CO_2 constraints increases from 1.5% in 2020 to 17.7% in 2050.

Total cumulative CO_2 sequestration over the period 2010-2050 amounts to 38 gigatonnes of carbon, which account for only 4% of global CO_2 storage capacity. This means that after meeting the stringent constraint imposed on global energy-related CO_2 emissions in 2050 (i.e., in the second half of the century), CCS can play a significant role in reducing CO_2 emissions.

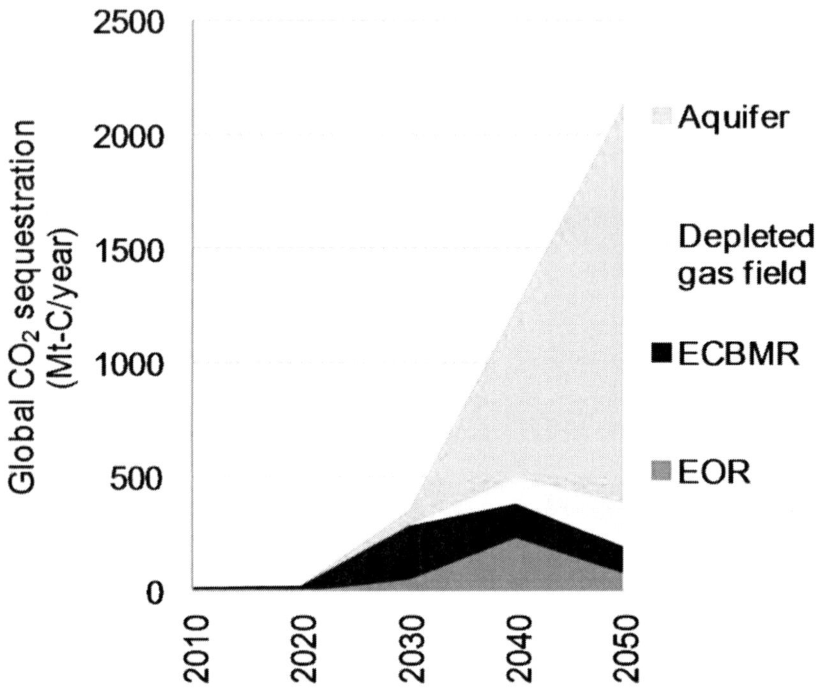

Figure 5. Global CO_2 sequestration by type of storage option.

As regards geologic CO_2 sequestration options, ECBMR is chosen first because very low, or even negative, costs can be achieved through offsetting revenues from the natural gas produced in ECBMR operations. This is also true for EOR: in the initial phase of CCS deployment, EOR operations are one of the pillars of global CO_2 sequestration. It can therefore be argued that CO_2 sequestration combined with enhanced hydrocarbon recovery creates an early opportunity for CCS.

In contrast, aquifers account for a dominant share of global CO_2 sequestration in the second half of the time horizon: their share in global CO_2 sequestration increases from 60.2% in 2040 to 82.0% in 2050. This is due to their abundant and widespread availability and their low cost (see Table 1 and Figure 4). It is noteworthy that all the four options have a visible participation in global CO_2 sequestration from 2030 onwards, when large-scale CCS deployment begins.

3.2. Cost-Optimal Pattern of CO_2 Sequestration and Transportation

Figures 6 and 7 illustrate the global pattern of geologic CO_2 sequestration and CO_2 transportation for 2030 and 2050, respectively. In 2030, large-scale CO_2 sequestration is carried out in a limited number of regions. More specifically, regions rich in coal resource base (e.g., the US, China, India, Australia, and South Africa) take the lead in CO_2 sequestration, reflecting the large share of ECBMR operations in global CO_2 sequestration at this point in time.

Although not shown here, CO_2 capture is implemented predominantly in coal-fueled plants in 2030 because the incremental cost of CO_2 capture is lower for them than for plants fueled by other carbonaceous feedstocks, which is due to economies of scale caused by the high carbon content of coal and the large scale of coal-fueled plants. This also explains why large-scale CCS deployment starts in these coal-rich regions. However, the capture of CO_2 from coal-fueled plants loses its share over time under the stringent CO_2 emissions reduction constraint. The main reason is that CO_2 emissions from coal-fueled plants with CCS are unacceptable under this constraint because coal has the highest carbon emission factor of all fossil fuels and because the efficiency of CO_2 capture for those plants is not 100% (about 90% - 95%).

Among others, EOR operations in regions with oil fields that are in decline (e.g., China, southeastern Asia, and West Africa) and aquifer storage in southeastern and western Asia play a role in global CO_2 sequestration in 2030. It is important to note that interregional CO_2 transportation is not chosen at all in 2030 because of the widespread distribution of geologic formations with large CO_2 storage capacity and the high cost of CO_2 transportation. In other words, it is cost-effective to sequester the CO_2 captured from a model region in geologic formations in that region.

It should be noted, however, that the realization of the CCS potential in today's developing countries actually depends on the effective transfer of technologies and funds from today's industrialized countries and on global efforts to reduce CO_2 emissions. Without them, CCS would not be deployed on such a large scale in today's developing countries.

As shown in Figure 7, the picture changes dramatically in 2050. At this point in time, geologic CO_2 sequestration is carried out all over the world. It should be emphasized that CCS is deployed fairly extensively in today's developing countries. In particular, Latin America, tropical central Africa, and southeastern Asia account for a large share of global CO_2 sequestration in 2050. This is because the strategy of negative CO_2 emissions provided by combining bioenergy conversion technologies with CCS is adopted in these regions rich in high-quality woody biomass resources to meet the stringent CO_2 emissions reduction constraint. On the contrary, the amount of CO_2 sequestration is tiny in Japan and Korea due to their very small CO_2 storage capacity.

Whereas the large amount of the captured CO_2 is stored in aquifers in many parts of the world, the other three options also make a non-negligible contribution to global CO_2 sequestration in 2050. Similar to the results for 2030, ECBMR is applied in coal-rich regions such as India, the US, Russia, and

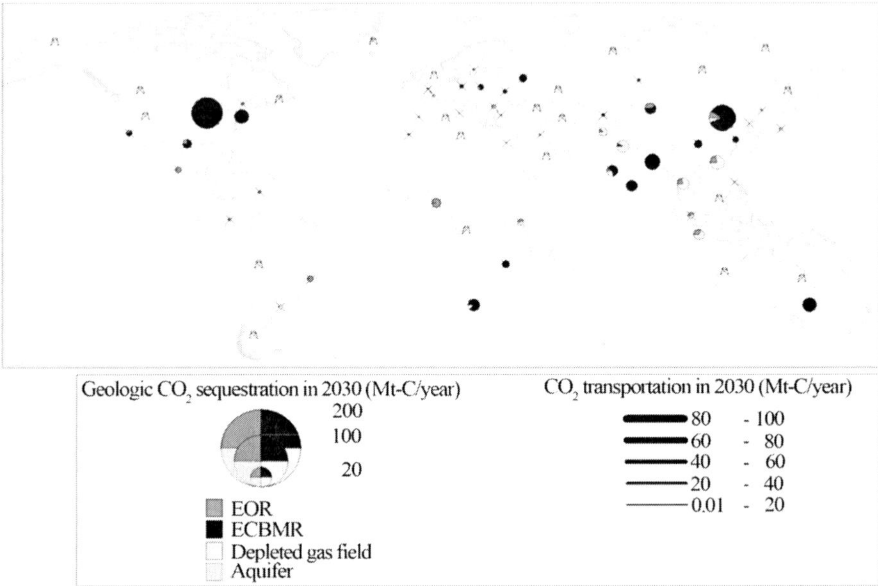

Figure 6. Global pattern of geologic CO_2 sequestration and CO_2 transportation in 2030[a]. [a]Same as footnote a in Figure 4.

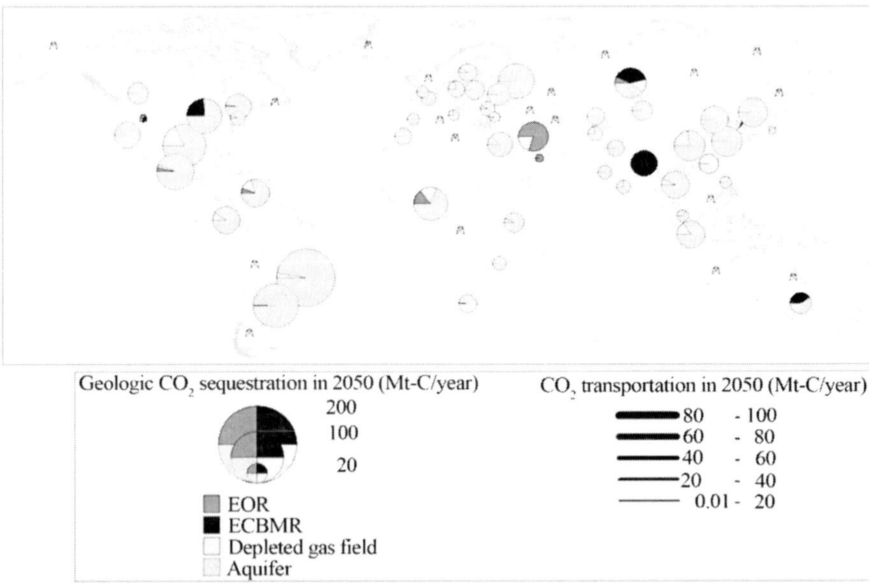

Figure 7. Global pattern of geologic CO_2 sequestration and CO_2 transportation in 2050[a]. [a]Same as footnote a in Figure 4.

Australia. On the other hand, EOR begins to be applied on a large scale in the Middle East, where giant oil fields exist, as oil production from them becomes increasingly mature. CO_2 sequestration in depleted gas fields is carried out on a large scale in regions where large quantities of conventional natural gas are produced (e.g., the US, the Middle East, and China).

Except for the route from Korea to the Far East in Russia, interregional CO_2 transportation is not chosen in 2050 for the same reason as was described for 2030. The captured CO_2 is interregionally transported only from Korea because it was estimated to have very small CO_2 storage capacity [11], because Korea has access to onshore geologic formations with sufficient CO_2 storage capacity within a relatively short distance, and because pipelines can make the short-distance transportation of CO_2 over land economical (see Figure 3). It can also be seen that the long-distance transportation of CO_2 by sea is unlikely to be done at least until 2050 due to the relatively close geographical proximity between CO_2 sources and sinks and to the high cost of shipping CO_2 by sea over such short distances (see Figure 3).

These results do not mean that CO_2 transportation will play a marginal role under the stringent CO_2 emissions reduction constraint. Rather, intraregional CO_2 transportation and CO_2 distribution networks built for this purpose will be a major component of a CCS system because there is to some extent a regional mismatch between CO_2 point sources and CO_2 sinks in most cases [2,13]. Taken together, great emphasis should be put on building infrastructure for short-distance CO_2 transportation over land to establish the cost-optimal CCS system.

Geologic CO_2 sequestration is hardly carried out in the 22 energy production regions of the model throughout the time horizon. This is because the interregional transportation of fossil fuels to the place of their use (i.e., the 48 energy production and consumption regions of the model) followed by their conversion to secondary energy combined with CCS is less costly than their conversion to secondary energy combined with CCS followed by the interregional transportation of secondary energy such as electricity and hydrogen.

4. CONCLUSIONS

Using the regionally disaggregated global energy system model with a detailed treatment of the whole chain of CCS, this paper has derived the cost-optimal global pattern of CO_2 sequestration in regional detail over the period 2010-2050 under the target of halving global energy-related CO_2 emissions in 2050 compared to the 2005 level. Also, prospects for interregional CO_2 transportation have been discussed under this target. Four major findings were obtained.

First, ECBMR will become a key early CO_2 sequestration opportunity because it can generate revenues that offset all or part of the costs of CO_2 capture and transportation. Accordingly, coal-rich regions such as the US, China, India, Australia, and South Africa will take the lead in CO_2 sequestration. For the same reason, EOR will also play a role in global CO_2 sequestration from the initial stage of CCS deployment, which may be applied mainly in China, southeastern Asia, and West Africa in 2030 and mainly in the Middle East in 2050.

Second, over time, CO_2 sequestration activities will become more spatially dispersed and an increasing amount of the captured CO_2 will be stored in aquifers in many parts of the world due to their abundant and widespread availability and their low cost. It is cost-effective to deploy CCS extensively in today's developing countries, implying the importance of the effective transfer of technologies and funds from today's industrialized countries. According to the results for 2050, CO_2 sequestration is carried out in the majority of world regions and the share of aquifers in global CO_2 sequestration reaches 82.0%.

Third, all the four options will make a visible contribution to global CO_2 sequestration. The captured CO_2 may be stored in depleted gas fields mainly in the US, the Middle East, and China.

Fourth, the interregional, long-distance transportation of CO_2 is unlikely to be chosen throughout the time horizon. This is due to the widespread distribution of geologic formations with large CO_2 storage capacity and the high cost of CO_2 transportation.

Readers, however, should be aware of a limitation of this study. As pointed out by many authors, regional differences in CO_2 policy actions and targets, CO_2 policy framework and incentives, and other political issues, which are not fully considered in the model, are likely to have a large impact on where CCS technologies are deployed in the "real world". Thus, the results of this study must be interpreted with caution. Nevertheless, this study is novel and significant in that the findings obtained from the regionally disaggregated results have important implications for the cost-optimal pattern of CCS deployment and the strategic planning of CCS infrastructure.

REFERENCES

1. IEA (2008) Energy technology perspectives 2008. IEA, Paris.
2. IEA (2004) Prospects for CO_2 capture and storage. IEA, Paris.
3. Takeshita, T. and Yamaji, K. (2008) Important roles of Fischer-Tropsch synfuels in the global energy future. Energy Policy, 36, 2791-2802.
4. Takeshita, T. (2012) Assessing the co-benefits of CO_2 mitigation on air pollutants emissions from road vehicles. Applied Energy, 97, 225-237.
5. IEA (2010) Energy technology perspectives 2010. IEA, Paris.

6. International Institute for Applied Systems Analysis (2007) GGI scenario database. http://www.iiasa.ac.at/Research/GGI/DB/
7. Rogner, H.-H. (2012) Energy resources and potentials. In: Johansson, T.B., Patwardhan, A., Nakicenovic, N. and Echeverri, L.G., Eds., Global Energy Assessment (GEA), Cambridge University Press, New York, 425-512.
8. Takeshita, T. (2009) A strategy for introducing modern bioenergy into developing Asia to avoid dangerous climate change. Applied Energy, 86, S222-S232.
9. Takeshita, T. (2011) Competitiveness, role, and impact of microalgal biodiesel in the global energy future. Applied Energy, 88, 3481-3491.
10. Riahi, K., Barreto, L., Rao, S. and Rubin, E.S. (2005) Towards fossil-based electricity systems with integrated CO_2 capture: Implications of an illustrative long-term technology policy. Proceedings of the 7th International Conference on Greenhouse Gas Control Technologies, Elsevier, Amsterdam, 921-929.
11. Research Institute of Innovative Technology for the Earth (2002) Study project for implementation planning of the New Earth program. New Energy and Industrial Technology Development Organization, Tokyo.
12. Takeshita, T. and Yamaji, K. (2006) Potential contribution of coal to the future global energy system. Environmental Economics and Policy Studies, 8, 55-88.
13. IPCC (2005) Special report on carbon dioxide capture and storage. Cambridge University Press, Cambridge.

Chapter 2

Numerical Simulation of CO_2 Sequestration in Large Saline Aquifers

Zheming Zhang[1] and Ramesh K. Agarwal[1]

[1] *Department of Mechanical Engineering and Materials Science, Washington University in St. Louis, St. Louis, USA*

1. INTRODUCTION

With heightened concerns over CO_2 emissions from pulverized-coal power plants, there has been major emphasis in recent years on the development of safe and economical geological carbon sequestration (GCS) technology. Although it is one of the most promising technologies to address globalwarming due to anthropogenic CO_2 emissions, the detailed mechanisms of GCS are not well understood. As a result, there remain many uncertainties in determining the sequestration capacity of the formation/reservoir and the safety of sequestered CO_2 due to leakage. These uncertainties arise due to lack of information about the detailed interior geometry of the formation and the heterogeneity in its geological properties, such as permeability and porosity, which influence the sequestration capacity and plume migration. Furthermore, the sequestration efficiency is highly dependent on the injection strategy, which includes injection rate, injection pressure, type of injection well employed and its orientation etc. The goal of GCS is to maximize the sequestration capacity and minimize the plume migration by optimizing the GCS operation before proceeding with its large-scale deployment. In this chapter, numerical simulations of GCS are conducted using the US Department of Energy (DOE) multi-phase flow solver TOUGH2 (Transport of Unsaturated Groundwater and Heat). A multi-objective optimization code based on genetic algorithm is also developed to optimize the

GCS operation for a given geological formation. It is described in Chapter titled," Optimization of CO2 Sequestration Saline Aquifers". Most of the studies are conducted for sequestration in a saline formation (aquifer). Large-scale GCS studies are conducted for the Mt.Simon, Frio and Utsira saline formations, for which some experimental data and computations performed by other investigators are available. These simulation studies provide important insights as to the key sources of uncertainties that can influence the accuracy of simulations.

2. SALINE AQUIFER GEOLOGICAL CARBON SEQUESTRATION (SAGCS)

Studies of GCS have suggested that various geological structures can serve as potential CO_2 storage sites. The major geological carbon sinks include the following structures: (1) conventional hydrocarbon reservoirs, (2) un-minable coal seams, (3) mature oil/gas reservoirs, and (4) deep saline formations. Among these candidates, this chapter focuses on carbon sequestration in saline aquifers considering the following facts.

- Concentrated locations of major sources of CO_2 (such as power plants) are close to existing saline aquifers.
- Geological survey has confirmed vast geological distribution of deep saline formations possibly suitable for GCS in the US and Canada.
- Preliminary estimates have suggested large storage capacity of the existing deep saline formations. The US DOE estimates an aggregate storage capacity of approximately 2102~20043 billion metric tons of CO_2 for SAGCS in US, which accounts for 80~90 percent of US overall GCS potential [1].
- Since most of the saline formations are located deep underground, i.e., at least 800 m below sea level, they provide great potential for secure long-term sequestration.
- Many surveys, research projects and commercial projects have already been conducted for SAGCS, making it attractive for further research and technical contributions.

A huge geographic distribution of deep saline aquifers in North America has been identified by the DOE. The DOE-estimated storage capacity for SAGCS takes into account more than 80 percent of the overall storage capacity of all possible GCS sites. In Table 1, the low-end capacity of 2102 billion metric ton of CO_2 is estimated under the condition that ineffective storage may occur due to improper and non-optimized sequestration approaches. On the other hand, the high-end capacity of 20043 billion metric ton of CO_2 is estimated under the conditions that most effective and optimal storage takes place. It can be seen that the high-end estimated capacity is over nine times the low-end estimated capacity. This large difference means it is important to deploy optimized reservoir engineering techniques for effective utilization of storage potential and successful GCS practice.

Table 1. Estimation of saline aquifer storage capacity of GCS for different regional carbon sequestration partnerships (RCSPs) [1]

RCSPs	Low-end capacity (billion metric tons)	High-end capacity (billion metric tons)
BSCSP (Big Sky Caron Sequestration Partnership)	98	1237
MGCS (Midwest Geological Sequestration Consortium)	11	158
MRCSP (Midwest Regional Carbon Sequestration Partnership)	95	123
PCOR (Plains CO2 Reduction Partnership)	174	511
SECARB (Southeast Regional Carbon Sequestration Partnership)	1376	14089
SWP (Southwest Regional Partnership on Carbon Sequestration)	266	2801
WESTCARB (West Coast Regional Carbon Sequestration Partnership)	82	1124
Total	2102	20043

3. THE APPROACH FOR NUMERICAL SIMULATIONS OF GCS

The large spatial extent (of the order of kilometers) and time duration (centuries) for CO_2 plume migration after injection makes the study of SAGCS very difficult at these large spatial and temporal macro-scales by using laboratory scale experiments, which can be conducted only on relatively small spatial and temporal scales varying from nanometers to a few meters and from nanoseconds to a few days/months. Conducting field tests in large-scale formations before the actual deployment takes place can be very expensive. However, numerical simulations using computation fluid dynamics (CFD) technology can be employed at industrial scale to determine the fate of injected CO_2 in a reservoir. With the development over the past four decades, CFD technology has now become mature and has been widely and successfully applied to various engineering problems. With the proper modeling of the storage formation and ground water transportation, CFD is capable of providing accurate enough analysis for quick estimation of reservoir performance at considerably lower cost. The governing equations of mass/energy transportation and numerical representations of the formation properties have been well explained in the TOUGH2 User's Guide [2],[3].

In a complex simulation like that of SAGCS, it is impractical to integrate all geophysical and geochemical effects into a single model while retaining acceptable computational efficiency. Therefore, careful examination of physical phenomenon of interest in SAGCS is essential to determine simplifications in modeling of features of interest [4],[5]. Another important benefit of numerical simulations is that one can investigate the effect of various injection parameters such as injection rate, injection duration, and injection well orientation and displacement on CO_2 storage efficiency and plume migration in a given reservoir. The advantage of numerical simulations makes it possible to perform optimization studies of these injection parameters to achieve the highest possible storage efficiency and minimum plume migration, as described in Chapter titled," Optimization of CO2 Sequestration Saline Aquifers". Such an optimization capability can be very beneficial in successful cost effective implementation of SAGCS on industrial scale. We have employed genetic algorithm in conjunction with numerical simulator TOUGH2 for optimization of SAGCS practice. Some salient features of the genetic algorithm and its integration with TOUGH2 are described in Chapter titled," Optimization of CO2 Sequestration Saline Aquifers".

4. SIMULATION CODE VALIDATION USING ANALYTICAL AND BENCHMARK SOLUTIONS

TOUGH2 is available as source files written in Fortran77. TOUGH2 does not provide graphical user interface (GUI) of any kind. All its input files and output results are in ASCII format. TOUGH2 has very high computing efficiency when executing large-scale simulations and it is very convenient for users to make modifications to the source code if needed. However, both the problem setup and result analysis capability in TOUGH2, such as mesh generation and contour map visualization, are not as comprehensive or straightforward as available in some newer commercial multiphase flow field simulators. For any complex problem, the modeling process tends to be tedious and error-prone. To address such deficiency, a third-party GUI for TOUGH2 named PetraSim was also installed on the same machine with TOUGH2. PetraSim preserves the original TOUGH2 binary files to execute simulations, while providing a smooth interface to a user-friendly computing environment. However, PetraSim in its original form lacks the compatibility for integrating a new optimization module and some recently developed equation-of-state modules.

Previous validation simulations on benchmark problems have shown that the simulation results obtained by TOUGH2 and PetraSim are identical [6],[7]. For our code validation purpose, we employed three widely used benchmark problems by GCS researchers worldwide. Simulations were conducted by both TOUGH2 and PetraSim. These three benchmark problems were first defined in the Workshop on Numerical Models for Carbon Dioxide Storage in Geological Formations at the University of Stuttgart, Germany [8],[9],[10],[11],[12]. We study the benchmark problem #1 and #3 using PetraSim, while benchmark problem #2 is simulated using the original version of TOUGH2 because of the limit on the availability of the needed equation-of-state module in PetraSim. When simulation is performed using TOUGH2, necessary post-processing programs such as Tecplot are employed for visualization and analysis.

4.1. Simulation of in situ CO_2 migration and comparison with analytical solution

As a first step towards code validation, simulations for CO_2 plume migration in an ideal simplified reservoir are performed. The analytical solution for this case is available [13], [14], [15] and is obtained as:

$$\frac{b_{ag}(r,t)}{B} = \frac{1}{\mu_b - \mu_{ag}}\left[\sqrt{\frac{\mu_b \mu_{ag} V(t)}{B\pi\phi r^2}} - \mu_{ag}\right] \tag{1}$$

where t is the time elapsed since the inception of injection, r is the distance (radius) from the injection well, $b_{ag}(r,t)$ is the plume thickness as a function of r and t, B is the total thickness of the reservoir, φ is the porosity of the reservoir, μ is the dynamic viscosity, subscripts *ag* and *b* stand for the injected CO_2 and brine respectively, and $V(t)=\int Q dt$ is the volume of the injected CO_2 within time t. For a horizontal reservoir, setting $b_{ag}(r,t) = 0$ yields Eq. (2), which gives a quick evaluation of the maximum CO_2 plume migration as:

$$R_{\max} = \sqrt{\frac{\mu_b V(t)}{\mu_{ag} B\pi\phi}} \tag{2}$$

In numerical simulations, a hypothetical deep saline reservoir of thickness of 100 m is assumed. A cylindrical computational domain is considered as shown in Figure 1. Generic hydro-geological properties are used. CO_2 injection rate is set at 1 kg per year for ten years. Detailed model parameters used in the simulations are summarized in Table 1. CO_2 plume migration in each year is computed by the simulation and compared with the analytical solution given by Eq. (2).

The simulation time is ten years and the CO_2 migration within the aquifer is computed for each of the ten years. In Figure 2, the CO_2 plume after 1, 4, 7 and 10 years since the inception of injection is shown.

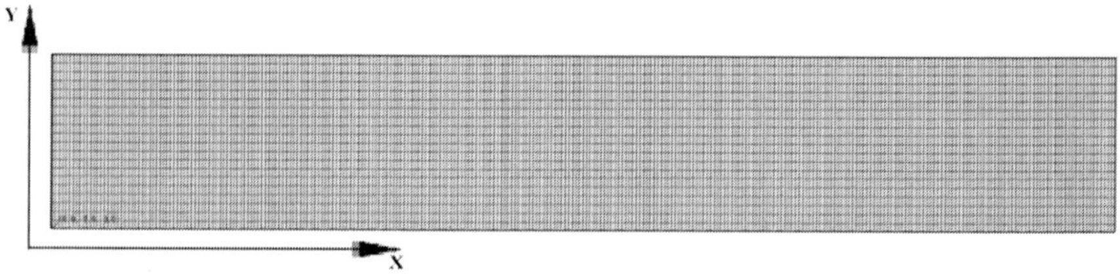

Figure 1. Computational domain and mesh of a generic cylindrical aquifer

Table 2. Geometry parameters and hydro-geological properties of the generic saline aquifer

Geometry	100 m in thickness; 3000 m in radius
Permeability	Isotropic, 100 mDarcy
Porosity	0.3
Temperature	20°C
Pressure	6.4 MPa
CO_2 Density	789.96 kg/m³
CO_2 Viscosity	0.0000712905 Pa•s
Brine Density	1029.69 kg/m³
Brine Viscosity	0.001488427 Pa•s
Relative Permeability	Linear
Brine Residual Saturation	0
CO_2 Residual Saturation	0
Capillary Pressure	None
Injection Rate	1 kg/s
Boundary Condition	Open boundary
Domain Discretization	300 × 20

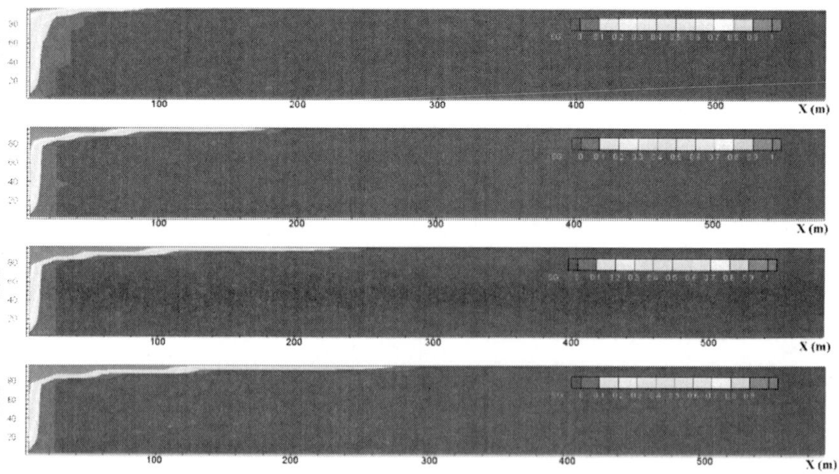

Figure 2. CO_2 plume after 1, 4, 7 and 10 years after injection

As seen in Figure 2, the injected CO_2 migrates upwards very rapidly and then prominently migrates underneath the caprock. A typical plume shape is already identifiable after one year of injection, when the farthest CO_2 migration reaches about 100 m from the injection well. In the following nine years, in situ CO_2 keeps migrating outwards and spreads to 300 m after the tenth year of injection. Physically, such a large radial migration of in situ CO_2 is caused by gravity separation. The horizontal extent of the plume can be analytically calculated by utilizing Eq. (2). Taking the necessary values of reservoir and fluid properties from Table 2, the horizontal extent of the plume predicted by Eq. (2) for the first 10 years since injection is summarized in Table 3. The horizontal extent of the plume given by TOUGH2 simulations is also summarized in Table 3 for comparison with the analytical solution.

Table 3. Maximum CO_2 migration underneath the caprock given by the analytical solution and TOUGH2 simulation

Year	Maximum Migration based on Numerical Simulation (A)	Maximum Migration based on Analytical Solution (B)	Deviation ((A-B)/B)
1	100.75 m	95.58 m	0.054090814
2	140.49 m	135.17 m	0.039357846
3	168.37 m	165.55 m	0.017034129
4	191.34 m	191.16 m	0.00094162
5	211.27 m	213.72 m	-0.011463597
6	229.23 m	234.12 m	-0.020886725
7	235.34 m	252.88 m	-0.069360962
8	260.81 m	270.34 m	-0.035251905
9	275.70 m	286.74 m	-0.038501779
10	289.36 m	302.25 m	-0.042646816

As shown in Table 3, TOUGH2 simulations successfully predict the maximum CO_2 plume migration underneath the caprock with excellent agreement with the analytical solution given by Bachu, Nordbotten and Celia [13],[14],[15]. The insignificant difference between the numerical and analytical solutions can be explained by the fact that CO_2 dissolution is accounted for in TOUGH2 while it is neglected in the derivation of the analytical solution. Since CO_2 dissolution is governed by the contact area between CO_2 and the ambient brine, the rate of CO_2 dissolution into brine should gradually increase over time as larger contact area becomes available as the CO_2 plume spreads. Nevertheless, Table 3 validates TOUGH2 as an accurate simulation tool for predicting the migration of in situ CO_2.

A schematic of the CO_2 plume flow is shown in Figure 3. Although based on a simplified analytical model, Figure 3 shows in situ CO_2 migration due to the combined pressure-driven Darcy flow and the buoyancy-drive CO_2

transport.With the injection well located on the left side of Figure 3, the CO_2 plume can be identified as consisting of two distinct regions. The first region is a smaller region on the left, adjacent to the injection well, marked as region (1) in Figure 3. In this region, CO_2 is distributed uniformly through the entire period of the injection interval. This implies strong hydrodynamic force caused by the pressure difference between the pressurized injection well and the unaffected aquifer. Within this region, lateral pressure gradient dominates the movement of CO_2 and Darcy flow occurs, causing CO_2 to migrate more radially through the aquifer. The second region is marked as region (2) in Figure 3, where the CO_2 plume fully develops. In this region, buoyancy due to the density difference between CO_2 and brine becomes dominant and drives the upward movement of CO_2 along with lateral migration. In this region, the vertical movement of CO_2 becomes dominant and results in plume flow. Since it is a phenomenon of fundamental concern in SAGCS, understanding the development of plume flow is critical for the success of SAGCS. The size of the two regions in Figure 3 can vary depending on the properties of the actual aquifer, but under most conditions, region (2) becomes dominant in size, which influences the safety and efficiency of SAGCS operations. Therefore, every effort should be made either to increase the size of region (1) or decrease the size of region (2) for successful and desirable implementation of SAGCS.

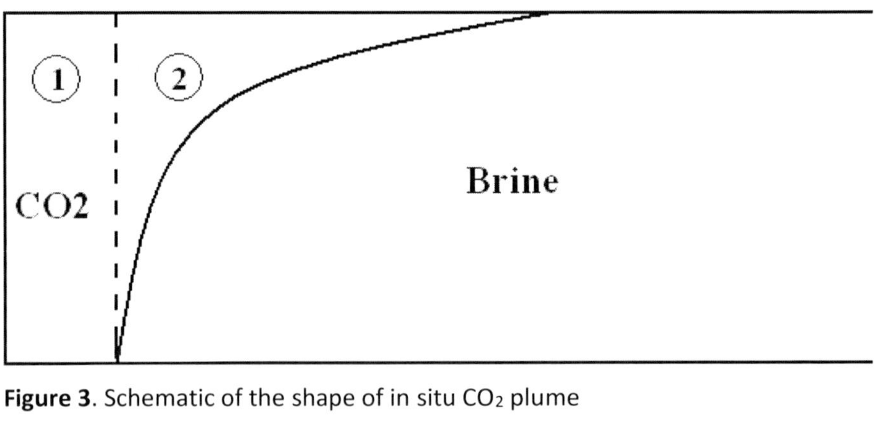

Figure 3. Schematic of the shape of in situ CO_2 plume

4.2. Simulation of benchmark problem #1 – CO_2 plume evolution and leakage through an abandoned well

A three-layered formation is modeled for the first benchmark problem [9]. CO_2 is injected into the deeper aquifer, shown schematically in Figure 4. It spreads in the aquifer and then rises up to a shallower aquifer upon reaching a leaky well. Quantification of the leakage rate, which depends on CO_2 plume evolution and the pressure buildup in the aquifer, is the main objective of this benchmark simulation. Figure 4 shows the schematic of the problem description by providing a cross-section of the formation.

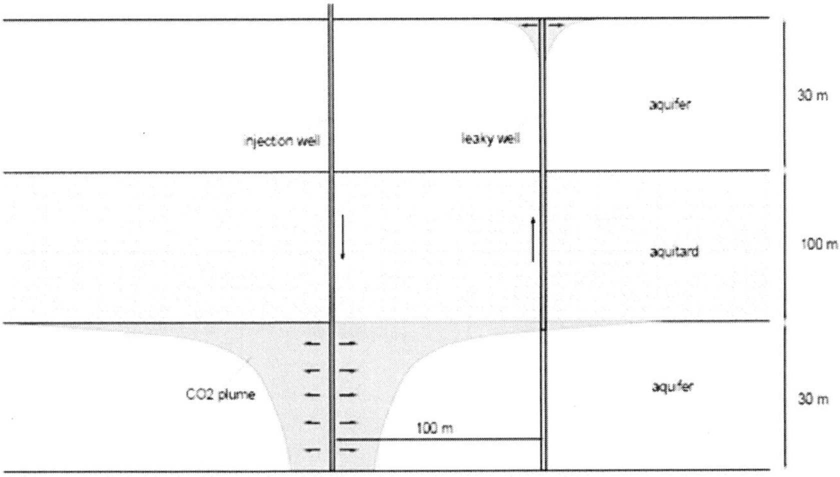

Figure 4. Schematic of benchmark problem #1 (cross-sectional view) [9]

The three layers in Figure 4 are identified as one aquitard layer and two (saline) aquifer layers. The lower aquifer layer is assumed to be 3000m below the ground surface. Typical saline aquifer conditions and hydrogeological properties, such as temperature, salinity, permeability, are assigned to the aquifer layers. The aquitard is assumed to be impermeable to both saline and CO_2, so it is considered an ideal geological seal to flow transportation. An "abandoned well" fully penetrating the three layers is located 100 m away from the CO_2 injection well. It can be either a crack in the formation or a physical abandoned well, which served as a pathway for upward CO_2 migration. Supercritical CO_2 is injected only into the lower aquifer through the injection well. Being less dense than brine, injected supercritical CO_2 gradually migrates to the ceiling of the lower aquifer and forms a plume. The formation and migration of the plume depends on the aquifer's geometric and hydrogeological properties. Table 4 summarizes the geometric properties of the aquifer in benchmark problem #1. It should be noted that the actual geometry of the injection well and abandoned well is circular, with a radius of 0.15 m. Since the use of an unstructured grid is not supported by PetraSim, an approximation to the circular geometry is made. Maintaining an identical cross-sectional area, the original circular injection well and the leaky well are replaced by wells of square cross-section with dimension of 0.266 m × 0.266 m.

Such an approximation is acceptable since the details of the flow pattern inside the wells are not critical in achieving the objective of this simulation.

Table 4. Geometry parameters for benchmark problem #1

Domain Dimension	1000 m × 1000 m × 160 m
Aquifer Depth	2840 m ~ 3000 m
Aquifer Thickness	30 m
Aquitard Thickness	100 m
Distance between Injection Well and Leaky Well	100 m
Injection & Leaky Well Geometry	0.266 m × 0.266 m

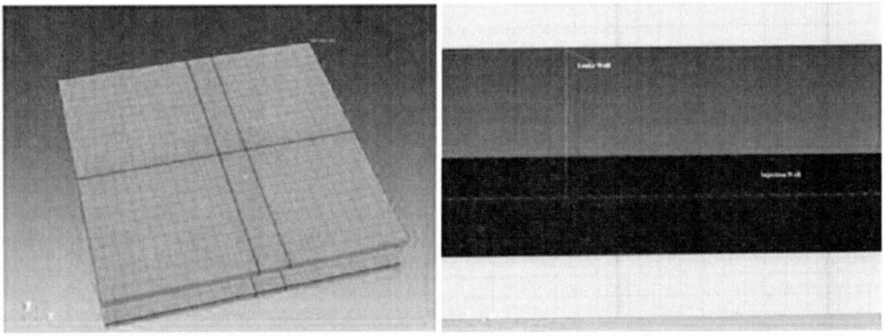

Figure 5. Entire computational domain (left) and the zoomed-in-view (right) for benchmark problem #1

Figure 5 shows the simulation domain and the structured mesh inside the domain.

To accurately model the small dimensions of the wells and accurately capture the CO_2 leakage rate, the mesh is highly refined in the neighborhood of the injection and leakage wells, as can be seen in the zoomed-in-view in Figure 5. Since high CO_2 concentration is expected at the ceiling of the lower aquifer due to gravity segregation, the mesh in this part of the lower aquifer is also refined. The mesh in the upper aquifer is not refined since it does not affect the accuracy of simulations and results in less computational efforts. The upper aquifer is uniformly discretized with vertical discretization length of 10 m, since it is assumed that the leakage amount is small and the leakage plume shape is not of great interest. By establishing the simulation domain and the mesh in this manner, reasonably accurate results are obtained while keeping the computational effort relatively low. The hydrogeological properties of the simulation domain are summarized in Table 5.

Table 5. Hydrogeological parameters for benchmark problem #1

Aquifer Permeability	20 mDarcy
Leaky Path Permeability	1 Darcy
Porosity	0.15
Residual Brine Saturation	0.2
Residual CO_2 Saturation	0.05
Relative Permeability	Linear ($k_r = S$)
Capillary Pressure	Brooks-Corey
Entry Pressure	1.0×10^5 Pa
Brooks-Corey Parameter	2.0

Table 6. Simulation parameters for benchmark problem #1

Thermal Condition	Isothermal
Initial Condition (Temperature)	Geothermal gradient: 0.03 K/m, Initial value at 800 m: 34°C
Initial Condition (Pressure)	Pressure gradient: 1045 Pa/m, 30.86 MPa at 3000 m depth
Boundary Conditions	Fixed-state on lateral boundaries No mass flow on top and bottom boundaries
Initial CO_2 Mass Fraction	$X_{CO2} = 0$
Initial Salt Mass Fraction	$X_{sm} = 0.20$
Injection Rate	8.87 kg/s
Simulation End Time	1000 days

Other simulation parameters such as initial conditions and boundary conditions are summarized in Table 6.

With the properties and parameters summarized above, the numerical model of benchmark problem #1 is setup in PetraSim. A pre-injection simulation is carried out first with no injection of CO_2 to achieve equilibrium condition under gravity. The equilibrium state is then implemented as an initial condition for the subsequent simulation with CO_2 injection. The equilibrium simulation is critical to provide the simulation with CO_2 injection with realistic initial conditions; this is a prerequisite procedure for all the simulations reported in this chapter. For this benchmark problem, it takes about five minutes of CPU time on the workstation for the simulation to finish. The leakage flux, pressure perturbation, and CO_2 saturation distribution throughout the aquifer after 80 days of CO_2 injection are examined and compared with the simulations of other investigators [12]. The leakage flux is a non-dimensional quantity defined as the ratio of CO_2 leakage rate to CO_2 injection rate. Detailed comparisons using various simulation codes are shown in Figure 6 and are summarized in Table 7.

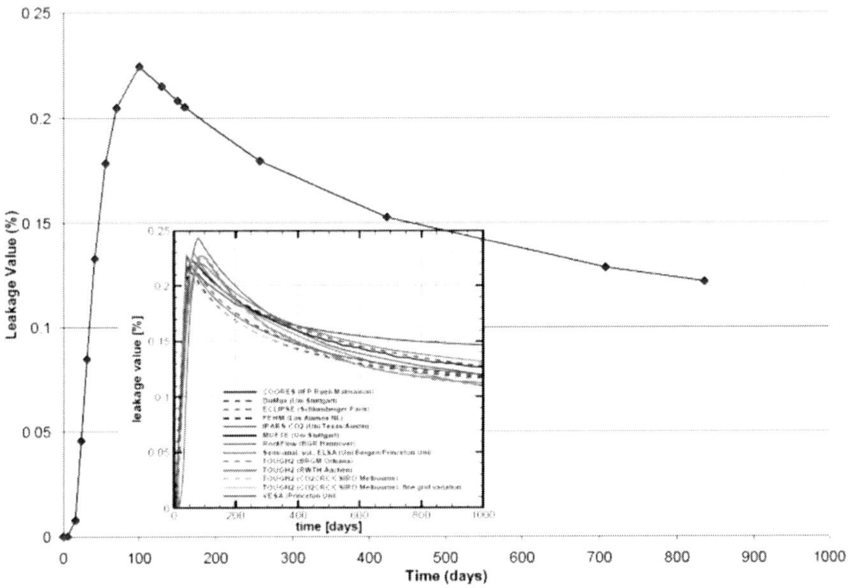

Figure 6. CO_2 leakage flux value obtained with WUSTL-TOUGH2 and other simulation codes

Table 7. Simulation results and comparisons with other codes for benchmark problem #1

	Max. Leakage	Time at Max. Leakage	Leakage at 1000th Day
TOUGH2 (WUSTL)	0.225 %	100 days	0.115 %
TOUGH2 (BRGM)	0.226 %	93 days	0.110 %

In Figure 6, our result (WUSTL-TOUGH2) is shown by the large graph, while comparisons with other simulation codes are shown in the inner box.

As additional comparisons, the pressure perturbation and CO_2 saturation distribution after 80 days of injection are also computed and compared with those from the MUFTE numerical solver [12]. Excellent agreement is obtained, as shown in Figure 7 and Figure 8.

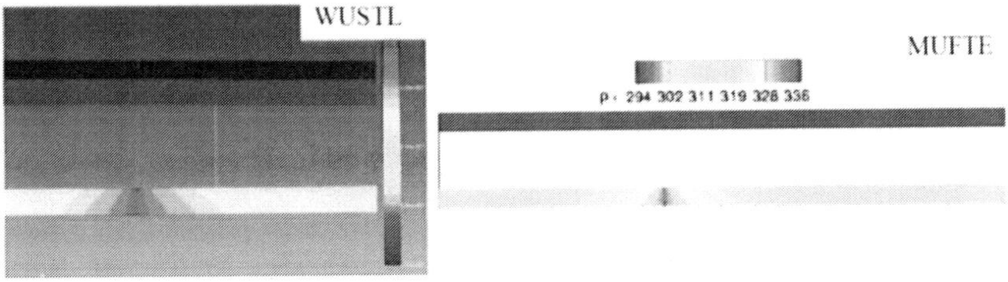

Figure 7. Pressure perturbation within the aquifer after 80 days of injection (left: WUSTL-TOUGH2; right: MUFTE)

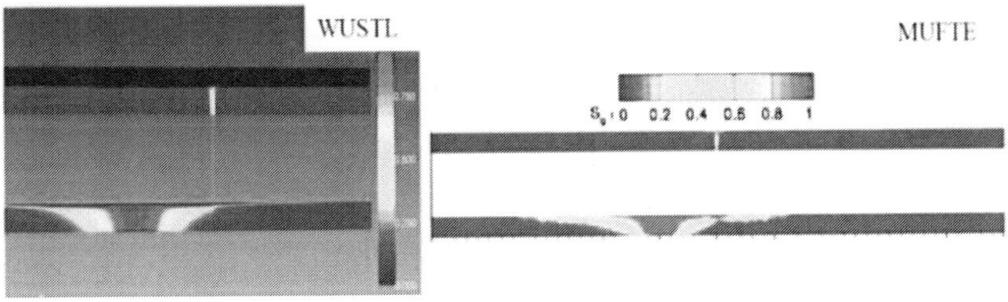

Figure 8. CO_2 distribution within the aquifer after 80 days of injection (left: WUSTL-TOUGH2; right: MUFTE)

As seen in Figure 8, the CO_2 plume is nicely captured in the simulation.

The simulation of benchmark problem #1 is very instructive. Three conclusions can be made: 1) Small variations among the results from different numerical simulators with different users are un-avoidable. Such variations are expected because some parameters are intentionally left unspecified. 2) Our results are in satisfactory agreement with the results of other investigators. 3) The most important CO_2 behavior under reservoir condition, i.e., the plume flow, is well captured and understood by the simulations. This simulation and others for benchmark problems #2 and #3 not only validate our numerical solver but also provide insights needed for the optimization studies reported in Chapter titled," Optimization of CO2 Sequestration Saline Aquifers".

4.3. Simulation of benchmark problem #2 – enhanced CH_4 recovery in combination with CO_2 sequestration in depleted gas reservoirs

For decades, the oil and gas industry has been using a reservoir engineering technique to increase the oil/gas production from mature reservoirs, known as enhanced oil/gas recovery (EOR/EGR). As the original formation fluid (oil or natural gas) gets extracted, pressure in the reservoir gradually decreases. Such de-pressurization process makes it increasingly difficult to maintain the desired production rate. The reservoir needs to be re-pressurized to mitigate the drop of oil/gas production. One of the means to do this is to inject CO_2 into the mature reservoir. With void space being occupied by the injected CO_2, remaining oil/gas is pushed out of the reservoir. Meanwhile, the depleted reservoir becomes an ideal carbon sink for long-term storage. EOR/EGR with CO_2 sequestration, also known as CSEOR/CSEGR, has been frequently used by the industry due to its strong economic merits.

In benchmark problem #2, a five-spot pattern domainis considered for modeling. The five-spot pattern is a common configuration for oil/gas production. A schematic of the reservoir is shown in Figure 3.

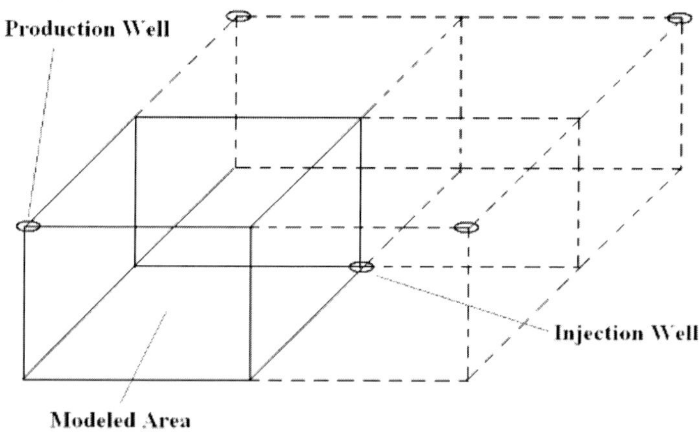

Figure 9. Schematic of the 3D five-spot pattern for benchmark problem #2

Natural gas is produced at the four upper corners of the reservoir, while CO_2 is injected in the middle of the domain at the bottom-most part. This is a direct result of CO_2 being heavier than CH_4 under the reservoir conditions. Injection of CO_2 at the bottom avoids gas mixing and creates better sweep efficiency. The main goal of this benchmark simulation is to identify the gas recovery factor, defined as the ratio of enhanced CH_4 production to the original remaining CH_4 amount until the shutdown of the production well. Additionally, the time until production shut-down, which is defined as the moment when the production contains up to 20% of CO_2 by mass, needs to be determined.

Due to symmetry, only a quarter of the domain is modeled, as shown in Figure 10 as the volume bounded by the solid lines. The dimension of the modeled domain is 201.19 m ×201.19 m with thickness of 45.72 m. Due to the relatively strong diffusion, the discretization length has a strong influence on the gas mixing [10].It is therefore strictly specified as 4.572 m for both vertical and horizontal direction.Figure 10 shows the CFD model and its mesh in the quarter computational domain of the five-spot reservoir.

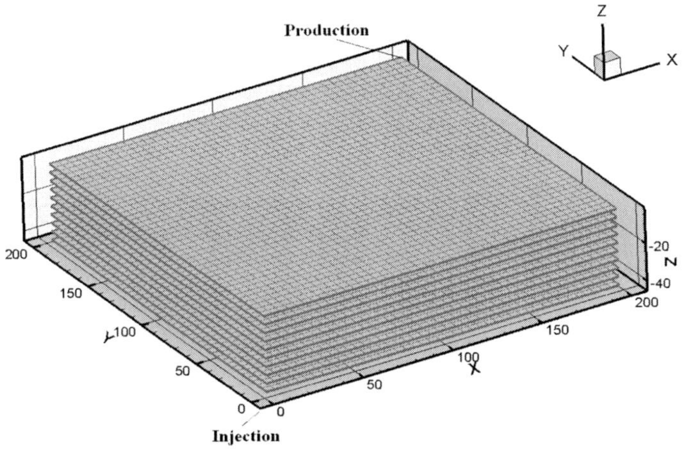

Figure 10. CFD model and its mesh in the quarter of the 5-spot domain for benchmark problem #2

Table 8. Hydrogeological properties of the domain for benchmark problem #2

Permeability	Horizontal: 50 mDarcy, Vertical: 5 mDarcy
Porosity	0.23
Residual Brine Saturation	0
Relative Permeability	Liquid: immobile, Gas: linear ($k_r = S$)
Capillary Pressure	None

Table 9. Simulation parameters for benchmark problem #2

Thermal Condition	Isothermal
Initial Condition (Temperature)	66.7 °C
Initial Condition (Pressure)	3.55 MPa
Boundary Conditions	No mass flow at all boundaries; Constant pressure at CH_4 production well
Initial CO_2 Mass Fraction	$X_{CO_2} = 0$
Initial CH_4 Mass Fraction	$X_{sm} = 1$
Injection Rate	0.1 kg/s until shut-down

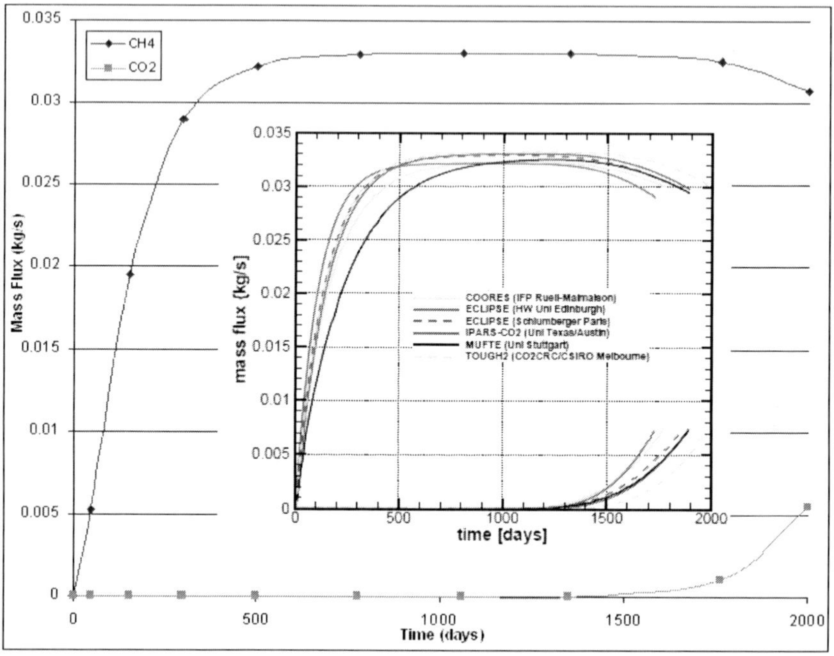

Figure 11. History of enhanced CH_4 recovery for benchmark problem #2

The hydrogeological properties assigned to the model are summarized in Table 8.

Initial conditions and boundary conditions and some other parameters of the domain are given in Table 9.

The termination of the simulation depends solely on the mass fraction of CO_2 in the reservoir. It takes about 30 minutes of CPU time to run 2,000 days of simulation before major CO_2 contamination occurs. The recovery factor, production shut-down time, pressure and CO_2 saturation distribution in the reservoir are investigated and compared with the results of other investigators.

In Figure 11, our results using TOUGH2 are shown as the large graph, while results of simulations from other investigators are shown in the inner box. Table 10 provides comparisons for recovery factor and production well shut-down time with other investigators' simulations [12].

Table 10. Comparisons of recovery factor and production shut-down time

	Recovery Factor	Production Well Shut-down Time
TOUGH2 (WUSTL)	61.4%	2063 days
TOUGH2 (CO_2/CRC)	58%	1987 days
MUFTE (U. Stuttgart)	53%	1894 days
IPARS (U. Texas)	55%	1891 days

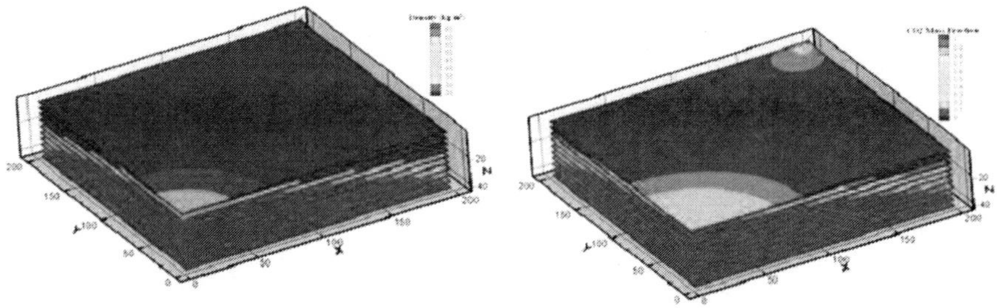

Figure 12. (a) Density profile (b) CO_2 mass fraction profile at production shut-down for benchmark problem #2

To visualize how the displacement process of CH_4 by CO_2 works, the density and CO_2 mass fraction profiles at production shut-down are examined in Figure 12.

Figure 11, Table 10 and Figure 12 lead to the following four conclusions: 1) Small variations in the results among different simulations with different users are unavoidable. Such variations are expected because some parameters are intentionally left unspecified. 2) Our results are in satisfactory agreement with those of other investigators. 3) It can be seen that the injected CO_2 migrates from the near lower corner to the far upper corner in a semi-spherical fashion. Unlike SAGCS, in situ CO_2 tends to sink to the bottom of the reservoir. This indicates strong gravity segregation caused by the density difference. 4) Production gas contamination caused by upward movement of CO_2 occurs at the production well despite the gravity segregation. This is due to the strong convective flow near the production well and mass diffusion.

4.4. Simulation of benchmark problem #3 – CO_2 injection in a heterogeneous geological formation

Accurate estimation of in situ CO_2 dissolution into the ambient brine is another important aspect of SAGCS simulation, since CO_2 becomes securely sequestrated once dissolved. Overestimation of CO_2 dissolution could lead to underestimating the potential leakage; on the other hand, underestimation of CO_2 dissolution would result in inefficient utilization of the formation's storage potential. In the meantime, it is instructive to model a realistic geological reservoir with heterogeneous hydrogeological properties for more realistic estimation of CO_2 storage capacity. In benchmark problem #3, part of the Johanson formation off the Norwegian coast is modeled for SAGCS [11]. The Johanson formation is a highly heterogeneous formation, especially in its porosity and permeability, as shown in Figure 13. CO_2 is injected in the middle of the modeled formation at 50 m from the bottom. The injection lasts for 25 years before it is shut down, and the total simulation time is 50 years. The goal of this benchmark study is to identify the amount of dissolved CO_2, the amount of CO_2 still in gaseous phase, and how these amounts evolve with respect to time. This study is very instructive to understand the dissolution process of injected CO_2 under reservoir conditions.

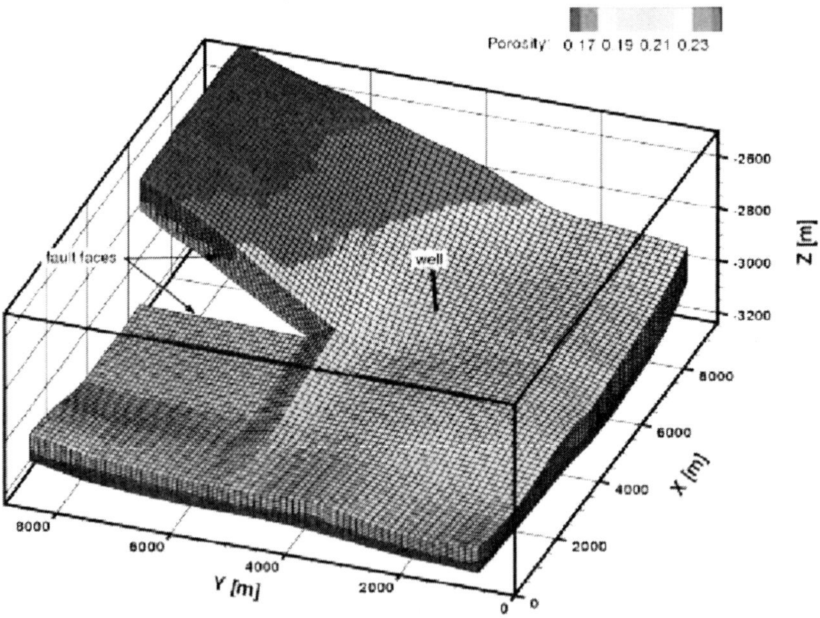

Figure 13. Johanson formation's porosity heterogeneity for benchmark problem #3[11]

The dimension of the modeled portion of the Johanson formation is 9600 m × 8900 m with varying thickness between 50 m to 150 m. The injection well is located at (x = 5440 m, y = 3300 m) over the bottom 50 m of the formation. The coordinates of vertices of 54756 hexahedral cells in Cartesian system have been provided for geometry construction [11]. Figure 14 shows the final CFD model of the Johanson formation.

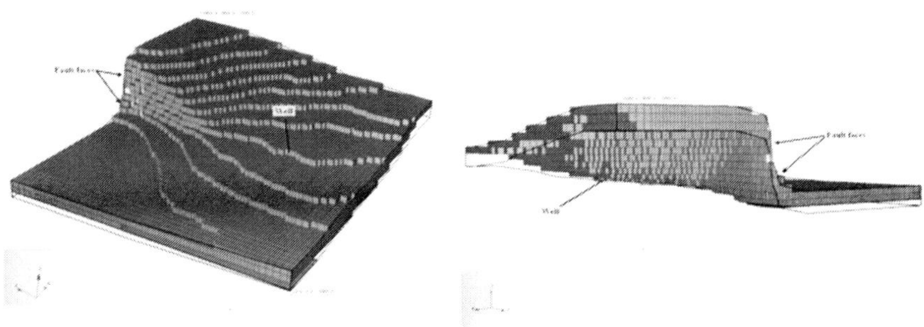

Figure 14. (a) Front view and (b) Rear view of the modeled Johanson formation

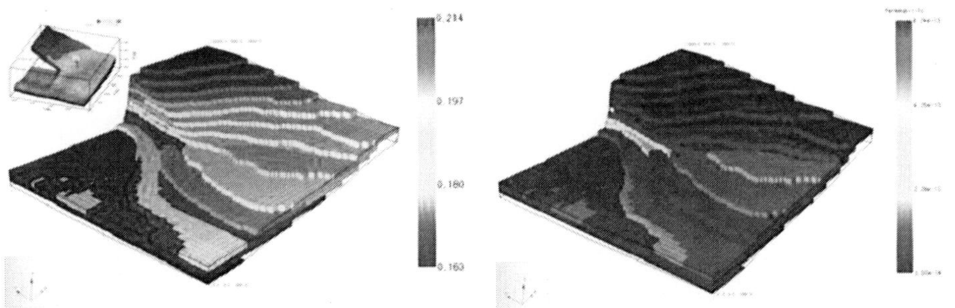

Figure 15. (a) Porosity and (b) Permeability of the modeled Johanson formation

The porosity and permeability of the modeled Johanson formation areshown in Figure 15.

Hydrogeological properties of modeled Johanson formation are summarized in Table 11.

Initial conditions, boundary conditions and other parameters of the modeled Johanson formation are summarized in Table 12.

Both gaseous and aqueous CO_2 accumulations after 50 years are considered as benchmark criteria for making comparisons with the simulations of other investigators. In Figure 16 our results using TOUGH2 are shown as the large graph, while results from other simulations are shown in the inner box.

Table 13 provides additional quantitative comparisons [12].

Table 11. Hydrogeological properties of the modeled Johanson formation

Permeability	Varies
Porosity	Varies
Residual Brine Saturation	0.2
Residual CO_2 Saturation	0.05
Relative Permeability	Brooks-Corey
Capillary Pressure	Brooks-Corey
Entry Pressure	1.0×10^4 Pa
Brooks-Corey Parameter	2.0

Table 12. Simulation parameters for the modeled Johanson formation

Thermal Condition	**Isothermal**
Initial Temperature	0.03 °C/m; 100 °C at 3000 m depth
Initial Pressure	1075 Pa/m, 30.86 MPa at 3000 m depth
Boundary Conditions	Fixed-state on lateral boundaries No mass flow on fault, top and bottom boundaries
Initial CO_2 Mass Fraction	$X_{CO2} = 0$
Initial Salt Mass Fraction	$X_{sm} = 0.1$
Injection Rate	15 kg/s (for 25 years), 0 kg/s thereafter
Discretization	Number of computational grids: 18804 Non-uniform for *x*-, *y*-, *z*-directions

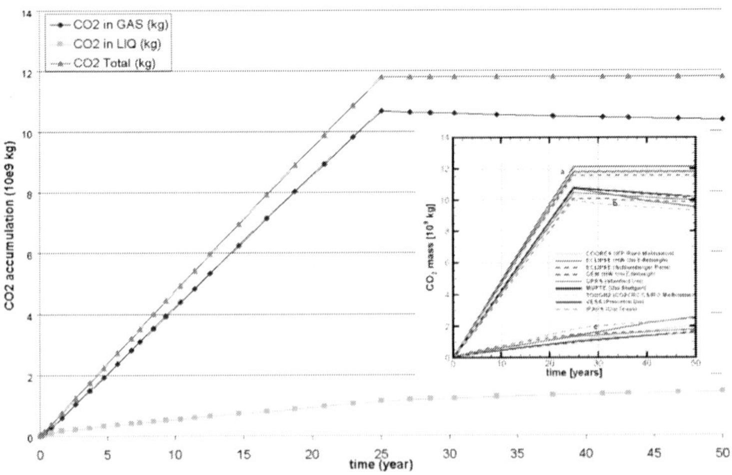

Figure 16. Gaseous and aqueous CO_2 accumulations for 50 years

Table 13. Comparisons of gaseous and aqueous CO_2 accumulations after 50 years

	Gashouse CO_2 at 50th Year	Dissolved CO_2 at 50th Year
TOUGH2 (WUSTL)	87.9%	12.1%
TOUGH2 (CO_2/CRC)	86.5%	13.5%
IPARS (U. Texas)	79.1%	20.9%

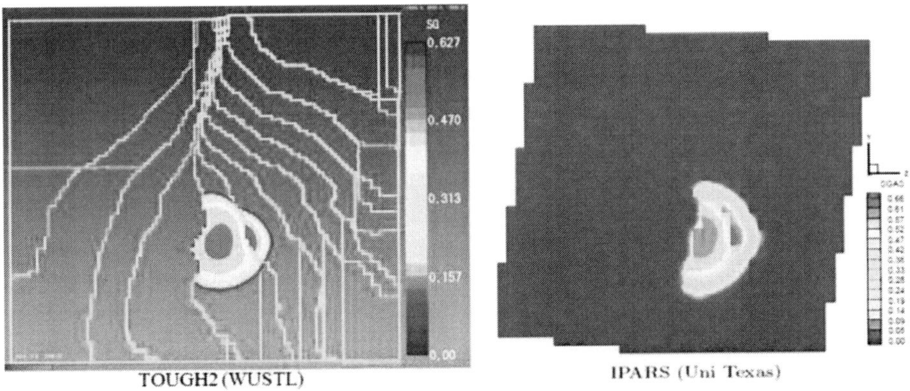

Figure 17. CO_2 saturation in the formation after 50 years, plan view

A comparison of the CO_2 migration after 50 years is given in Figure 17.

Figure 16, Table 13, and Figure 17 lead to four conclusions: 1) Small variations in the results among different simulations with different users are unavoidable. Such variations are expected because some parameters are intentionally left unspecified. 2) Our results are in satisfactory agreement with the results of other investigators. 3) CO_2 dissolution into the ambient porewater is a process that takes place very slowly. 4) The greater slope of aqueous CO_2 during the first 25 years (when injection continues) implies enhanced carbon dissolution due to convection during the first 25 years.

4.5. Conclusions of benchmark problems

The benchmark simulations presented in this chapter demonstrate that the TOUGH2 numerical simulator is capable of producing accurate and consistent results for various types of problems related to GCS. These simulations allow us to conduct simulations of large-scale SAGCS in identified saline formations with confidence, and proceed towards the development of a numerical optimization module for TOUGH2 and perform optimization designs of innovative reservoir engineering techniques for enhanced SAGCS safety and storage efficiency, as reported in Chapter titled," Optimization of CO2 Sequestration Saline Aquifers".

5. SIMULATIONS OF GCS IN IDENTIFIED LARGE SCALE SALINE AQUIFERS

Accurate large-scale simulations of existing (completed/continuing) SAGCS projects for identified known aquifers are crucial to create confidence in the future deployment of SAGCS projects. Although detailed history-matching simulations of existing SAGCS projects are challenging due to various uncertainties, e.g., in the reservoir topography and hydrogeology, the simulations can still provide informative insights into several aspects of SAGCS, such as the variance in multiphase flow properties, integrity of the geological seals, and the mechanism of CO_2 trapping. Such insights are essential for better understanding of the nature of SAGCS and its best practices for deployment. Detailed history-matching simulations have always been an important part in the SAGCS research activity. In the following sections, the results of SAGCS simulations for three large-scale identified aquifers are described.

5.1. SAGCS simulation for Mt.Simon formation

Located in the Illinois basin, the Mt.Simon sandstone formationis a huge saline aquifer that covers most of Illinois, southwestern Indiana, southern Ohio and western Kentucky. The estimated storage capacity of the Mt.Simon formation ranges from 27000 to 109000 million tons of CO_2[16],[17]. The Midwest Geological Sequestration Consortium (MGSC) is the regional consortium conducting studies of the possibility of large-scale GCS throughout the Illinois basin. ADM-Decatur GCS Project and FutureGen 2.0 Project are the two best-known SAGCS projects being currently carried out in theMt.Simon formation.

The depth of the Mt.Simon formation varies significantly throughout its coverage [18],[19]. In the southern part, it reaches as deep as 4300 m below mean sea level (MSL) while it increases to 80 m below MSL in the north. Consequently, a south-north geological slope of approximately 8 m/km has been estimated. The thickness of the Mt.Simon formation also changes significantly.A maximum thickness of 800 m in the north has been measured while it diminishes to zero further to the south. Other than the variance in topography, analysis of rock samples has suggested strong anisotropy in the formation's hydrogeological properties, with porosity ranging from 0.062 to 0.2 and permeability ranging from 5 mDarcyto 1000 mDarcy. Low-permeable Eau

Claire shale, which sits above the Mt.Simon formation, serves as the caprock. Except for some small regions near the Mississippi River, Eau Claire shale is very thick (more than 90 m) throughout most of the Illinois basin. The security of SAGCS over Mt.Simon formation is therefore assured by the continuous coverage of Eau Claire shale. A Precambrian granite formation stretches beneath the Mt.Simon saline aquifer.

A recent geological survey has suggested an area in the center of the Mt.Simon formation to be the core injection area – an area in which future storage sites are likely to be located. This core injection area is indicated as the area compressed by the white boundary in Figure 12, along with the elevation information of the Mt.Simon formation. As can be seen from Figure 12, both the ADM and FutureGen 2.0 projects are located in the core injection area.

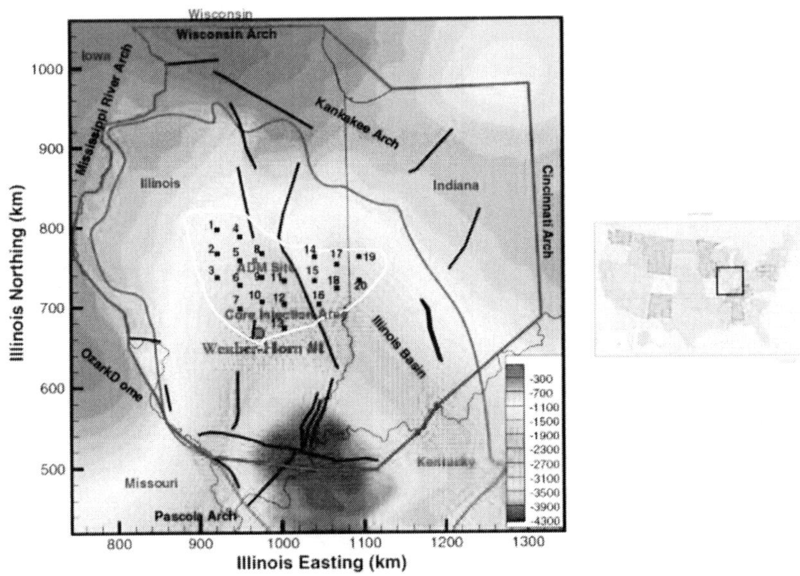

Figure 18. Core injection area and elevation of Mt.Simon sandstone

Mt. Simon sandstone is a typical stratified saline formation. According to the geological survey, strong anisotropy in porosity, permeability and capillary pressure exists through the entire depth of the formation. Based on variance of porosity, the Mt. Simon formation can be distinguished as four subunits, namely an upper unit with sandstone and shale tidally influenced and deposited, a middle unit with relatively clean sandstone, an Arkosic unit with highly porous and permeable sandstone, and a lower unit with decreased porosity and permeability. The high porosity and permeability of the Arkosic unit makes it an ideal candidate for the injection site. When modeling, these four subunits of Mt.Simon are further divided into 24 layers, each of which has a layer-averaged porosity and permeability value [16],[19]. With the consideration of data availability, a candidate site for future sequestration project, the Weaver-Horn #1 well (WH #1 well is shown as the red dot in Figure 18), has been chosen for our simulation study. The detailed well log of WH #1 is summarized in Table 14[20]. It is desired to model the anisotropy of hydrogeological properties as accurately as possible to capture its effect on in situ CO_2 transport. It should be noted that the lower unit of the Mt.Simon formation is not considered in the model due to its absence near WH #1 well. Both Eau Claire shale and Precambrian granite are modeled as impermeable formations.

Table 14. Porosity, permeability and characteristic capillary pressure of the 24 layers of Mt.Simon at injection site WH #1 [20]

Sub-Unit	Sublayer	Layer Depth (m)	Mean Porosity	Mean Permeability (mDarcy)	Characteristic Capillary Pressure (bar)
	1	2140 – 2150	0.061	5	0.37
	2	2150 – 2182	0.109	300	0.06
	3	2182 – 2197	0.074	10	0.28
Upper Unit	4	2197 – 2203	0.083	3.6	0.4875
	5	2203 – 2230	0.195	110	0.1
	6	2230 – 2232	0.071	1.1	0.8
	7	2232 – 2280	0.13	210	0.083
	8	2280 – 2322	0.083	5.4	0.4125
	9	2322 – 2331	0.24	150	0.0875
	10	2331 – 2340	0.088	8	0.35
	11	2340 – 2350	0.156	800	0.095
	12	2350 – 2370	0.25	80	0.125
	13	2370 – 2378	0.163	900	0.095
Middle Unit	14	2378 – 2385	0.195	105	0.1007
	15	2385 – 2399	0.163	800	0.05
	16	2399 – 2406	0.136	72	0.1167
	17	2406 – 2412	0.156	700	0.05
	18	2412 – 2424	0.129	160	0.09
	19	2424 – 2430	0.161	850	0.05
	20	2430 – 2462	0.128	60	0.15
	21	2462 – 2500	0.202	1000	0.05
Arkosic Unit	22	2500 – 2502	0.14	190	0.09
	23	2502 – 2537	0.151	1000	0.04

A cylindrical model of the Mt.Simon formation is constructed. For thermal condition,the model uses calculated values with a thermal gradient of 9.2°C/km. The reservoir pressure is assumed to be hydrostatic pressure with a gradient of about 10.8 MPa/km from the ground surface. Salinity is assumed to increase with the depth, starting from 235 mg/L at 450 m below ground surface with a gradient of 12.8 mg/L per meter in depth. A north-south geological slope of 8 m/km is also considered in the model. A "no-flux" boundary condition is applied at the top and bottom of the model, representing the impermeable upper and lower bounding formations. A "fixed-state" boundary condition is imposed at the lateral boundary to represent an essentially "open" reservoir. The permeability and porosity of the 24 sublayers can be seen in Figure 19.

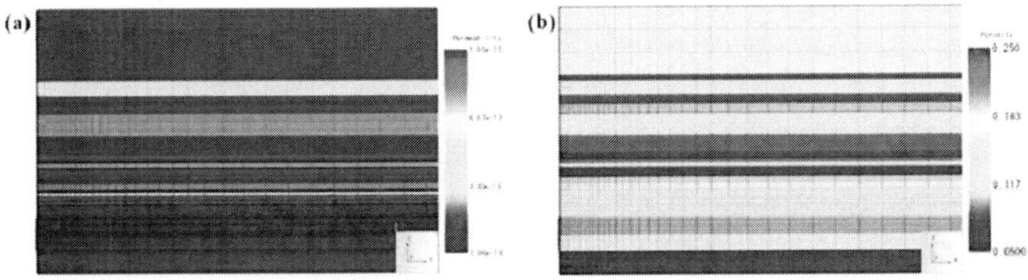

Figure 19. (a) Permeability, (b) porosity, and computational mesh of the 24 sublayers of the Mt.Simon formation model at WH #1 well

Due to the relatively high porosity and permeability, CO_2 injection is assigned at the bottom Arkosic unit (bottom three sublayers). The injection rate is assigned to be 5 million tons per year and injection lasts for 50 years. CO_2 footprint at 5th year, 25th year and 50th year since the beginning of injection is examined.

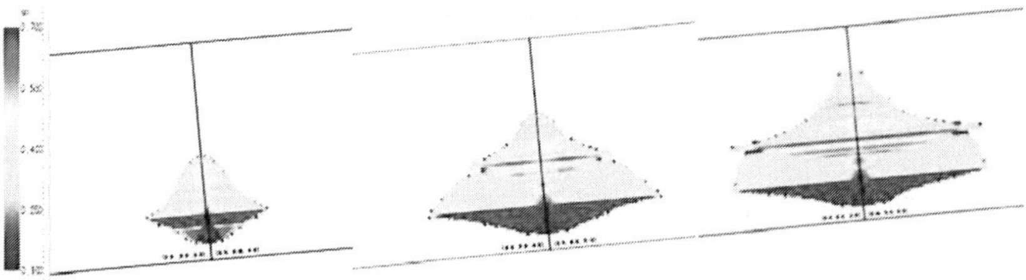

Figure 20. Saturation of gaseous CO_2 at (a) 5th (b) 25th and (c) 50th year of injection

As seen in Figure 20, the CO_2 plume evolves with a complex spatial pattern during the 50 years of injection. Within the Arkosic unit where the injector is located, extensive lateral migration with relatively higher concentration of gaseous CO_2 is observed. In the overlying sublayers, however, a strong secondary sealing effect that retards the vertical migration of gaseous CO_2 is observed, as can be seen from the pyramid-shaped sub-plume. Detailed analysis of secondary sealing effect is done as follows. As shown in Figure 20, the injected CO_2 migrates laterally away from the injector within the highly permeable Arkosic unit in the first 5 years after the start of injection. Simultaneously, buoyancy also leads to upward movement of CO_2 until it encounters the immediately overlying low-permeability sublayer (sublayer #20). The low permeability of sublayer #20 directly results in higher capillary pressure experienced by mobile CO_2, and thus a stronger vertical pressure gradient is required for mobile CO_2 to penetrate sublayer #20. When the capillary pressure is greater than the phase pressure of CO_2, sublayer #20 appears to be "impermeable" to the underlying CO_2 plume. Consequently, gaseous CO_2 accumulates under this layer and continues spreading out laterally, finally reaching a maximum extent of approximately 3000 m. Meanwhile, the increased CO_2 column under sublayer #20 brings up its phase pressure. Once the phase pressure of CO_2 exceeds the entry pressure of sublayer #20, mobile CO_2 breaks the capillary barrier of its overlying layer and starts to penetrate it. Such accumulation-penetration-breakthrough behavior of gaseous CO_2 occurs each time the upward migrating CO_2 encounters an overlying sublayer with lower permeability. Because the high capillary entry pressure of the overlying layer temporarily prevents CO_2 from migrating upwards,this phenomenon is identified as the secondary sealing effect. As can be seen from Figure 20, this effect is a very effective means to retard the upward migration of in situ CO_2. Its contribution makes gaseous CO_2 barely reach the Eau Clair shale even after 50 years of injection.

5.2. SAGCS simulation of Frio formation

The SAGCS pilot project for the Frio deep saline formation near the Mexican Gulf Coast is the subject of study in this simulation. The Frio project has two characteristics that make it attractive for numerical study. First, it is a completed pilot project with detailed field data available; secondly, hysteresis information of relative permeability and capillary pressure has been obtained from the core sample of the Frio saline formation. The hysteresis effect is an important factor in obtaining accurate estimation of CO_2 migration and dissolution for full-term SAGCS simulation. A full-term simulation refers to a simulation that investigates the fate of in situ CO_2 through the entire life cycle of a SAGCS project, which consists of both injection and post-injection periods.

The Frio SAGCS pilot project was conducted at the South Liberty oil field operated by Texas American Resources in Dayton, Texas. Starting from October 4, 2004, 1600 tons of CO_2 was injected into the Frio formation about 1500 m below the ground surface within 10 days. The Frio formation consists of brine-bearing sandstone with high permeability beneath the Gulf Coast. It is a relatively thin sandstone layer only 23 m thick. A steep geological slope of 16° from south to north has been identified for Frio formation [21]. The Frio pilot project employed one injection well and one observation well about 33 m to its north. Other than the conventional pre-injection geological surveys, laboratory analysis of core samples suggested the hysteresis behavior of relative permeability and capillary pressure in the Frio formation. The hysteresis has been considered in our simulations.

The hydrogeological parameters of the modeled Frio formation are summarized in Table 15.

Table 15. Geometry and hydrogeological parameters for Frio formation

Permeability	Isotropic, 1 mDarcy
Porosity	0.28
Residual Brine Saturation	0.15
Residual CO_2 Saturation	0.2
Relative Permeability	van Genuchten-Mualem with hysteresis
Capillary Pressure	van Genuchten-Mualem with hysteresis
Initial Conditions	$P = 15.2$ MPa, $T = 59°C$
Initial CO_2 Mass Fraction	$X_{CO2} = 0$
Initial Salt Mass Fraction	$X_{sm} = 0.093$

Characteristics of capillary pressure and relative permeability have been obtained from mercury-injection laboratory experiments on core samples from Frio formation, given in Figure 21. Hysteresis in both capillary pressure and relative permeability can be clearly observed. Drainage (of pore-water) curves are marked red and imbibition (of pore-water) curves are marked blue. When multiple drainage-imbibition cycles occur, different imbibition curves represent different orders of drainage-imbibition cycles. The primary imbibition curve, i.e., when brine imbibition occurs for the first time, is depicted as a bold solid curve. Since only one drainage-imbibition cycle takes place when continuous CO_2 injection is imposed before it is permanently shut down, only the primary imbibition curve needs to be considered in the model.

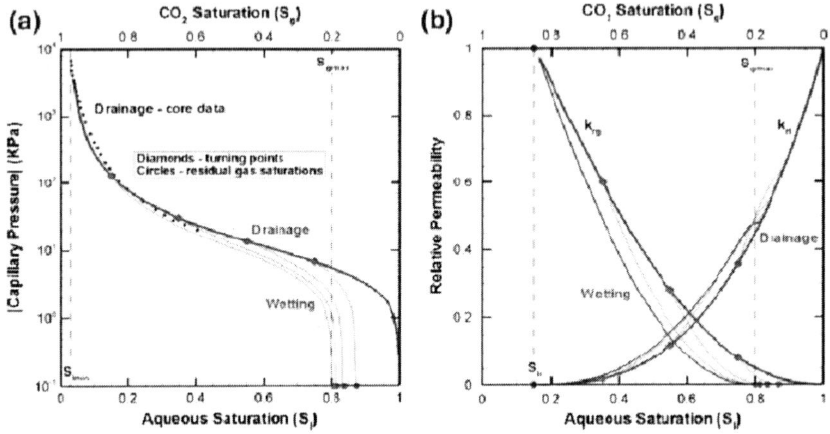

Figure 21. Capillary pressure and relative permeability characteristics of Frio formation[22],[23]

Following Doughty et al.'s work [22], a rectangular portion of Frio formation with dimension of 2500 m northwest-southwest, 800 m northeast-southeast, and 23 m thickness is modeled, as shown in Figure 22. The injection well is located at a point with coordinate (*x*=560 m, *y*=800 m) from the lower left corner of the computational domain. Although the formation is 23 m thick, injection takes place only over the first 8 m from the caprock. An observation well is located 33 m to the north. Because flow transport is most intense near the injection and observation wells, it is evolved in a computational domain of dimension 30 m × 30 m whose mesh is refined for accurate capture of the flowpattern. The injection and observation well locations, well depth, computational mesh, and north-south slope of the numerical model are all shown in Figure 22.

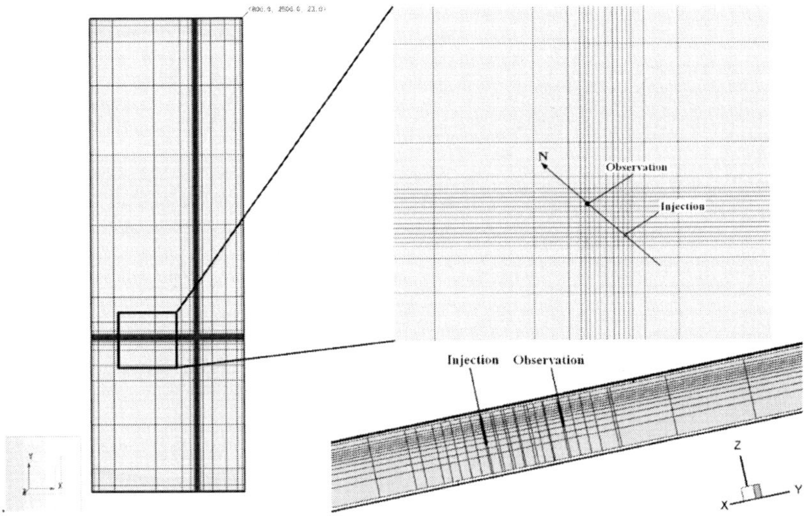

Figure 22. Model geometry and mesh in a portion of Frio formation and zoomed-in side view of the injection and observation wells

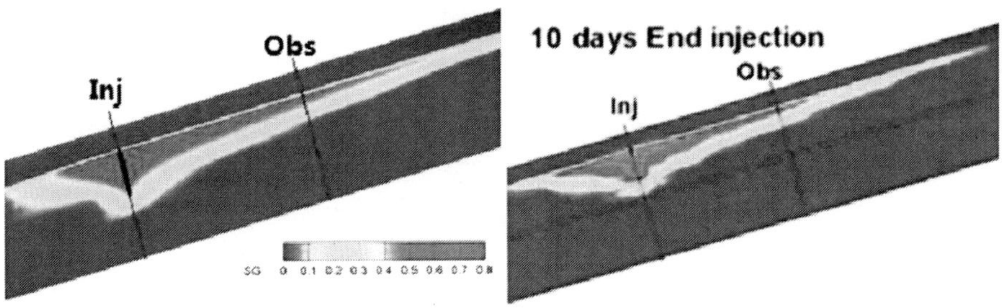

Figure 23. CO_2 footprint on 10th day when injection stops (comparison with Doughty et al.'s work[22])

The simulation time is set at 10 days to match the actual duration of injection. It takes approximately 12 hours of CPU time for the simulation to finish. The profiles of gaseous phase CO_2 at the end of injection in the vertical cross-section containing both injection and observation wells are shown in Figure 23. Doughty et al.'s result [22] is also shown in Figure 23 for comparison.

Additionally, Figure 24 shows the CO_2 saturation profiles at the injection and observation wells obtained by our simulation. They are compared with those given by Doughty et al. and the reservoir saturation tool (RST) logs [22]. The RST well logs are actual measurements in the field during the pilot project.

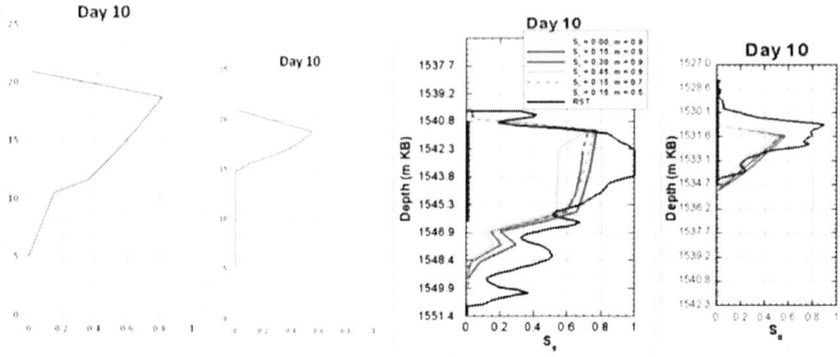

Figure 24. CO_2 saturation profiles given by the numerical simulations and the RST logs

As seen from Figure 23, the highly asymmetric CO_2 plume suggests a strong tendency of movement upwards in the direction of the geological slope. Unlike the case of Mt. Simon SAGCS, the CO_2 plume in the Frio project is shaped like an inverted pyramid, which implies the lack of secondary sealing effect. Both the asymmetric migration and inverted pyramid-shaped plume indicate strong evidence of the dominant role of gravity segregation in determining the in situ CO_2 migration. Considering the relatively short-term injection (10 days) for the Frio SAGCS project, this implies that in situ CO_2 migrates mostly convectively. Furthermore, it demonstrates that the poor permeability caprock layer above the injection serves quite well as the CO_2 barrier.

Comparing our results with those from Doughty et al. [22], the following conclusions can be drawn: (1) Overall our results are in good agreement with those of Doughty et al. for the plume shape, tendency of plume migration induced by the slope, distance of plume migration, and gaseous CO_2 saturation, etc. (2) Discrepancy still exists at the detailed simulation level. The results show that in our simulations, CO_2 saturation at the injection well reaches a maximum of 0.8 by the 10th day of injection. Although being consistent with Doughty et al.'s work, it differs from the field data. Results from the RST measurement suggest a CO_2 saturation value of 1.0, i.e., dry-out of brine, occurs adjacent to the injection well. The occurrence of brine dry-out is fairly common near the injection well due to the strong pressure gradient. However, the absence of brine dry-out in both our and Doughty et al.'s simulations can be explained by the designated brine residual saturation value. In our TOUGH2 simulations, a brine residual saturation value of 0.15 is pre-assigned to the entire computational grid including the grid near the injection wells. Since residual saturation describes the minimum saturation value of a certain phase being displaced, it means that at least 15% of the pore space will remain occupied by brine regardless of the pressure gradient. A direct result is the capped CO_2 saturation value at 0.85 and the absence of brine dry-out. (3) Our simulation shows quicker decrease in gas saturation during the injection interval. In Doughty et al.'s work, the gas saturation only drops from 0.8 to 0.65 for the upper 5 m of injection depth. In contrast, it drops from 0.8 to 0.4 in our simulation. This implies stronger buoyancy in our simulation, and thus results in a steeper inclined CO_2-brine interface. This also explains the slight overshoot in the plume migration to the north in our simulation.

5.3. SAGCS simulation for Utsira formation

The Sleipner project near the Norwegian coast on the North Sea is probably the most prestigious, important and successful SAGCS demonstration so far. It has the most complete topographic description, industrial-scale injection amount, and long-term monitoring data. Nevertheless, great uncertainties still exist for accurate reservoir-scale simulation of the Sleipner SAGCS project. Simulation studies of this project can provide helpful insights in understanding the transport behavior of in situ CO_2 and the reservoir performance.

Starting from 1996, the Sleipner field in the North Sea has been the host of the world's first commercial SAGCS project. CO_2 is captured from the gas mixture produced from a nearby deeper natural gas reservoir. To date, approximately 1 million tons of supercritical CO_2 has been sequestered annually. Utsira saline aquifer is the target formation for permanent carbon sequestration for the Sleipner SAGCS project. The Utsira formation is located at a depth of 800 m – 1100 m from the seabed with thickness of about 200 m – 250 m. The injection site is located at the southern portion of the Utsira formation. A 250 m – 330 m thick shale layer known as the Nordland Formation serves as the caprock, and core testing has suggested its potential for bearing a CO_2 column of at least 100 m but perhaps up to 400 m (depending on the in situ conditions). It is estimated that the Utsira formation has permeability of about 1– 8 mDarcy, porosity of about 0.35 – 0.4, and temperature of about 34°C – 37°C. It is also estimated that the reservoir bears hydrostatic pressure from its overburden formations. Similar to the Mt.Simon formation, the Utsira formation is also highly stratified, consisting of sublayers with high-permeability sandstone and low-permeability shale. Therefore, it is expected that the secondary sealing effect will occur. Figure 25 shows a 2-D seismic image taken in 2008 revealing the CO_2 plume in the Utsira formation. Multiple layers can be distinctly identified from the seismic image.

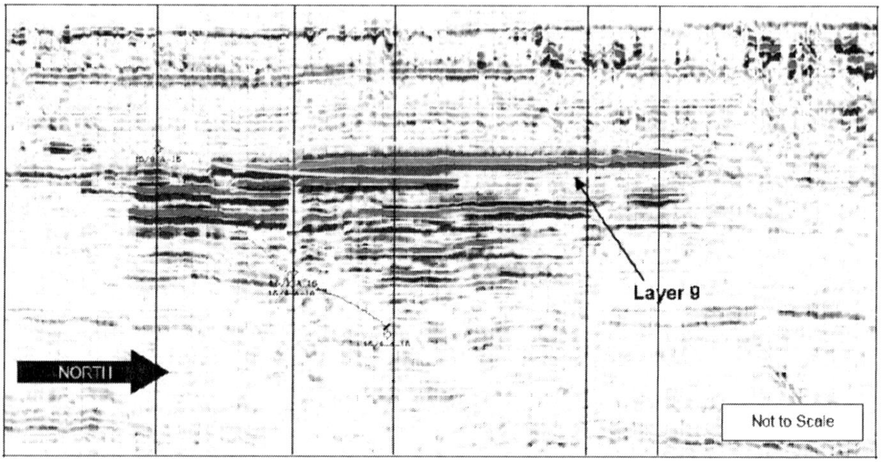

Figure 25. eismic image of Utsira formation after 9 years of injection, S-N cross-section [**24**]

Two numerical models have been constructed to study the Sleipner SAGCS project. The first model is a generalized axisymmetric layered model for estimating the ballpark migration of in situ CO_2. The purpose of this simulation is to determine the secondary sealing effect and gain an overview of the plume migration within the Utsira formation. The second model describers a 48 km^2 area of detailed topmost sandstone layer (marked as Layer #9 in Figure 26). Layer #9 is of particular interest regarding the safety of the sequestration project, as it is the layer within which the highest concentration of gaseous CO_2 exists and the most significant plume migration occurs. Detailed topography of Layer #9 is shown in model #2, making it a complicated 3D model. The 3D Layer #9 model is introduced to investigate the effect of actual topography on in situ CO_2 migration, while avoiding intensive computational effort associated with full 3D modeling and simulation of the entire Utsira formation.

5.3.1. Model #1 – Generalized stratified model of the Utsira formation

A pre-injection geological survey unveiled the layered structure of the Utsira formation. The majority of the formation can be identified as an eight-layered structure, but one extra layer needs to be added to the structure near the injection site due to the existence of a sand wedge. Therefore, a cylindrical domain with nine alternating sandstone and shale layers is constructed as shown in Figure 26. According to the seismic survey, it is assumed that all four shale layers have identical thickness of 5 m, four shallower sandstone layers have identical thickness of 25 m, and the bottom sandstone layer has a thickness of 60 m. This adds up to a total 180 m thickness for the modeled Utsira formation. The lateral radius of the generalized cylindrical model reaches 100 km, which is about the same as the actual extent of the southern part of the Utsira formation. According to Audigane et al. [25], all sandstone layers have identical and isotropic hydrogeological properties, as do the shale layers. Figure 26 shows the layered structure and computational mesh of the modeled Utsira formation as well as the location of CO_2 injection. The permeability and porosity are 3 mDarcy and 0.42 for the sand layer and 10 mDarcy and 0.1025 for the shale layer. A temperature of 37°C and pressure of 11 MPa are applied as the pre-injection conditions. van Genuchten-Mualem functions are used to describe both relative permeability and capillary pressure. CO_2 injection of 30 kg/s is assigned as a point source at the middle of the bottom-most sand layer.

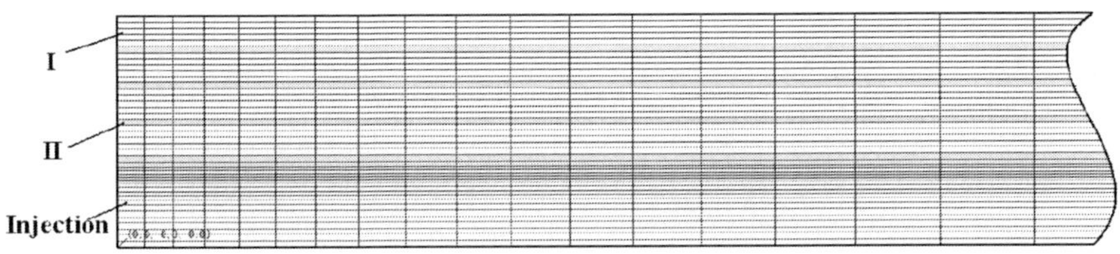

Figure 26. Computational mesh and layered structure of the generalized 9-layered model of the Utsira formation

The simulation time is set at 15 years and the CO_2 plume profile is examined for each year. Figure 27 shows the cross-sectional view of gaseous CO_2 in the reservoir for ten consecutive years since the start of injection.

The results shown in Figure 27 provide evidence of strong secondary sealing effect for migration of in situ CO_2. Similar to the case of Mt. Simon SAGCS, the injected CO_2 first migrates upwards driven by buoyancy until it reaches the first shale layer. Due to the low permeability and high capillary entry pressure, CO_2 is confined by this shale layer and is forced to migrate radially. Simultaneously, CO_2 concentration builds up beneath the shale layer and finally breaks through the capillary barrier upon sustaining sufficient CO_2 column height. The accumulation-penetration-breakthrough takes place each time the CO_2 plume encounters a new shale layer and forms an inverted pyramid shaped sub-plume, as documented clearly by the first and second year plume shapes in Figure 27. Due to the secondary sealing effect, in situ CO_2 has very limited contact with the caprock of the Utsira formation by the third year of injection.

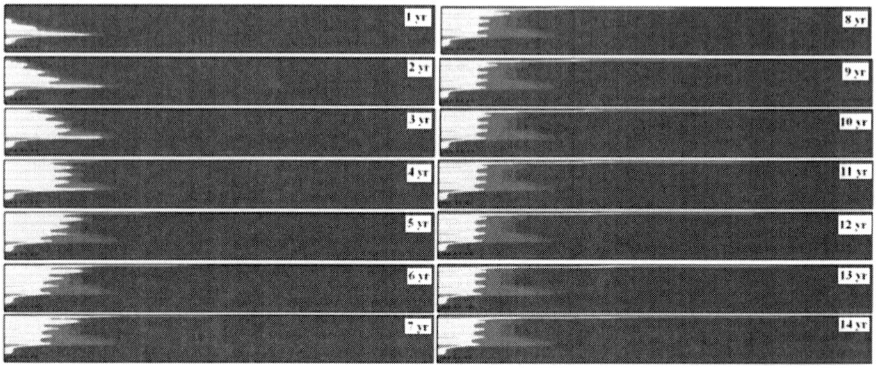

Figure 27. In situ CO_2 distribution for 15 years of injection

Additionally, ten-year CO_2 flux analysis has been made for the topmost sandstone layer (Layer #9) since it is critical to identify the accumulation of CO_2 underneath the caprock. As shown in Figure 30, excellent agreement between our simulation and the seismic amplitudes analysis [26] is observed, suggesting the overall accuracy of our model despite some discrepancy at the detailed level. The flux analysis shown in Figure 28 also implies that the accumulation rate of CO_2 in the topmost sandstone layer tends to increase until it becomes stabilized. This fact can be explained by mechanism of secondary sealing effect.

5.3.2. Model #2 – Detailed 3D model of the Utsira Layer #9 formation

In situ CO_2 possesses strong potential to migrate upward due to buoyancy, and thus accumulates under the caprock unless capillary barrier is compromised. Previous experience has demonstrated that the accumulation of CO_2 under the caprock occurs in a relatively short period compared to the entire time span of SAGCS projects, and it is a major concern for storage security. Therefore, it is critical for a SAGCS project to identify the accumulation of CO_2 and its tendency for migration underneath the caprock. With such information available, precautionary treatments can be deployed to avoid potential leakage. The Utsira formation near the injection site has been identified as a nine-layer structure as shown in Figure 29. The topmost sandstone layer, Layer #9, is of most interest since it has the highest concentration of gaseous CO_2 and has direct contact with the overlying caprock formation. Seismic surveys have shown striking growth of CO_2 accumulation in Layer #9 between 1999 and 2006 as shown in Figure 29 [26].

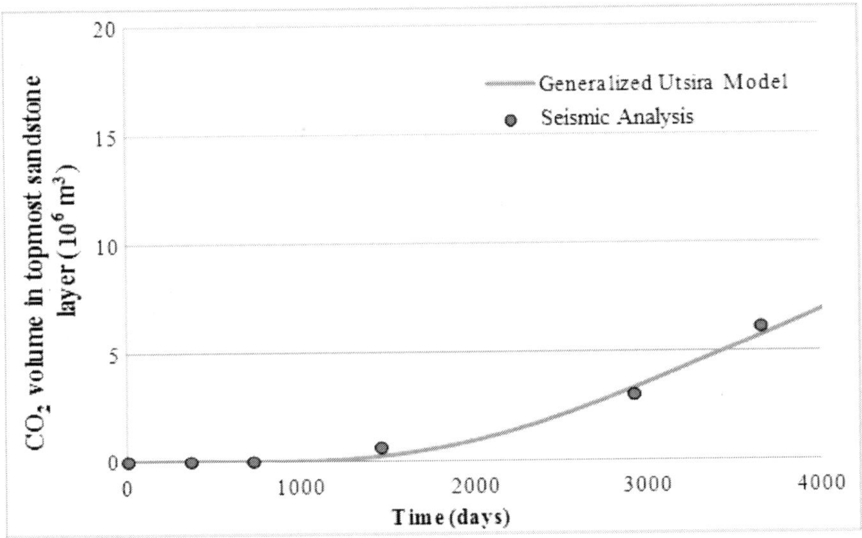

Figure 28. Gaseous CO_2 accumulation in the topmost sandstone layer

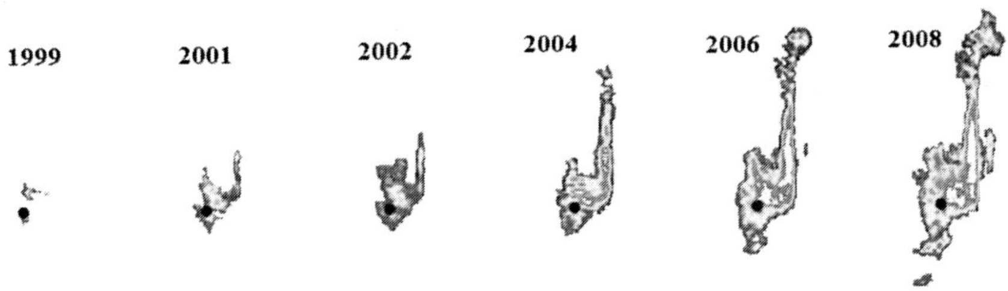

Figure 29. Amplitude maps of Layer #9 from 1999 to 2008, from Singh et al. [26]

The black dot in Figure 29 marks the location of the injection well, which is roughly 200 m under Layer #9. Two distinct local CO_2 accumulations appeared after about three years of injection (recall that injection began in 1996), indicating CO_2 began to accumulate under the caprock. However, CO_2 migration in Layer #9 was not symmetric due to the topography of the caprock. The northward migration of initially impacted CO_2, seen as the "body" of the plume in Figure 29, implies a local topographic dome; a prominent north-tending migration, seen as the "finger" of the plume in Figure 30. It implies the spill of local structurally trapped CO_2 along a north-tending topographic ridge. CO_2 migration along the north-tending ridge was rather fast at about 1 m/day between 2001 and 2004 [24].

In order to examine the plume's evolution within the topmost layer more closely, a 3D model of Utsira Layer #9 is created with detailed topography. It should be noted that only Layer #9, not the entire depth, is modeled because of the following considerations. To ensure the accurate capture of topographic effect on plume shaping, a computational domain with considerable fine mesh resolution has to be modeled based on geological survey data. The computational effort and thus the feasibility of highly detailed model of the entire Utsira formation is very intensive and time consuming. Secondly, CO_2 has to breakthrough several layers of relatively low-permeability shale prior to reaching the topmost layer. While it is difficult to quantify the breakthrough of gaseous CO_2, the quantification of CO_2 feeding into the topmost layer (Layer #9) is rather reliable. Therefore, a model of only the topmost layer (Layer #9) could provide an ideal platform to investigate the effect of various parameters such as topography on the shaping of the CO_2 plume, and also could be used for optimization purpose to achieve high efficiency sequestration while maintaining an affordable computational effort and cost.

A reservoir model with dimension of 1600 m × 4900 m with varying thickness was constructed. It covers the portion of the Utsira formation where the plume, shown in Figure 31, is located. As mentioned earlier, the topography of this portion of the Utsira formation is accurately modeled based on seismic geological survey data (provided by Zhu and Lu of the University of Indiana [27]) with 50 m × 50 m mesh resolution. Because only Layer #9 is modeled, the thickness of the computational domain varies from 3.5 m to 26.3 m with an average thickness of 11.3 m. However, to accurately capture the accumulation and upward and lateral movement of CO_2, 37 layers are used along the thickness. The topmost layer and the bottom two layers are designated to represent the low permeability shale, while the 34 layers in the middle are assigned the properties of mudstone. In the 3D Layer #9 model, permeability anisotropy is considered with west-east permeability of 2 mDarcy, north-south permeability of 10 mDarcy, and vertical permeability of 200 mDarcy. A 3D overview of the Layer #9 model is shown in Figure 24.

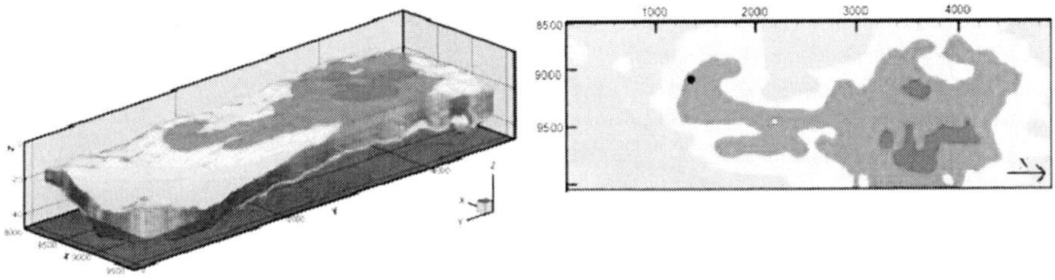

Figure 30. 3D overview and plan view of the 3D Layer #9 model of Utsira indicating feeder locations (black dot: main feeder; cyan square: secondary feeder)

Table 16. Hydrogeological properties of the Utsira Layer #9 model

Temperature	33 °C
Pressure	8.6 MPa
Total Utsira Formation Area	26100 km^2
Total Utsira Formation Thickness	50 m ~ 300 m
Layer#9 Area	1600 m × 4900 m
Layer#9 Thickness	3.5 m ~ 26.3 m
Shale Permeability	Horizontal: 0.001 mDarcy, Vertical: 0.0001 mDarcy
Mudstone Permeability	W-E: 2 mDarcy, N-S: 10 mDarcy, Vertical: 200 mDarcy
Utsira Porosity (shale/mudstone)	35.7 %
Residual CO_2 Saturation	0.02
Residual Brine Saturation	0.11
Relative Permeability Type	Corey/van Genuchten-Muller
Capillary Pressure	None
Porewater Salinity	3.3 %
North-south Geological Slope	8.2 m/km, 5.8 m/km
CO_2 Feeder Location	Main feeder: W-E: 516 m, N-S: 1210 m, bottom mudstone Secondary feeder: W-E: 925 m, N-S: 2250 m, bottom mudstone

Table 16 summarizes the hydrogeological properties of the Layer #9 model.

It should be noted that in the 3D Layer #9 model, the source of CO_2 is identified as "feeder" but not "injector" to emphasize that CO_2 is supplied from the lower aquifer through leakage pathways rather than by direct injection. Since the actual CO_2 injector is located at about 200 m under Layer #9, information on the injection rate recorded at the injector is not applicable for the CO_2 feeders in the Layer #9 model. To determine the CO_2 feeding rate to Layer #9, seismic surveys of CO_2 distribution are used to obtain its volume under in situ conditions, and then converted to mass flow rate. Information of CO_2 accumulative mass provided by Zhu and Lu [28] is summarized inTable 17.

Table 17. Accumulated CO_2 mass in Layer #9, 1999-2008

Year	Accumulated Mass (kg)	Yearly Feeding Mass (kg)	Feeding Rate (kg/s)
1999	0.00	0.00	0.00
2000	1.82×10^{7}	1.82×10^{7}	0.577
2001	5.52×10^{7}	3.70×10^{7}	1.17
2002	9.49×10^{7}	3.97×10^{7}	1.26
2003	1.45×10^{8}	5.01×10^{7}	1.59
2004	2.13×10^{8}	6.80×10^{7}	2.16
2005	3.07×10^{8}	9.40×10^{7}	2.98
2006	4.34×10^{8}	1.27×10^{8}	4.03
2007	6.03×10^{8}	1.69×10^{8}	5.36
2008	8.20×10^{8}	2.17×10^{8}	6.88

Table 17 gives the CO_2 accumulation in Layer #9 from 1999 to 2008. It can be seen that CO_2 feeding rate to Layer #9 keeps on increasing for the recorded nine years. Recalling the analysis of secondary sealing effect given for the previous cases, it is the pressure gradient between the gaseous CO_2 phase pressure at lower aquifer and the capillary pressure of the overlying shale layer that determines the breakthrough of CO_2 and its flow rate. When breakthrough first occurs, the pressure gradient just breaks the equilibrium state, resulting in relatively low breakthrough mass flux to Layer #9. However, as more CO_2 accumulates, the pressure gradient gradually increases and leads to increasing breakthrough mass flux, as depicted in Table 17. A nine-year average feeding rate of about 2.89 kg/s can be obtained from Table 17. Both the nine-year average value and the values in Table 17 have been considered in the simulations.

The significant north-tending plume finger is rather perplexing for regular pressure-gradient driven Darcy flow. Analysis suggests three possible explanations for the cause of the prominent north-tending CO_2 fingering along the ridge: (1) significantly higher permeability applied to the ridge; (2) existence of north-south geological slope which enhances the buoyancy-drive migration along the ridge; and (3) existence of a secondary (or even multiple) CO_2 pathways under the ridge. The hypothesis of significantly higher permeability at the ridge can be easily ruled out since no such evidence is obtained from the geological survey. It is still under debate whether geological slope should be considered when analyzing the CO_2 migration in the Utsira formation. Chadwick and Noy's work [24] suggested two possible values of geological slope based on the seismic images of the cross-section of the Utsira formation, which are 8.2 m/km and 5.8 m/km.

The simulation time is set at nine years, which corresponds to the injection period of 1999~2008. CO_2 plume migration at the topmost layer is examined for each year. Considering all the uncertainties mentioned above, a total of nine simulations are performed until good history-matching is obtained, as shown in Figure 31.

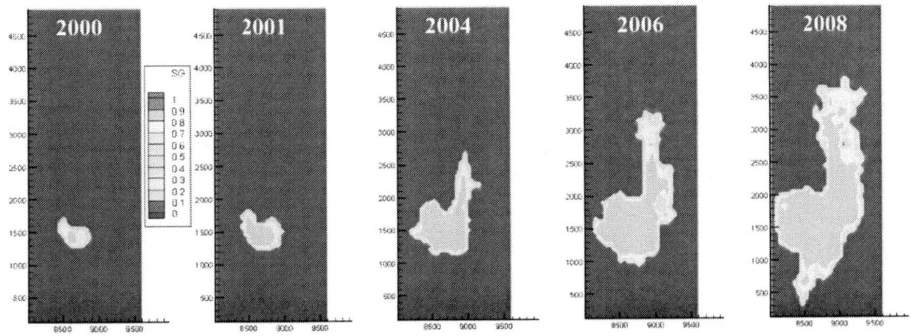

Figure 31. Simulated CO_2 migration in Layer #9, 2000 ~ 2008

Both the 2D generalized Utsira formation model and the 3D detailed Utsira Layer #9 model have generated satisfactory simulation results, as seen by history-matching. In summary, five major conclusions can be reached, as follows. First, the simulations show that the permeability anisotropy should be accurately modeled. Vertical-to-horizontal anisotropy close to 10:1 is needed to accurately capture the upward migration of CO_2. Horizontal anisotropy of 2:10 is needed to capture the northern spill of CO_2 into the north-tending ridge. Second, a secondary feeder is likely to exist directly under the north-tending ridge to generate sufficient plume migration along the ridge. This suggests multiple pathways for CO_2 breakthrough from the lower aquifer structure. Third, the fact that the injection gas is a CO_2-methane mixture is very important in modeling since the presence of 2% methane enhances the buoyancy. Fourth, it is critical that time-dependent feeding of CO_2 be modeled. This is consistent with the behavior of CO_2 path flow breaking the capillary pressure barrier, as noted before for the secondary-sealing effect in the case of the Mt.Simon formation. And finally, simulation results suggest strong mobility of gaseous CO_2 under the caprock (shale) without major leakage, implying that the caprock serves quite well as the non-permeable CO_2 barrier while exerting little resistance for the lateral flow of CO_2 underneath.

The simulation studies of the three actual large-scale identified deep saline aquifers conclude this chapter. These studies have provided important insights and best practices for obtaining accurate simulations of GCS using TOUGH2. In Chapter titled," Optimization of CO2 Sequestration Saline Aquifers", innovative reservoir techniques and their optimization, for more efficient and secured SAGCS operations, are discussed.

6. CONCLUDING REMARKS

In this chapter, some key factors relevant to saline aquifer geological carbon sequestration (SAGCS) have been investigated. First, numerical simulations have been performed for completed/ongoing SAGCS projects in three large scale identified saline formations using the DOE numerical simulator TOUGH2. Before performing these studies, TOUGH2 was validated against the available analytical solutions and the benchmark numerical test cases. The three case studies of SAGCS in large-scale saline formations have provided important insights into the reservoir performance and sequestration uncertainties.

7. ACKNOWLEDGEMENTS

Financial support for this work was provided by the Consortium for Clean Coal Utilization (CCCU) at Washington University in St. Louis, MO, USA.

REFERENCES

1. US Department of Energy. The 2012 United States Carbon Utilization and Storage Atlas – Fourth Edition; 2008.http://www.netl.doe.gov/technologies/carbon_seq/refshelf/atlasIV/ (accessed 24 May 2013).
2. Pruess K. TOUGH2: A General Numerical Simulator for Multiphase Fluid and Heat Flow, LawrenceBerkeley Laboratory Report LBL-29400, California: Berkeley; 1999.
3. Pruess K., Oldenburg C., and Moridis G. TOUGH2 User's Guide, Version 2.0 (revised), Lawrence Berkeley Laboratory Report LBL-43134, California, Berkeley; 2011.
4. Celia M.A. and Nordbotten J.M. Practical Modeling Approaches for Geological Storage of Carbon Dioxide. Underground Water 2009; 47(5) 627-638.
5. Celia M.A. and Nordbotten J.M. How Simple Can We Make Models for CO2 Injection, Migration, and Leakage? Energy Procedia 2001; 4 3857-3864.

6. PetraSim User Manual, Thunderhead Engineering website, http://www.thunderheadeng.com/downloads/petrasim/PetraSim-4-manual.pdf(accessed 24 May 2013).

7. PetraSim Example Manual, Thunderhead Engineering website, http://www.thunderheadeng.com/downloads/petrasim/4/examples/ PetraSimExamples.pdf(accessed 24 May 2013).

8. Workshop on Numerical Models for Carbon Dioxide Storage in Geological Formations, University of Stuttgart website, http:// www.iws.uni-stuttgart.de/CO2-workshop/(accessed 24 May 2013).

9. Ebigbo A., Nordbotten J.M., and Class H. Numerical Investigations of CO2 Sequestration in Geological Formations: Problem-Oriented Benchmarks, Problem 1, CO2Plume Evolution and Leakage through an Abandoned Well. R&D-Program of GEOTECHNOLOGIEN website, http://www.hydrosys.uni-stuttgart.de/co2-workshop/ Problem1. Pdf (accessed 24 May 2013).

10. Ebigbo A., Nordbotten J.M., and Class H. Numerical Investigations of CO2 Sequestration in Geological Formations: Problem-Oriented Benchmarks, Problem 2, Enhanced CH4 Recovery in Combination with CO2Storage in Depleted Gas Reservoirs. R&D-Program of GEOTECHNOLOGIEN website, http://www.hydrosys.uni-stuttgart.de /co2-workshop/Problem2.pdf(accessed 24 May 2013).

11. Ebigbo A., Nordbotten J.M., Class H. and Eigestad G. Numerical Investigations of CO2 Sequestration in Geological Formations: Problem-Oriented Benchmarks, Problem 3, Estimation of the CO2 Storage Capacity of a Geological Formation. R&D-Program of GEOTECHNOLOGIEN website, http://www.hydrosys.uni-stuttgart.de/ co2-workshop/Problem3.pdf (accessed 24 May 2013).

12. Class H., Ebigbo A., Rainer H., Dahle H.K., Nordbotten J.M., Celia M.A., Audigane P., Darcis M., Ennis-King J., and Fan Y. A Benchmark Study on Problems Related to CO2 Storage in Geologic Formations. Computational Geosciences 2009; 13409-434.

13. Bachu S., Nordbotten J.M., and Celia M.A. Evaluation of the Spread of Acid-Gas Plumes Injected in Deep Saline Aquifers in Western Canada as an Analogue for CO2 Injection into Continental Sedimentary Basins.In: 7thInternational Conference on Greenhouse Gas Control Technologies, Vancouver, Canada, September 2004.

14. Celia M.A., Bachu S., Nordbotten J.M., Gasda S., and Dahle H. Quantitative Estimation of CO2 Leakage from Geological Storage: Analytical Models, Numerical Models, and Data Needs. In:

7thInternational Conference on Greenhouse Gas Control Technologies, Vancouver, Canada, September 2004.

15. Nordbotten J.M., Celia M.A., and Bachu S. Injection and Storage of CO2 in Deep Saline Aquifers: Analytical Solution for CO2 Plume Evolution during Injection. Transport in Porous Media 2005; 58(3) 339-360.

16. Leetaru H.E., Frailey S.M., Damico J., Finley R., McBride J.H., and Morse D.G. Developing a Geological Model for the Phase III (ADM) Saline Sequestration Validation Site. In: 7th Annual Conference on Carbon Capture and Sequestration, Pittsburgh, US,May 2008.

17. Barnes D.A., Bacon D.H., and Kelley S.R. Geological Sequestration of Carbon Dioxide in the Cambrian Mount Simon Sandstone: Regional Storage Capacity, Site Characterization, and Large Scale Injection Feasibility, Michigan Basin. Environmental Geosciences 2009; 16(3) 163-183.

18. Midwest Regional Carbon Sequestration Partnership, CO2 Injection Test in the Cambrian-Age Mt. Simon Formation, Battelle Final Report, April 29, 2011.http://addap.tk/userdata/phase_II_reports/report_only/east_bend_report_final-mainreport.pdf (accessed 24 May 2013).

19. Zhou Q., Birkholzer J.T., Mehnert E., Lin Y., and Zhang K. Modeling Basin- and Plume-Scale Processes of CO2 Storage for Full-Scale Deployment. Ground Water 2010; 48(4) 494–514.

20. Zhang Z. and Agarwal R. Numerical Simulation of Geological Carbon Sequestration in Saline Aquifers - three case studies. In: 12th Annual Conference on Carbon Capture and Sequestration, May 13-16 2013, Pittsburgh, PA, USA.

21. Hovorka S.D., Benson S.M., Doughty C., Freifeld B.M., Sakurai S., Daley T.M., Kharaka Y.K., Holtz M.H., Trautz R.C., Nance H.S., Meyer L.R., and Knauss K.G. Measuring Permanence of CO2 Storage in Saline Formations—the Frio Experiment. Environmental Geosciences 2006; 13(2) 105–121.

22. Doughty C., Freifeld B.M., and Trautz R.C. Site Characterization for CO2 Geologic Storage and Vice Versa: The Frio Brine Pilot, Texas, USA as A Case Study. Environmental Geology 2008; 54 1635–1656.

23. Bachu S. and Bennion B. Effects of In-situ Conditions on Relative Permeability Characteristics of CO2-Brine Systems. Environmental Geology 2007; 54 1635–1656.

24. Chadwick R.A. and Noy D.J. History- Matching Flow Simulations and Time-Lapse Seismic Data from the Sleipner CO2 Plume. In: Proceedings of the 7th Petroleum Geology Conference, 2010; 7 1171–1182

25. Audigane P., Gaus I., Czernichowski-Lauriol I., Pruess K., and Xu T. Two-Dimensional Reactive Transport Modeling of CO2 Injection in a Saline Aquifer at the Sleipner Site. American Journal of Science 2007; 307 974-1008.

26. Singh V., Cavanagh A., Hansen H., Nazarian B., Iding M., and Ringrose P. Reservoir Modeling of CO2 Plume Behavior Calibrated Against Monitoring Data from Sleipner, Norway. Society of Petroleum Engineers 2010; 134891-PP.

27. Zhu C. CO2-Water-Rock Interaction in Geological Carbon Sequestration, Seminar presentation at Washington University in St. Louis, 2011.

28. Zhu C. and Lu P. Personal Communication, Department of Geological Sciences, University of Indiana, 2012.

Chapter 3

CO_2 Utilization: A Process Systems Engineering Vision

Ofélia de Queiroz F. Araújo[1], José Luiz de Medeiros[1] and Rita Maria B. Alves[2, 3]

[1] *Federal University of Rio de Janeiro, Brasil*

[2] *BRASKEM S.A., Brasil*

[3] *University of São Paulo, Brasil*

1. INTRODUCTION

The development of economies results in increased energy consumptions, as observed in Figure 1. In the coming decades, the supply of such expanding demand will remain based on fossil fuels technologies. Expanding the share of renewable energy (*e.g.*, biofuels in the case of transportation fuels) would require massive investments in creating a new infrastructure, which would eventually raise the standards to a new economic order entirely based on renewables in the near future. On the other hand, the current scenario involves the announcement of large proven reserves of non-conventional gas and oil and expansion of installed infrastructure of production and refining.

Fossil fuel based energy is recognized as carbon emitter. The challenge of the century is thus to expand energy supply in a carbon-constrained economy. According to the World Economic Forum (King, 2010), no truly low-carbon technology will be able to penetrate the mass market in the short term. The use of installed fossil processing infrastructure, with co-processing of biomass and fossil feedstock and capturing and utilization of emitted CO_2 is the "escape route" for a moderate transition from the present to a long-term sustainable future. In this context, putting a price on carbon will gradually build the road to a greener tomorrow. Meanwhile, bio-based products are a realistic supplement to fossil-based products, but they cannot mitigate the rising demand for fossil fuels.

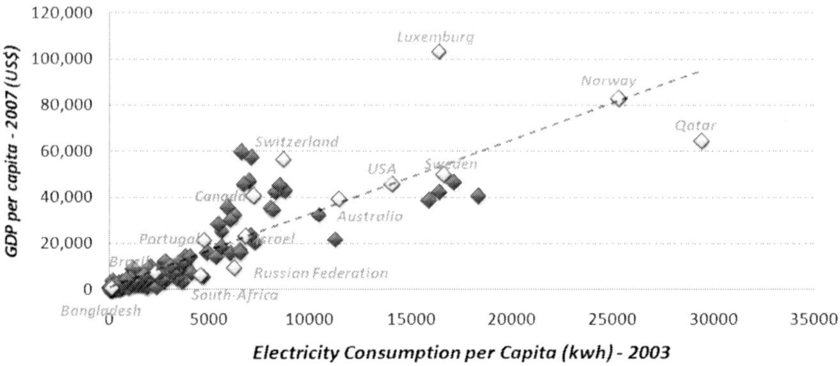

Figure 1. Economic development and electricity consumption. [graph constructed with data available at hdr.undp.org]

According to an IEA Technology Roadmap, the manufacture of only 18 chemicals account for 80% of the energy demand in the chemical industry and 75% of its greenhouse gas (GHG) emissions. The study concludes that, "in the short to medium term (to 2025), steady progress in implementing incremental improvements and deploying best practice technologies (BPT) could provide substantial energy savings and emissions reductions compared to business as usual". "A step change in the sector's energy consumption and GHG emissions would require the development of 'game changer' technologies, such as sustainable biomass feedstocks and hydrogen from renewable energy sources which have not yet reached commercial maturity." (IEA, 2013)

With this prospect, CO_2 utilization in the short term should allow parallel production routes based on BPT, driven by emission-capture-utilization synergies. In this sense, production and conversion of synthesis gas (SYNGAS) exhibits the highest potential for medium term of commercial success. Nevertheless, it is worth noting that, while the utilization of CO_2 has potential to reduce greenhouse gas emissions into the atmosphere, CO_2 has disadvantages as a chemical reactant due to its relative significant chemical inertness. This inertness is the underlying reason why CO_2 has broad industrial application as solvent (supercritical CO_2), as fire and pest extinguisher, and as a non-toxic amusement additive in the food industry. From the standpoint of building a low-carbon economy, each potential use of CO_2 as reactant has a customarily high energy requirement, entailing associated energy-related GHG emissions. Obviously, such GHG emissions should not exceed the yield of chemical conversion of CO_2. Reverse water gas shift (RWGS) and dry reforming to yield SYNGAS and CO_2 hydrogenation to methanol are the most prominent CO_2 conversion alternatives to high volume chemical commodities. On the other hand, it is reasonable to expect that CO_2 conversion and biomass-based processing alternatives will coexist for a while with fossil-based technologies. Figure 2 illustrates the concept of co-existing fossil and biomass feedstock, having SYNGAS generation as the integration phase. In this scenario, SYNGAS can be obtained from coexisting fossil and green feedstock via gasification (of biomass, coal and heavy residues), steam reform and dry (CO_2) reform of natural gas, which, in turn, will also coexist with downstream nonconventional conversion routes – e.g., Fischer-Tropsch (FT) and methanol to olefins (MTO) – to fuels and chemical intermediates to supply the installed petrochemical industries.

In countries with large bioresources as well as significant oil and natural gas production infrastructure (e.g., USA, Canada, Mexico, China, Russia, Australia, Argentina and Brazil), fossil feedstock (mainly nonconventional oil and gas) will present the greatest challenge to biomass. In Brazil, for instance, gas supply capacity in Santos Basin increased from 600 thousand m^3/d, in 2009, to 22.2 million m^3/d in 2013 (PETROBRAS, 2013). The identified risks associated to such expansion are, beside depletion of natural resources, increased CO_2 emissions.

Therefore, CO_2 capture, transportation and utilization must be included in the scene since CO_2 stands conceptually as a renewable feedstock. In this sense, Figure 3 illustrates CO_2 emissions capture and utilization associated to the production and refining of fossil fuel. It is worth noting that, in addition to the variety of alternative technologies, other factors influence the conception of capture and utilization of CO_2 for the production of chemicals.

Rostrup-Nielsen and Christiansen (2011) present four recognizable trends in next generation successful plants for producing chemical commodities: (a) location of cheap raw materials; (b) economy of scale; (c) highly integrated process plants; and (d) low CO_2 footprint (t CO_2/t product) Consequently, the production chain and the associated GHG emissions are complex and require a system view for optimal decision making on technologies for production of energy and chemicals taking into account the CO_2 perspective.

Chemical Process Systems Engineering (cPSE) deals with the set of basic unit operations involved in turning raw materials into products via chemical and bio-chemical processes. In a wider definition, PSE is concerned with the optimization of decision-making process for creation and operation of chemical supply chains. Such integrated framework embraces product and process design, manufacture and distribution of chemical products with multiple and conflicting technical, economic, energetic, environmental and social objectives.

With this vision, c-PSE of industrial use and reuse of CO_2 supply chain is assessed in a life cycle approach of technological alternatives involving integration of CO_2 capture (separation), CO_2 transportation and CO_2 chemical/biochemical utilization. Apart the conventional utilizations of CO_2 listed above, there are three new potential chains of CO_2 utilizations, namely:

(i) production of chemicals via chemical conversion (CC); (ii) enhanced oil recovery (EOR) and carbon geological storage (CGS); and (iii) conversion via algae and microalgae to biomass or biochemical conversion (BC).

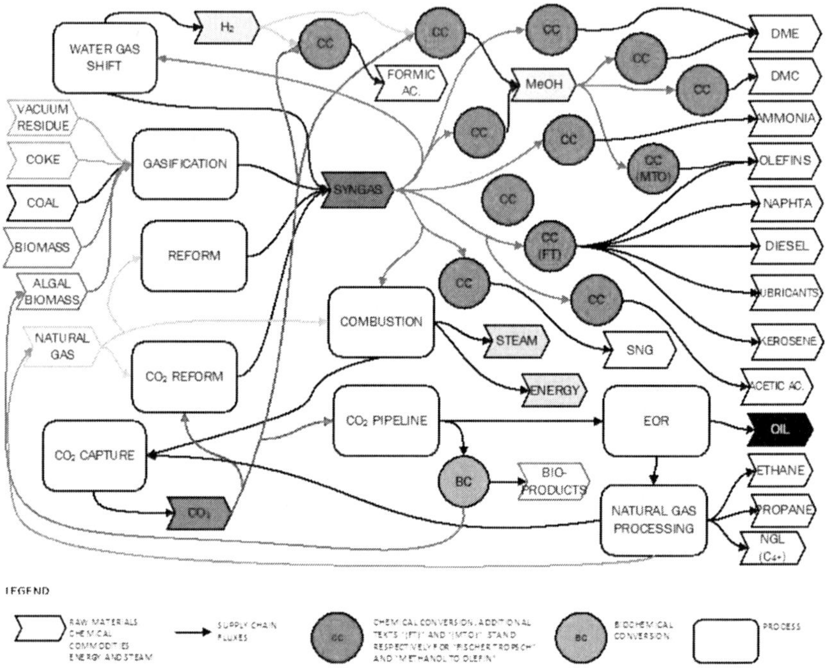

Figure 2. Coexistence of fossil and biomass feedstock uses: CO_2 capture and utilization dropped into existing production infrastructure

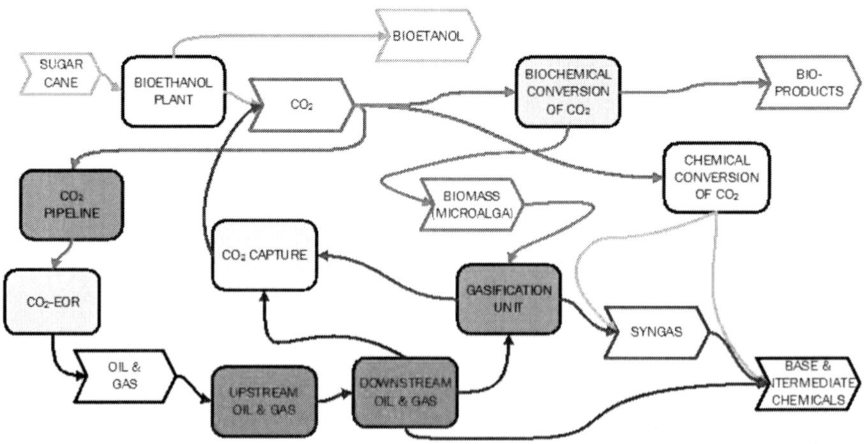

Figure 3. CO_2 capture and utilization in the oil & gas supply chain

EOR and CGS constitute, under present conditions, the only CO_2 processing chain that has some steps ready to be put into operation at high scales. These encompass some separation technologies for CO_2 post-combustion capture, CO_2 compression and CO_2 transportation via long pipelines. EOR and CGS can be reinforced if oxy-fuel technologies finally attain industrial maturity, which may occur within the next 5 decades. Pre-combustion technologies are also potential powerful contributors to EOR-CGS systems, which should be fully developed within the next 5 decades. Pre-combustion relies on some separation technologies also present in post-combustion alternatives.

BC via solar photosynthesis is a promising but still incipient package of technologies that have yet to be proven feasible in large scales. The main drawback is the impressive footprint and liquid hold-up of algae cultivation and processing plants, which have to comply with the high dispersion and periodicity of solar light and small concentration of biomass in growing medium (always below 10g/L). Besides, the use of non-solar light energy is totally out of question due to the photosynthesis efficiency limitations of green life forms (always below 10%, i.e., only a maximum of 10% of the incident light, already with appropriate wavelength, is biochemically usable to bio-convert CO_2).

CC context encompasses novel approaches extending the technology scenario for CO_2 supply chain via chemical conversion to benign, stable compounds for long-term storage or to value-added products like plastics, chemical intermediates and oxygenated octane enhancers like DMC – dimethyl-carbonate (Souza et al., 2013) for reuse.

The cPSE vision of CO_2 treats routes of CC and BC (and BC integrated with CC, i.e., BCC) to useable products and fuels, not as substitute, but rather as a complement to EOR and CGS. The present text is oriented to analyze CO_2 as a profitable feedstock, i.e., its potential use as feedstock to chemicals and fuels within economic applications, besides its relevant use in EOR. CGS is not considered an economic application; instead, it constitutes only a plausible, secure, destination of the excess of carbon, i.e., all CO_2 that has to be discarded because it is not dispatched to any economic use.

Among the routes for CO_2 utilization, BC is a natural choice as photosynthesis yields biomass, allowing the production of bioproducts, biofuels and chemicals through downstream CC processing routes. In fact, biomass gasification is the most flexible technology for dropping into conventional downstream CC routes. This integrated option configures BCC – biochemical and chemical conversion of CO_2.

The beneficial use of alkaline wastes or metallic ions to convert CO_2 via neutralization of alkaline wastes, or reaction of CO_2 with metallic ions to form less soluble carbonates that can be removed from produced water (oil & gas industry) is also a relevant CC application of CO_2. Lastly, this study presents the

EOR use of CO_2, i.e., by injection into depleting oil or gas fields to maximize hydrocarbon production.

Finally, analysis of routes that undertake CO_2 reduction must take into account the life cycle of the processes in order to assess whether additional CO_2 production occurs beyond the amount abated from atmospheric emissions. This is precisely the case of CC of CO_2 into fuels and chemicals that always requires high energy input, normally derived from burning fossil fuels, entailing further associated GHG emissions. In this context, new or mature cPSE solutions should comply with the triple objective of sustainability, namely: economically feasible, environmentally benign and socially beneficial, in a supply chain approach. In connection with this, the chapter presents the CO_2 Capture Cycle as well as promising alternatives of its reutilization.

2. THE THERMODYNAMICS OF PURE CO_2

The phase behavior of pure CO_2 (Figure 4) exhibits particularities when compared with common light species of natural gas (NG) like CH_4, C_2H_6 and C_3H_8. In general grounds, CO_2 has a fluid phase behavior similar to ethane (C_2H_6) with a very similar Critical Point (CP) temperature. The pronounced distinctive characteristic is its high Triple Point (TP) (Figure 4) temperature comparatively with light hydrocarbons. This means that solid CO_2 (dry ice) can be easily encountered below -56.6°C and above 5.2bar, if the original processing stream is rich enough in CO_2. This freeze-out of solid is not observed with the light hydrocarbon species unless below -182°C, which is about 20°C below the lowest temperature that can be achieved in LNG processing (i.e., the Normal Boiling Point of CH_4). In other words, without CO_2, the coldest NG processing (LNG plant) does not have solid formation. The phase behavior of pure CO_2 (Figure 4) is characterized by two larger (and infinite) continents corresponding to gas and solid states, and a finite intermediate liquid continent extending between the TP and CP temperatures. The three continents are two-dimensional (2D) objects due to the Phase Rule, which stipulates two degrees of freedom for a pure species at one-phase condition. One-dimensional (1D) equilibrium boundaries – SLE, SVE and VLE lines – are positioned between two neighboring continents. Their 1D nature is also a consequence of the phase Rule, which stipulates one degree of freedom for a pure species at a two-phase condition. SLE – solid-liquid (or solid-fluid) equilibrium line – is an endless line extending from the TP to indefinitely high pressures, characterized by the coexistence of solid and liquid CO_2. SVE – solid-vapor equilibrium line – is characterized by coexistence of solid and gas CO_2 ending at the TP and lower bounded by the absolute 0K. VLE – vapor-liquid equilibrium line – is a finite locus between TP and CP where liquid and gas CO_2 coexist. Just above the liquid continent there is a somewhat indefinite supercritical fluid zone (SCF). The SCF is a fusion of gas and liquid behaviors at high pressures above the CP pressure and high temperatures above the CP temperature. The SCF is characterized by high densities and high compressibility.

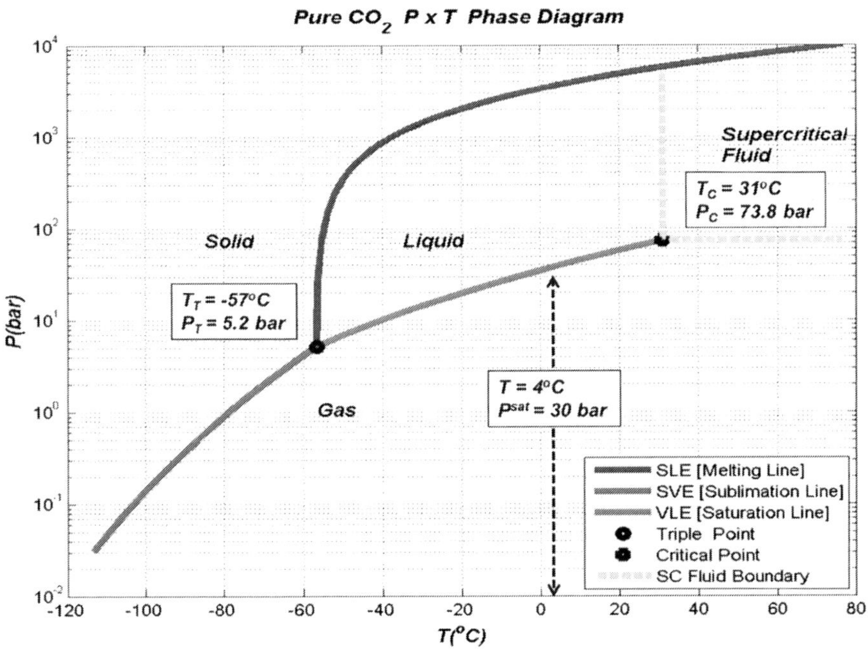

Figure 4. Phase behavior of pure CO_2 [T_C, P_C critical point; T_T, P_T triple point]

The contact point for the three continents is the Triple Point TP, an invariant three-phase point (zero degrees of freedom) by the Phase Rule. The VLE locus ends at the Critical Point CP where the differences between liquid and vapor vanish. The CP is also an invariant point with zero degrees of freedom, but, contrary to the TP, the CP is a single phase point satisfying two extra criticality conditions. The phase behavior of CO_2 depicted in Figure 4 can be enriched if density (kg/m^3) is put on the third axis as shown in Figure 5. In Figure 5, densities of gas phase are calculated via the Peng-Robinson (PR) equation of state (EOS), whereas liquid and solid phase densities (saturated or not) are calculated via the correlations presented by Span and Wagner (1996) and Trusler (2011), respectively. In Figure 5, the saturation lines SLE, SVE and VLE are also depicted, where the VLE line is presented with its gas and liquid branches merging at the CP. Figure 5 reveals that CO_2 can be found with densities well above the density of water (1000 kg/m^3), either as a solid or as high pressure dense liquid or SCF fluid.

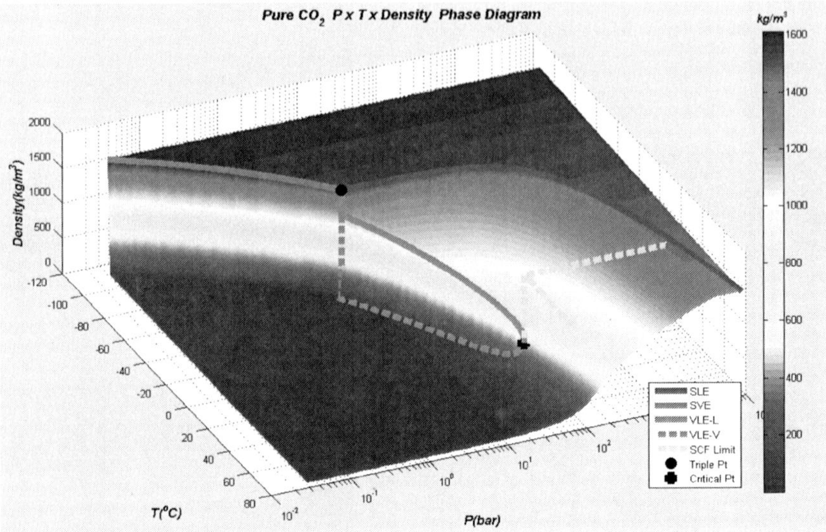

Figure 5. Density x Pressure x Temperature of Pure CO_2 [with TP and CP points, VLE, SVE & SLE lines and SCF domain]

The PR-EOS is a very simple thermodynamic relationship that can be used to predict gas and liquid properties of pure CO_2 and of CO_2 rich mixtures. The PR-EOS is certainly not the most accurate state relationship to address the fluid properties of CO_2 (Genesis, 2011; see also the EOS presented by Span and Wagner, 1996), but it does represent the best compromise between simplicity of use and accuracy for CO_2 rich systems without water, either in single fluid phase or in two coexisting fluid phases (VLE), at low or high pressures and densities (Li, 2008; Li and Yan, 2009; Genesis, 2011). According to Genesis (2011), the PR-EOS produces errors for compressibility factor (Z), molar enthalpy (H) and sound speed (C) below 1% for CO_2 subcritical gas phases and between 5% and 20% for subcritical liquid phases. Near the CP, the errors in Z, H, C can be higher but the uncertainties of experimental values are also higher. In the SCF domain, the PR errors for Z, H, C fall between 1% and 15%. In all the aforementioned cases, errors are relative to Span-Wagner EOS for fluid CO_2. The PR-EOS also seems adequate to describe the supercritical fluid (SCF) domain of CO_2 and its rich mixtures near critical transitions. This means that PR-EOS can be used to address any property (density, enthalpy, entropy, exergy, etc) of liquid and vapor phases of CO_2 and its mixtures with NG species under moderate errors, which are compatible with engineering applications (Li, 2008 and Li and Yan, 2009; Genesis, 2011). The PR-EOS is presented in Eq. (1), while its classical mixing rules follow in Eqs. (2) and (3). PR-EOS component parameters are given in Eq. (4) to (6) where Tc_i, Pc_i, ω_i and nc are, respectively, critical temperature, critical pressure, acentric factor for component i, and the number of components. K_{ij} represents the binary interaction parameter (BIP) for species *i* and *j* which is symmetric, can be used as zero in some cases, and is not necessary for pure CO_2. In Eqs. (1) to (3), *V, T,P,R* and N_i represent volume (m^3), temperature (K), pressure (bar), ideal gas constant ($8.314.10^{-5}$ bar.m^3/mol.K) and the mol number of species *i*.

$$P=\frac{NRT}{V-Nb}-\frac{N^2a}{V^2+UVNb+W(Nb)^2}\quad,\ U=2,\ \ W=-1 \tag{1}$$

$$Nb=\sum_{i=1}^{nc} N_i b_i \tag{2}$$

$$N^2a=\sum_i^{nc}\sum_j^{nc} N_iN_j\sqrt{a_i}\sqrt{a_j}\sqrt{\Phi_i(T)}\sqrt{\Phi_j(T)}\left(1-K_{ij}\right)\ ,\ (K_{ij}=K_{ji}) \tag{3}$$

$$a_i=0.45724\frac{R^2Tc_i^2}{Pc_i}\ ,\ \ b_i=0.07780\frac{RTc_i}{Pc_i} \tag{4}$$

$$\Phi_i(T)=\left(1+g(\omega_i)\left(1-\sqrt{T/Tc_i}\right)\right)^2 \tag{5}$$

$$g(\omega_i)=0.37464+1.54226\omega_i-0.26992\omega_i^2 \tag{6}$$

If dimensionless terms in Eq. (7) are used in Eq. (1), the classic *Z* cubic form results in Eq. (8):

$$Z=\frac{PV}{NRT}\ ,\ B=\frac{PNb}{NRT}\ ,\ A=\frac{PN^2a}{(NRT)^2} \tag{7}$$

$$Z^3 - (1 + B - UB)Z^2 + (A + WB^2 - UB - UB^2)Z - AB - WB^2 - WB^3 = 0$$

(8)

3. THE CO_2 CAPTURE CYCLE

According to Oi (2010), CO_2 removal from process streams at an industrial scale has occurred since about 1930, mainly from natural gas and from industrial gases at high pressures for ammonia and methanol production. Like any separation unit in any kind of process, the CO_2 capture technology has to be judiciously chosen as it may undermine the profitability, controllability, safety, and simplicity of the plant. Nevertheless, even when properly selected, separations usually raise concerns like heat and mechanical energy consumption, increased utility use, carbon emission, chemicals demands, size, weight, footprint, construction restraints, operational hazards, etc.

The CO_2 capture cycle encompasses two or three main unit operations, which have to separate CO_2 from the gas mixture and send it to an appropriate destination. First, there is the CO_2 transfer across the gas phase into the medium that contains the binding material: a solvent or an adsorbent or a selective barrier. Second, there is (or not) the regeneration of the binding medium with concomitant CO_2 release. Third, there is the compression and cooling of the captured CO_2 because it has to be handled at high density or as a liquid.

Technologies for CO_2 capture from gas streams include chemical absorption (e.g., aqueous ethanolamines and aqueous K_2CO_3), physical absorption (e.g., propylene carbonate, selexol and rectisol), physical adsorption, membrane permeators, membrane contactors, cryogenic distillation and hybrid technologies (e.g., membrane permeator followed by ethanolamine absorption). Among those, the most relevant technologies include membrane permeation and chemical absorption with aqueous ethanolamines, the later standing as the most mature CO_2 capture technology from gas streams.

3.1. Chemical absorption with aqueous alkanolamines

With regards to the solvent, the advantages of the chemical absorption of acid gases – CO_2 and H_2S – by aqueous alkanolamines are well-known: the former are weak acids and the later are weak alkali, such that they reversibly bind at low temperature and high acid gas fugacity and subsequently untie at higher temperatures and low fugacity, leading to efficient acid gas stripping, which regenerates the solvent. The relevant variable in the liquid phase is the solvent loading (in mol of acid gas per mol of amine), which expresses the degree of conversion of amine in the solvent. Typically, the loading assumes values in the range 0 to 1.2 mol/mol amine.

AGWA (Acid Gas, Water and Amines) systems is a convenient denomination (de Medeiros et al., 2013a; de Medeiros et al., 2013b) of such reactive vapor-liquid equilibrium (RVLE) systems containing Acid Gas, Water and Amines. Amines are understood to be the common alkanolamines like monoethanolamine (MEA, MW=61), diethanolamine (DEA, MW=105), methyl-diethanolamine (MDEA, MW=119) and 1-amino-2-propanol (AMP, MW=75).

MEA is a benchmark co-solvent for CO_2 capture, with: (i) satisfactory absorption capacity; (ii) fast kinetics; (iii) miscible with water in all proportions, and (iv) low cost. On the other hand, MEA is problematic in terms of solvent regeneration, because it exhibits: (i) the highest energy load per unit of stripped gas; (ii) corrosion and chemical/thermal degradation concerns; (iii) non-negligible evaporation losses due to its low boiling point; and (iv) high reactivity entailing low H_2S/CO_2 selectivity. As a secondary amine, DEA is less reactive than MEA but it has the following comparative advantages: (i) lower energy per unit of stripped gas; (ii) more resistant to degradation; (iii) less corrosive; and (iv) less volatile. MDEA exhibits the lowest reactivity with CO_2 and the greatest resistance to degradation. When compared to DEA and MEA, MDEA presents the following advantages: (i) the highest equilibrium capacity of acid gas absorption (in the case of CO_2, nearly two times the capacity of primary amine); (ii) lowest regeneration cost (does not form stable products with CO_2); (iii) high H_2S/CO_2 selectivity thanks to its low reactivity with CO_2; (iv) lowest enthalpies of reaction and lowest regeneration heat per unit of stripped gas; and (v) negligible losses due to very low vapor pressure. Alkanolamines with steric hindrance like AMP show reduced carbamate stability and the methyl adjacent to the amine group may affect absorption capacity and/or its rate. AMP exhibits absorption capacity and stripping heat similar to MDEA, but a faster reaction with CO_2 during the capture step (Medeiros et al., 2013b).

AGWA chemical reactions are really three-reactant transformations in the liquid phase with 1:1 amine to acid gas mol ratio. Water is necessary at high mol ratio to amine (8:1 or 7:1), otherwise the reaction simply does not evolve sufficiently, leaving non-solvated amine unconverted (low loading). This is the reason why MEA concentration in the solvent is upper bounded at 30%w/w in water (or 11.2%mol MEA + 88.8%mol H_2O), and 50%w/w in the case of MDEA (or 13.1%mol MDEA + 86.9%mol H_2O). Figure 6 presents a schematic of the chemical absorption system for CO_2/H_2S capture. AGWA absorption reactions take place in the colder higher pressure absorption column, whereas in the hotter lower pressure stripper column, AGWA reactions are reverted, thus releasing free acid gas and regenerating amine at the bottom. The main consumption of heat occurs in the reboiler in the bottom of the stripper column, where water is vaporized breaking the liquid phase association of acid gas, water and amine. Consequently, typical heat consumptions of MEA strippers lay between 167kJ and 200kJ per mol of stripped CO_2 (or between 3.8 GJ and 4.5 GJ/tonne of stripped CO_2). These figures are impressive high values, equaling 4 to 5 times the molar heat of vaporization of water per mol of stripped CO_2.

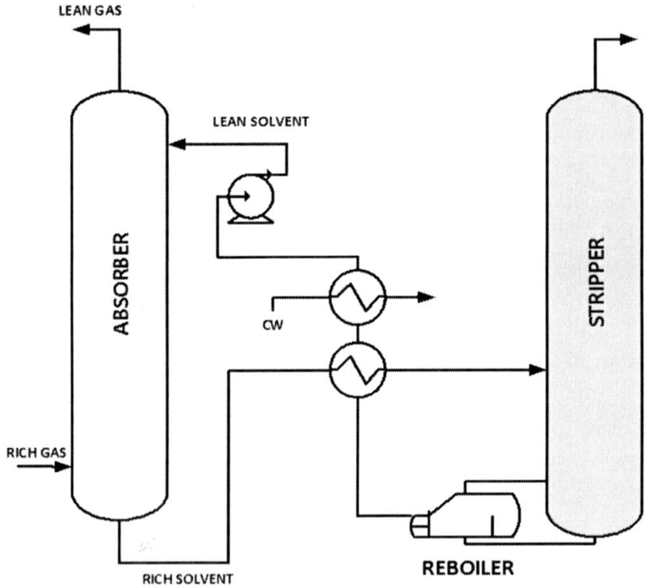

Figure 6. CO_2 capture by chemical absorption

The literature presents several modeling approaches for absorption and stripping with AGWA systems. The most common approach involves cumulatively ionic species within ideal solution, ideal gas vapor phase, reversible chemical kinetics and rate-based interfacial mass transfer (de Medeiros et al., 2013b). This kind of approach is classical and is more adequate to low pressure and to dilute AGWA systems as in CO_2 capture from combustion gases. For high capacity and high pressure AGWA with rich CO_2 natural gas systems, high loadings and high heat effects may appear. For such AGWA systems de Medeiros et al. (2013a, 2013b) proposed a molecular Chemical Theory approach where molecular complex species are formed in the liquid phase via chemical equilibrium reactions having as reactants real AGWA species CO_2, H_2S, H_2O, MEA, DEA, MDEA and AMP. Each complex species is created by reacting three real species: an acid gas, water and an amine, as shown in Eq. (9). These complexes are reversibly created during the absorption step and destroyed during the stripping step. The main advantages of the molecular chemical theory of de Medeiros et al. (2013b) are:

i. Theory is a Reactive Vapor-Liquid Equilibrium (RVLE) framework that employs only VLE AGWA data for model tuning via implicit statistical procedures. Model can be tuned with low or high pressure data, appropriate either for the absorption column (low temperature, high pressure) or for the stripping column (high temperature, low pressure).

ii. AGWA RVLE is, in fact, an ionic scenario dominated by weak ions and several unknown solvated complexes. Thus, the limited knowledge to handle all the possible ions is circumvented via such nonionic, nonvolatile complexes.

iii. Model tuning requires only AGWA VLE data, much more available than non-equilibrium counterparts.

iv. The nonionic assumption is proposed to mimic the weakness of AGWA electrolytes, all created by incomplete dissociations.

v. All thermodynamics properties (of both vapor and liquid phases) in the formalism of de Medeiros et al. (2013b) are calculated via cubic EOS (Soave-Redlich-Kwong - SRK and Peng-Robinson - PR), because all species considered are purely molecular.

vi. The formalism can be used with both acid gases CO_2 and H_2S and with blends of alkanolamines, which are used for conjugating desirable qualities like good reactivity of MEA with lower costs of regeneration and corrosion of MDEA. Such blends are promising in terms of gains relatively to individual amines.

The AGWA formalism of de Medeiros et al. (2013b) was calibrated with VLE AGWA data from Literature. These data configure a large database of AGWA-VLE with 1331 runs shown in Figure 7. This database includes several runs with alkanolamines (MEA, DEA, MDEA and AMP), water and two acid gases or solutes (CO_2 and H_2S) at pressures ranging from 0.1 bar to 30 bar and temperatures from 25°C to 140°C. The calibration parameters correspond to chemical equilibrium constants belonging to chemical reactions that convert acid gas, water and amine into molecular complex species in liquid phase as shown in Eqs. (9). These chemical reactions are chemical equilibrium (ChE) reactions, which evolve to the right when CO_2 and H_2S are absorbed by the solvent and to the left when CO_2 and H_2S are stripped from the solvent by the action of heat and low pressure. In other words, the set of chemical equations in Eq. (9) can reproduce both phenomena absorption and solvent regeneration. This is a physically sound approach that allows, among other things, to estimate heat effects that occur in these processes like the release of heat during gas absorption and the absorption of heat during stripping of CO_2 and H_2S.

$$
\begin{aligned}
&CO_2(g)+H_2O(g)+MEA(g)\xleftrightarrow{K_1}CO_2-H_2O-MEA(Liq)\\
&CO_2(g)+H_2O(g)+MDEA(g)\xleftrightarrow{K_2}CO_2-H_2O-MDEA(Liq)\\
&CO_2(g)+H_2O(g)+DEA(g)\xleftrightarrow{K_3}CO_2-H_2O-DEA(Liq)\\
&CO_2(g)+H_2O(g)+AMP(g)\xleftrightarrow{K_4}CO_2-H_2O-AMP(Liq)\\
&H_2S(g)+H_2O(g)+MEA(g)\xleftrightarrow{K_5}H_2S-H_2O-MEA(Liq)\\
&H_2S(g)+H_2O(g)+MDEA(g)\xleftrightarrow{K_6}H_2S-H_2O-MDEA(Liq)\\
&H_2S(g)+H_2O(g)+DEA(g)\xleftrightarrow{K_7}H_2S-H_2O-DEA(Liq)\\
&H_2S(g)+H_2O(g)+AMP(g)\xleftrightarrow{K_8}H_2S-H_2O-AMP(Liq)
\end{aligned}
\tag{9}
$$

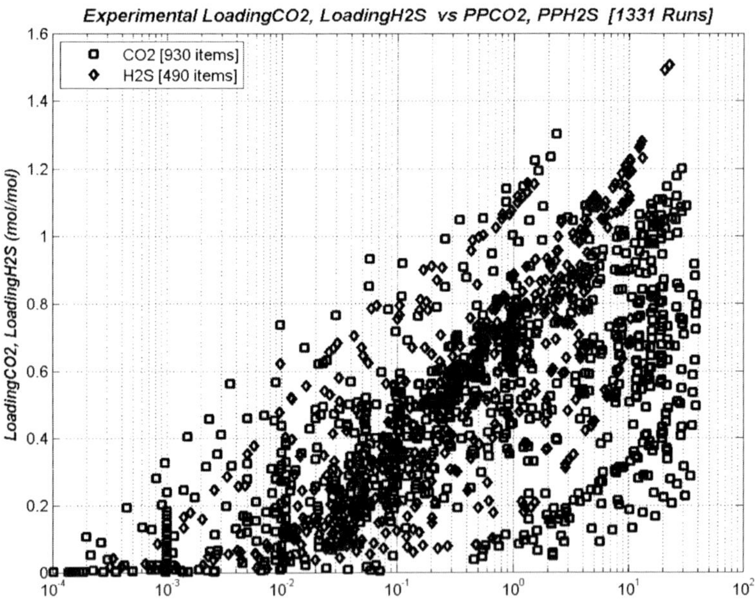

Figure 7. AGWA Database used to estimate parameters in the Chemical Theory Model of de Medeiros et al. (2013b) [PPCO2, PPH2S – partial pressures (bar) of CO_2 and H_2S]

The parameter estimation of the AGWA model of de Medeiros et al. (2013b) involves estimating ChE constants for the chemical reactions in Eq. (9) at selected temperatures. Thus, several experimental AGWA VLE points at a chosen temperature are extracted from the AGWA database in Figure 7 and are processed via a maximum likelihood algorithm to estimate the ChE constants. The estimation algorithm also assures that the set of nonlinear constraints of each selected experiment is satisfied. The set of constraints of a given experiment is shown in Table 1 with Eqs. (10) to (16) including: (i) Real species – CO_2, H_2S, H_2O, MEA, DEA, MDEA, AMP – mass balance (RMB); (ii) mol fraction normalization for each phase (SXY); (iii) definition partial pressures of solutes (acid gases) (PPS); (iv) definition of loadings of solutes (LDG); (v) VLE of Real species (VLE); (vi) ChE of Complex formation in Eq. (9) (CHE). The nomenclature used in Table 1 is shown in Table 2, including the matrix of stoichiometric coefficients of Eq. (9) in Eq. (10).

Figures 8 and 9 depict some results after calibration of the AGWA model of de Medeiros et al. (2013b) with the database of AGWA experiments in Figure 7. Both figures refer to parameter estimation using 320 experimental AGWA data points at 40°C. Figure 8 presents experimental versus predicted loadings (mol/mol amine) of solutes CO_2 and H_2S at 40°C, whereas Figure 9 presents experimental versus predicted partial pressures (bar) of solutes CO_2 and H_2S at 40°C. As can be observed, there is a satisfactory agreement between experimental values and predicted counterparts.

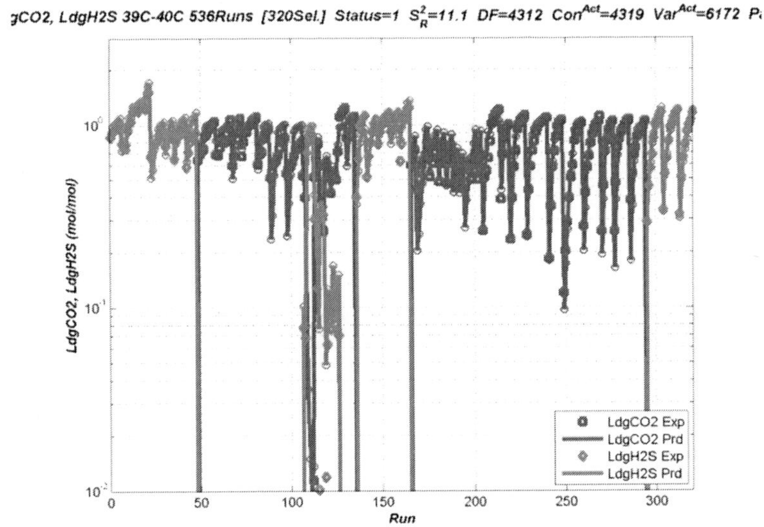

Figure 8. Experimental vs predicted Loadings of CO_2 (blue) and H_2S (red) after calibrating the AGWA model at 40°C with 320 data points (de Medeiros et al., 2013b) [LdgCO2, LdgH2S – Loadings of CO_2 and H_2S (mol/mol amine)]

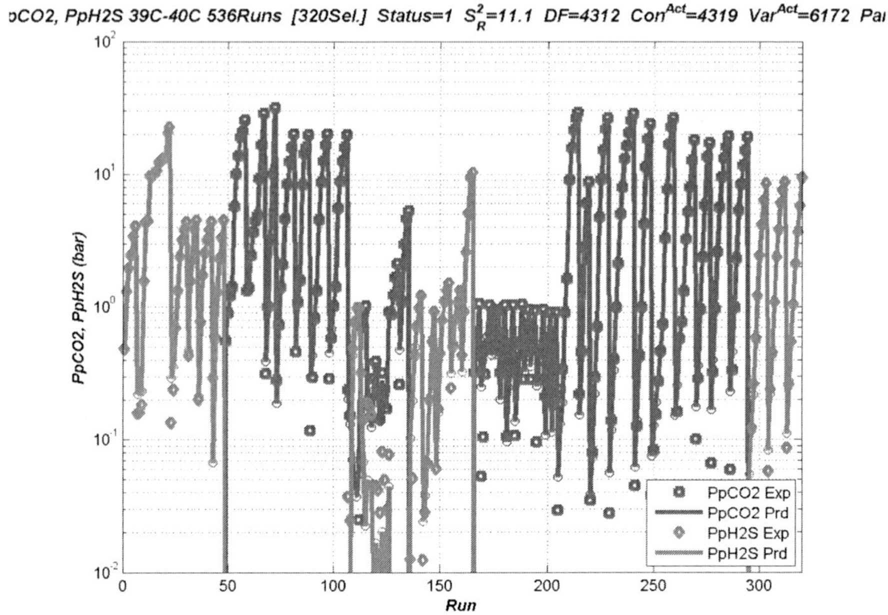

Figure 9. Experimental vs predicted partial pressures of CO_2 (blue) and H_2S (red) after calibrating the AGWA model at 40°C with 320 data points (de Medeiros et al., 2013b) [PpCO2, PpH2S – Partial pressures of CO_2 and H_2S (bar)]

Table 1. Set of Constraints of AGWA VLE Experiments [real species (*n=7*): CO_2, H_2S, H_2O, MEA, DEA, MDEA, AMP], [solutes (*ns=2*): CO_2, H_2S], [Complexes (*nr=8*): see Eq. (9)]

Constraints [Label]	Expression	Eq.
n Real Species Balances [RMB]	$\underline{N} + L\underline{\underline{\Pi}}\,\underline{X}_C - L\underline{X} - V\underline{Y} = \underline{0}$	I
Normalizations [SXY]	$\sum_{j=1}^{n} X_j + \sum_{k=1}^{n_C} X_{Ck} - 1 = 0$	II
	$\sum_{j=1}^{n} Y_j - 1 = 0$	III
n_S Solute Partial Pressures [PPS]	$\underline{P}_S - P\underline{\underline{S}}_S\underline{Y} = \underline{0}$	V
n_S Solute Loadings [LDG]	$(\underline{1}^T \underline{\underline{S}}_A \underline{N})\underline{\alpha}_S - \underline{\underline{S}}_S(\underline{N} - V\underline{Y}) = \underline{0}$	VI
n VLE of Real Species [VLE]	$\ln \underline{\hat{f}}^V - \ln \underline{\hat{f}}^L = \underline{0}$	VII
n_r Reactions ChE [CHE]	$\ln \underline{X}_C + \underline{\underline{\Pi}}^T \ln \underline{\hat{f}}^L - \ln \underline{K}(T) = \underline{0}$	VIII

Table 2. Nomenclature used in Table 1

Symbol	Definition
$\underline{\alpha}_S$	Vector of solute loadings (mol/mol of amine)
$\underline{\hat{f}}^V, \underline{\hat{f}}^L$	Vectors of Real species fugacities in vapor and liquid phases (bar)
$\underline{K}(T)$	Vector of ChE constants of Eq. (9)
L	Liquid phase mole number
n_S, n, n_r	Numbers of solutes (=2), Real species (=7) and Complex species (=8)
$\underline{N}$	Vector of total mole number of Real species in the experiment
$\underline{P}_S$	Vector of solute partial pressures (bar)
P	Pressure (bar)
$\underline{\underline{\Pi}}$	Matrix of stoichiometric coefficients of Eq. (9), shown in Eq. (17).
$\underline{\underline{S}}_A, \underline{\underline{S}}_S$	Selection matrices for amines and solutes with sizes 4 X 7 and 2 X 7
V	Vapor phase mole number
$\underline{X}, \underline{Y}$	Vectors of mol fractions of Real species in liquid and vapor phases
$\underline{X}_C$	Vector of mol fractions of Complex species in liquid phase

$$\underline{\underline{\Pi}} = \begin{bmatrix} -1 & -1 & -1 & -1 & 0 & 0 & 0 & 0 \\ 0 & 0 & 0 & 0 & -1 & -1 & -1 & -1 \\ -1 & -1 & -1 & -1 & -1 & -1 & -1 & -1 \\ -1 & 0 & 0 & 0 & -1 & 0 & 0 & 0 \\ 0 & -1 & 0 & 0 & 0 & -1 & 0 & 0 \\ 0 & 0 & -1 & 0 & 0 & 0 & -1 & 0 \\ 0 & 0 & 0 & -1 & 0 & 0 & 0 & -1 \end{bmatrix} \quad (10)$$

Predicted values are calculated solving the set of constraints of the AGWA model in Eqs. (I) to (VIII) with all thermodynamic properties (e.g. vapor and liquid fugacities) estimated with SRK-EOS. The estimated parameters are the eight ChE constants in Eq. (9) and (VIII) at the corresponding temperature of 40°C.

The huge heat consumption of MEA strippers (167 kJ to 200kJ/mol of stripped CO_2) forces investigation on more efficient alternatives of solvent regeneration. Wagener and Rochelle (2010) recognize as a “monumental task” reducing energy penalty of CO_2 capture from coal-fired power plants (approximately 30%). They presented an analysis of various stripper configurations, concluding that increasing complexity improves performance at the cost of higher capital and operational expenditures, i.e., an optimal scheme should exist. The alternative configurations include simple stripping with vapor recompression, multi-pressure, double matrix, split product, internal exchange, and flashing feed. Wagener and Rochelle (2010) concluded that operating with multiple pressure levels reduces the energy requirement as "equivalent work" (including reboiler duty, pumping and heat exchangers) in 33.6 kJ/molCO_2 captured, with an optimal lean loading of 0.375 mol/mol amine. Moreover, they claim that the arrangement benefits from stripping at high pressure, whilst improves reversibility when returning to atmospheric conditions for the absorber. Wagner and Rochelle (2011) revisited several configurations with varying levels of complexity and reported that an inter-heated column and a simple stripper required 33.4 kJ/mol CO_2 and 35.0 kJ/mol CO_2 of equivalent work, respectively, at their optimum lean loadings.

3.2. Membrane permeators

The first membrane modules were developed as planar films. However, such arrangement has a low ratio of membrane transfer area per equipment volume. Presently, the majority of modules for CO_2 gas separation are manufactured in hollow-fiber or spiral wound configurations. Figure 10 shows a schematic of a hollow-fiber membrane permeator module and the selective transport of molecules across the hollow-fiber membrane.

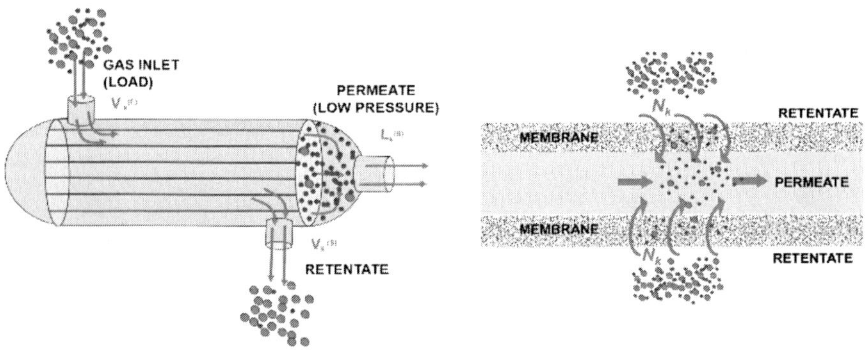

Figure 10. Hollow-fiber membrane permeator and the transfer through a hollow-fiber

The membrane acts selectively against gas diffusion from the high pressure side (the retentate) to the low pressure side (the permeate) creating separation. The membrane – either in hollow-fiber or spiral wound configurations – is composed by two polymeric structures: a permselective dense (nonporous) skin over a thicker microporous substrate. Inside the permeator gas coming from the high pressure retentate stream solubilizes into the dense skin and diffuses to the low pressure side creating an almost adiabatic expansion with consequent fall of temperature via Joule-Thomson effect. Such fall of temperature may create problems to the proper functioning of the module due to eventual condensation of less volatile species that are retained and accumulate in the retentate. This condensation is undesirable and can damage the membrane plastic material.

Ho and Wiley (2005) evaluated the economic performance of membrane separation for low pressure (flue gas applications) and high pressure (e.g., natural gas processing). The authors concluded that the highest share in capital cost was due to the compression phase (around 80%) while membrane and the respective shells exhibited 10% of the investment. For natural gas processing, however, membrane has the largest share of CAPEX (about 62%), as the compression is an existing stage of natural gas processing. Moreover, Ho and Wiley also report an alternative multi-stage configuration in order to obtain a CO_2 rich permeate in a second membrane stage, which demands a recompression step corresponding to 30% of the total CAPEX.

In oil and natural gas deep-water offshore rigs, an issue in the processing of large volumes of NG with high CO_2/CH_4 ratio *vis-à-vis* climate concerns, involves large capacity CO_2 separators, whose targets are exportation of saleable NG from the retentate product via long pipelines and feeding enhanced oil recovery (EOR) systems with hyperpressurized CO_2 from the permeate product. In this scenario, robustness, angular indifference, modularity, and compactness also influence the selection of a separation technology. In this context, membrane permeation batteries are usually favored against the more traditional absorption columns with amine solvents (Medeiros et al., 2013a). On the other hand, large permeation batteries also have their own shortcomings, mostly related to the permselective dense (nonporous) skin over the microporous substrate. The permselective skin forces the existence of a high ΔP and exhibits low fluxes which means low capacity per unit area and high consumption of power for permeate recompression and/or permeate recycles. It also has CO_2/CH_4 limited selectivity and demands feed necessarily dew-point conditioned (to prevent gas condensation in the retentate provoked by the decrease of temperature associated with permeation) and continuous inspection looking for membrane bursts.

A model for hollow-fiber permeators can be built by writing mass, momentum and energy conservation principia for both permeate and retentate. Additionally, an appropriate model for trans-membrane flux transport is also necessary. Typically, trans-membrane flux models are written as products of a permeancy coefficient and a driving force term. Each species has a permeancy coefficient for a given membrane and given conditions of temperature, pressure and gas composition, but, in general, permeances are assumed independent of gas composition and pressure. The driving force term is usually expressed as a difference of component fugacities between the retentate and permeate sides. In this context and assuming a co-current compressible flow of both permeate and retentate, Nakao et al. (2009) proposed a stationary hollow-fiber membrane permeator rigorous model which can be used to predict CO_2 separation from CO_2 rich NG. The model of Nakao et al. (2009) builds a spatial 1-dimensional axial description where species mass balances, energy and momentum balances are rigorously written for both permeate and retentate. All permeate and retentate thermodynamic properties (enthalpies, densities, fugacities, etc) are rigorously calculated via PR-EOS, Eq. (1). The model is too involved to be discussed here in detail, but the clarity and usefulness of it results are worth presenting in the context of CO_2 capture. Figure 11 depicts a natural gas feed named GBS2 (from Basin of Santos, Brazil) with 2.79554 $MMNm^3/d$ at 53bar and 38°C with, initially, 25.32%mol CO_2 and 55.64%mol CH_4. Figure 11 depicts the VLE locus of GBS2 and the heat capacity (*Cp* in kJ/mol.K) map versus *T X P* with vivid identification of the critical and supercritical neighborhoods due to the second-order transition across the critical state that is revealed by second order properties like *Cp*. In Figure 11 [A], the gas feed is located at 28°C above its dew boundary (blue). GBS2 passes through a battery of 40 horizontal modules (0.2m X 10m each) of cellulose-acetate hollow-fibers (0.5mm ID), with CO_2 and CH_4 permeances respectively of $1.27.10^{-8}$ and 4.4. 10^{-10} $mol/s.m^2.Pa$ and $1854m^2$ of permeation area per module. Permeate pressure is 4bar and the external temperature is 27°C with a heat transfer coefficient of $5W/m^2.K$. Resulting profiles in the axial flow direction (10m long) for one module are shown in Figure 12. Permeate and retentate initiate contact at axial position *X=0m* and cease contact at *X=10m*. The background along the axis of flow is painted in four colors for discrimination among the four quarters of a typical module. Figure 12 [A] depicts retentate profiles of %mol CO_2 and CH_4 showing a CO_2 decrease from 25.62% to 9.3%mol and a CH_4 increase from 55.64% to 65.5%mol. Figure 12 [B] shows that the final permeate recoveries of CO_2 and CH_4 are 72% and 11.5%. Figure 12 [C] depicts permeate and retentate temperature profiles revealing a drop of retentate temperature from 38°C to 29°C, insufficient for condensation (Figure 11 [A]).

3.3. Gas-liquid contactors

Gas–liquid contactors (GLC) constitute a new and versatile kind of membrane operation for CO_2 removal from NG. A GLC unit admits a gas stream which is contacted with a liquid stream (the solvent stream) through a polymeric micro-porous membrane. The solvent phase is an aqueous solution of ethanolamines (e.g. MEA and/or MDEA) that can absorb CO_2 as occurs in an absorption column. But the difference here is that the liquid and gas phases really do not mix as they do in direct contact devices like packing towers. Assuming that the GLC is manufactured with hollow-fiber membranes, the solvent phase flows in the hollow-fiber inner space while the gas phase flows in the outside shell space, but there are also configurations where the roles of liquid and gas phases are inverted. A module of hollow-fiber GLC with gas flowing in the shell side and with solvent flowing in the inner space of the hollow-fibers is sketched in Figure 13. As can be seen, the module resembles closely a shell and tube indirect contact heat exchanger.

GLC membrane operation can outperform common nonporous permeators in terms of capacity per unit area while sustaining high CO_2/CH_4 selectivity. High fluxes are possible in GLC because the membrane does not have a dense skin in order to be selective. The underlying reason is that selectivity is imposed by the solvent in the inner (permeate) side cutting the necessity of high ΔP across the membrane. As the reader can see, terms permeate and retentate are also used here despite some improperness, and refer, respectively, to the inner solvent flow and to the gas flow that was not transferred to the inner solvent.

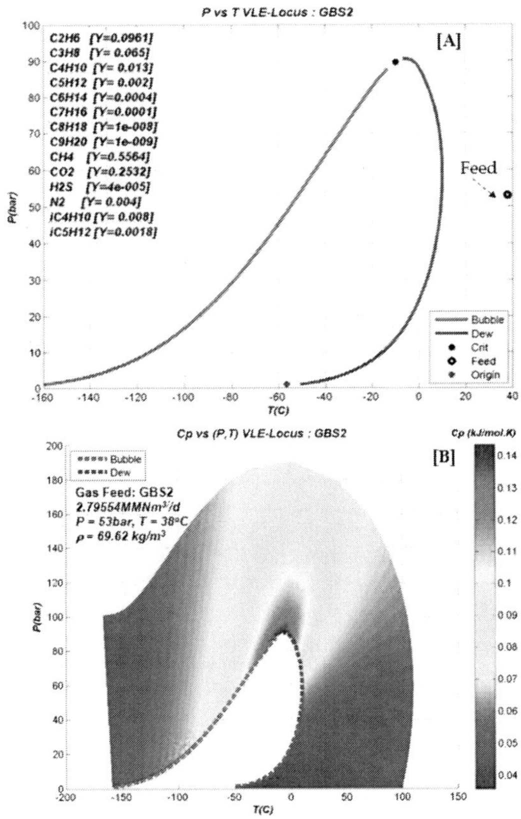

Figure 11. GBS2 a CO_2 rich natural gas (25.32%CO_2,55.64%CH_4,9.61%C_2H_6, 9.43%C_3^+) for Hollow-Fiber Permeator: [A] VLE Locus *P*(bar) X *T*(°C); [B] C_P(kJ/mol.K) vs *P*(bar) X *T*(°C)

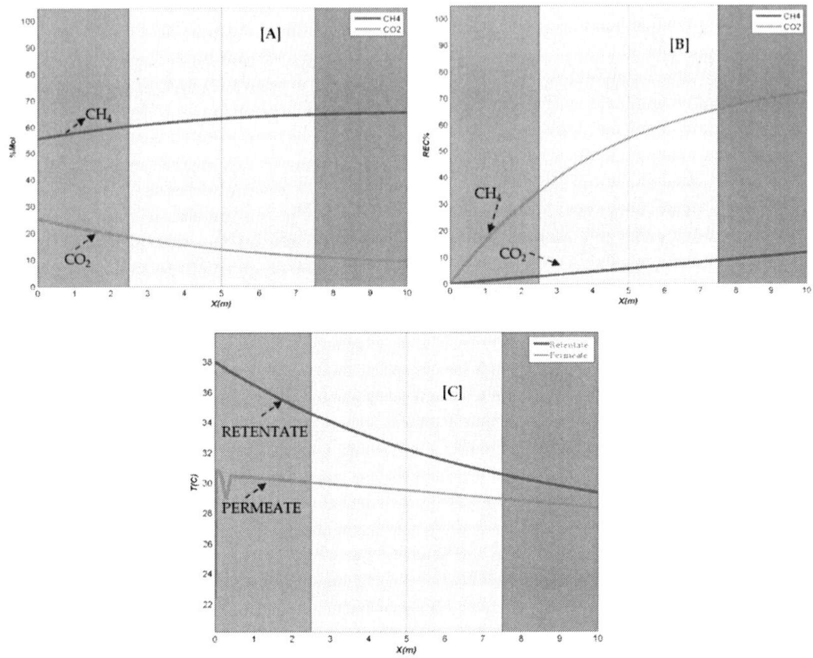

Figure 12. Profiles in a hollow-fiber module in co-current flow with feed GBS2 in Fig. 3.6 (25.32% CO_2 & 55.64% CH_4): [A] CO_2 & CH_4 %mol in retentate; [B] CO_2 & CH_4 permeate % recoveries; [C] retentate and permeate temperatures.

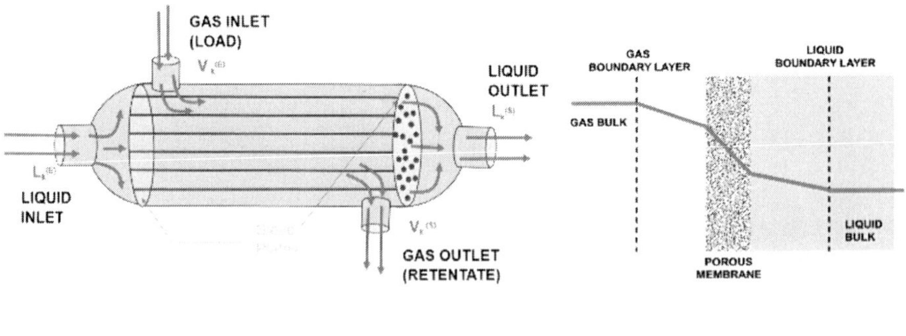

Figure 13. Hollow-fiber gas-liquid contactor with solvent in the inner space of membrane

GLC can separate CO_2 from NG offering the advantages of both membrane and absorption technologies, but leaving behind the respective shortcomings like the low flux coexisting with high difference of pressure between the membrane sides and the flooding concerns coexisting with dependence on gravity in the case of absorption columns. GLC combines polymeric membrane separation and chemical absorption, using a physically and chemically active solvent for selective CO_2 removal. However, the new aspect is that there is a membrane standing as a physical barrier against the unnecessary mixing of gas and liquid phases. Other advantages of GLC are: (i) it allows independent manipulation of liquid and gas flows; (ii) offers larger area of gas–liquid interface; (iii) modularity ; (iv) it exhibits flexibility to increase/decrease operational scales; (v) no dew-point conditioning of the gas feed is necessary; and (v) angular indifference allowing horizontal or vertical operational arrangements. Figure 14 sketches a typical process flowsheet for operation of a GLC unit capturing CO_2 from a NG feed.

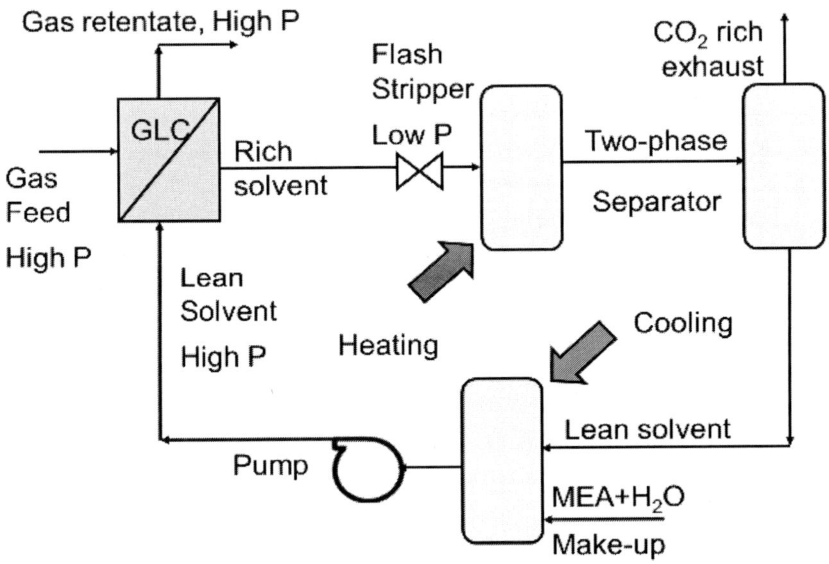

Figure 14. Typical process flowsheet for CO_2 capture with gas-liquid contactor (GLC)

Here the GLC operates t high NG pressure, but the rich solvent has to be regenerated in a low pressure stripper where CO_2 is released by the action of heating and low pressure as occurs in the second regeneration column shown in Figure 6.

A model for stationary gas-liquid hollow-fiber contactor was proposed by de Medeiros et al. (2013a) for separating CO_2 from CO_2 rich natural gas. This model is based on the AGWA theory discussed in the sub-section 3.1 (de Medeiros et al., 2013b). The model assumes a hollow-fiber contactor with co-current compressible flows of both permeate and retentate, where the permeate corresponds to the inner space inside the hollow-fibers. Permeate and retentate are separated by the membrane. The permeate is supposed in two-phase flow because the trans-membrane flux of methane will support the maintenance of a gas phase in the inner permeate flow. The permeate is, in fact, a two-phase reactive flow in continuous reactive vapor-liquid equilibrium (RVLE) because CO_2 is in reactive vapor-liquid equilibrium with water and amine via a set of chemical equations similar to Eq. (9) (without the H_2S chemical equations if the GLC is designed for CO_2 capture only).

The model of de Medeiros et al. (2013a) is based on 1-dimensional axial geometry of the GLC. Species mass balances and momentum/energy balances are written for both RVLE permeate and gas retentate. All properties of gas and liquid phases are calculated with PR-EOS. The trans-membrane flux terms are written as products of permeancy coefficients and a driving force term. Each species has a permeancy coefficient for a given membrane and given temperature conditions. The driving force term is expressed as a difference of component fugacities between the retentate and RVLE permeate sides. The major difficulty encountered in the GLC model is to represent the RVLE two-phase flow in the inner membrane space. This is accomplished by solving the AGWA VLE model (de Medeiros et al., 2012b) described in Table 1, along the entire path of the permeate in the inner space of the hollow-fibers. As an example, consider the equilibrium map in Figure 15.

This map was calculated assuming that 1 mol of natural gas (with 16.7%mol CO_2 +82.3%mol CH_4) is contacted with 1.2 mol of liquid solvent containing 14.5%mol MEA +14.5%mol MDEA +71%mol H_2O. Only the chemical reactions of CO_2 absorption by MEA (Eq. 9a) and MDEA (Eq. 9b) are considered. Each point in Figure 15 represents the resulting equilibrium vapor phase mol fraction of CO_2 (Y_{CO2}) versus T(°C) X P (bar) under reactive VLE. Clearly, the locations with low Y_{CO2} (blue) are dominated by absorption phenomena, while those with high Y_{CO2} (red) are dominated by stripping phenomena. This kind of inner RVLE solver is a key element in the construction of the GLC model of de Medeiros et al. (2013a).

3.4. The CO_2 transportation cycle

An efficient and reliable transportation system is required to displace enormous quantities of captured CO_2 to their destination site at appropriate geological formations that are capable to accommodate hundreds of billions or trillions of Nm^3 of CO_2 under stable and secure conditions. Clearly a not especially large size thermoelectric coal plant can produce something like 1 million of metric tonnes (1Mt) of CO_2 per year or about 500.10^6 Nm^3/y. Such huge capacities can only be attended by large pipeline systems operating with dense compressed fluid (liquid or dense supercritical CO_2), because the other existing alternatives – road, railroad and barge transport – simply cannot cope with dense fluid pipelines in terms of unitary cost and capacity of transportation.

In other words, despite of the existence of many options for transporting compressed (gas or liquid) CO_2 – including highway tankers, railway tankers, ships, and pipelines – it is evident that the impressive volumes that must be transported dictates that only pipelines working at high pressures, high densities and high capacities are suitable for this service. For instance, 2-3Mt/y of CO_2 have to be transported to dispose of the entire production of a single 500MW coal-fired power plant. This corresponds to transporting 230-350t/h of CO_2, just to service a single, medium-sized, client. Thus, only a network of large-scale pipelines could provide viable overland transport of such massive flow rates of CO_2. Presently, about 50Mt/y of CO_2 (equivalent to the output of 16 coal-fired power plants) are transported by 3100km of CO_2 pipelines, mainly for EOR processes in the USA and Canada (de Medeiros et al., 2008). The best example is the 808 km long, 30" diameter, Cortez Pipeline that transports 13Mt/y of CO_2 from highlands in Colorado to oilfields in Texas, USA.

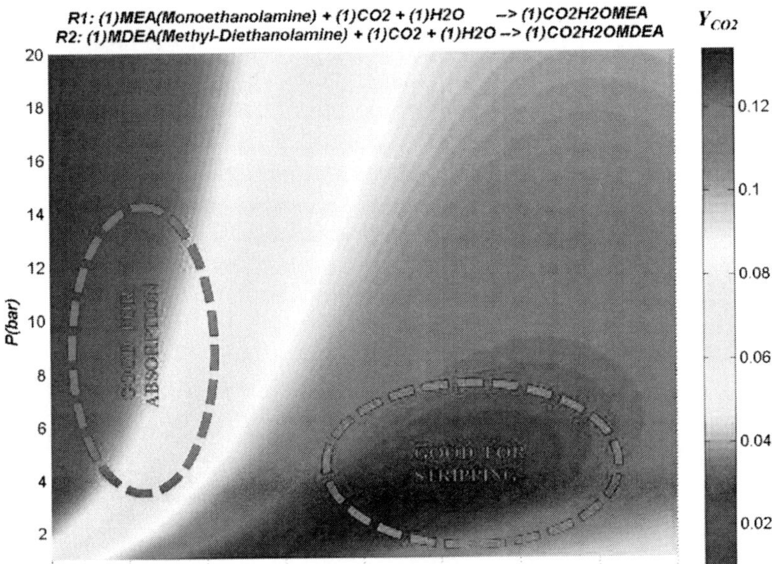

Figure 15. Equilibrium vapor phase mol fraction of CO_2 (Y_{CO2}) versus T(°C) X P (bar) under reactive VLE: 1mol of a 16.7%mol CO_2 natural gas is contacted with 1.2 mol of a liquid solvent with 14.5%mol MEA, 14.5%mol MDEA and 71%mol water.

Based on historical capital and O&M data for a 480-km long CO_2 pipeline without booster compressors, McCoy (2008) projected a fixed O&M coefficient of \$3,250/y/km for CO_2 pipelines. Considering a horizontal pipeline without appreciable elevation changes and an annualized fixed cost of 15% of capital, McCoy estimated the total unitary cost of CO_2 transportation as only \$1.16 per tonne of CO_2 per 100km. Based on a Monte Carlo sensitivity analysis, McCoy (2008) determined a range of \$0.75 to \$3.56 per tonne of CO_2 per 100km for this cost, recommending the median value of \$1.65 per tonne per 100km as a suitable estimate for investment decisions.

The design of CO_2 pipelines depends on reliable compressible flow models for dense compressible fluid near critical conditions. This model should account for thermal compressibility effects inside the fluid, i.e., temperature increases (decreases) during downhill (uphill) flow due to gravity compression (expansion).

In the same way, external heat transfer and elevation effects must be allowed. The extremely high compressibility of CO_2 near its critical state at 31°C (shown in Figures 4 and 5), leads to potential abrupt changes of velocity due to abrupt changes of density as the fluid compresses (decompresses) near the critical state. In other words, any candidate model for high capacity CO_2 pipelines must be able to calculate thermodynamic properties of dense supercritical CO_2 near its critical transition with accuracy. Such a CO_2 pipeline model has been proposed by de Medeiros et al. (2008) by solving rigorous species mass balances and energy/momentum balances along the pipeline with all thermodynamic properties given by PR – EOS or SRK – EOS.

This model is demonstrated in the simulation of a sub-sea CO_2 pipeline (Figures 16, 17 and 18) for transportation of 20 $MMSm^3/d$ of CO_2 from onshore plant to five EOR wellheads 2100m deep, 380km from the coast. The 20'' pipeline extends 380km from west to east and 320km from south to north with 700 km of length.

As seen in Figure 17, a problematic factor was inserted between the continental shelf and the continental slope, namely, a big rift 1500m deep lies in the pipeline route. Rifts are not common in Santos Basin, Brazil, but they exist in the Norway arctic coast (Pettersen, 2011).

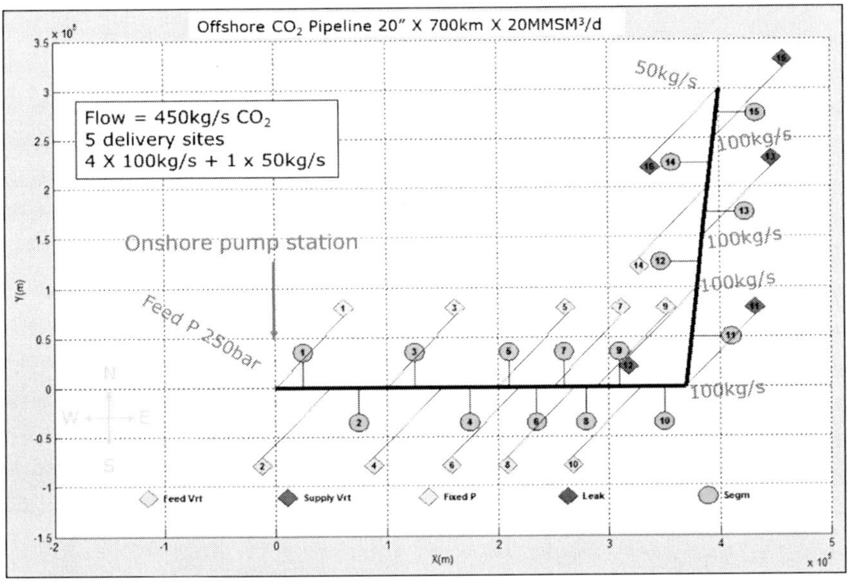

Figure 16. Hypothetical 20"X 700 km sub-sea pipeline for 20MMSm³/d of liquid CO_2 (≈450kg/s) at 250 bar from onshore plant to five offshore EOR wellheads (4X100kg/s + 1X50kg/s) 2100m deep, 380km from the coast.

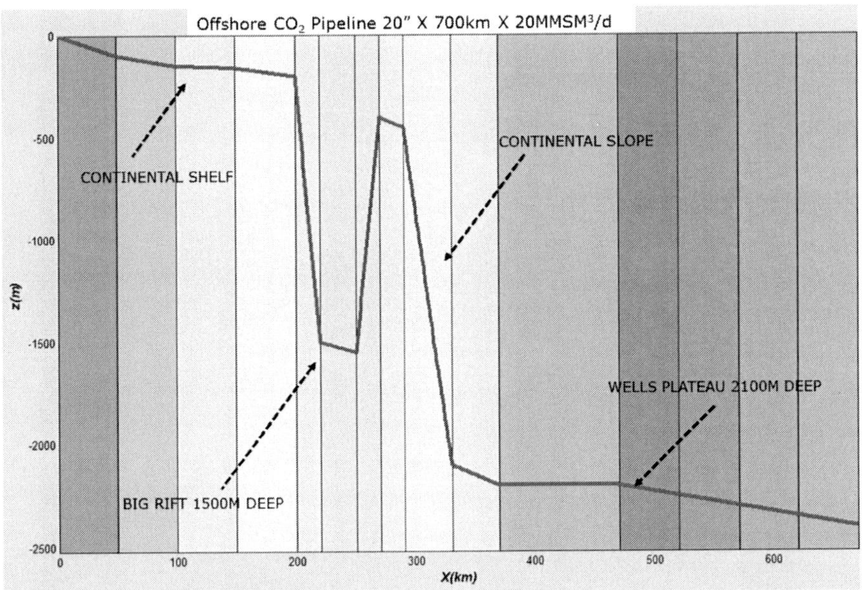

Figure 17. Elevation profile for the hypothetical sub-sea CO_2 pipeline from onshore plant to five EOR offshore wellheads 2100m deep, 380km from the coast.

4. THE CO_2 UTILIZATION CYCLE

Although CGS has been regarded worldwide as a mitigation technology, it deals with CO_2 as a waste with an energy and an economic penalty for its disposal (Armstrong, 2012). Rather than treating CO_2 as a waste, carbon dioxide utilization (CDU) recognizes it as a raw material in chemical process to produce high added-value carbon containing products. It is also worth noting that the CDU is a complementary technology to CGS, not a competing technology, adding value to a process and thus it may help balance the costs of CGS.

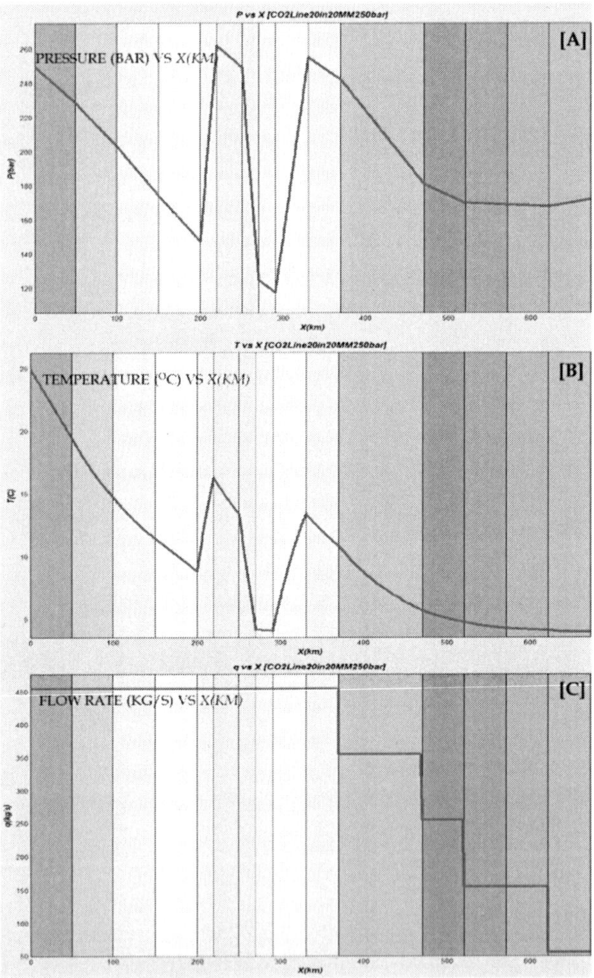

Figure 18. Calculated profiles for the hypothetical 20"X 700 km sub-sea pipeline with 20MMSm^3/d (≈ 450kg/s): [A] pressure (bar); [B] temperature (°C); [C] mass flow rate (kg/s)

While CO_2 has broad industrial application as solvent (supercritical CO_2, fire extinguishers) and in the food industry, it has disadvantages as a chemical raw material due to its low reactivity and few reactions are thermodynamically feasible. Furthermore, each potential use of CO_2 as reactant has an energy requirement that needs to be determined and must not exceed the CO_2 utilized and, although the utilization of CO_2 has been subject of research since before 1970´s, there is much research still needed for CO_2 activation. Moreover, the utilization of CO_2 to cause an effective reduction in its emission into the atmosphere, must observe certain guidelines (Aresta, 2010): (i) the new process must reduce the overall CO_2 emissions; (ii) it mustbe less energy - and material - intensive with respect to the on–stream processes that it aims to replace; (iii) the new process must reduce the overall CO_2 emissions; (iv) it must employ safer and more eco-friendly working conditions; (v) it needs to be able to operate on a commercial scale and (vi) it must be economically viable.

According to Song (2006), the global market for CO_2 is estimated to be \$3.2 billion/year in 2003. Carbon markets across the world were valued at 96 billion euros (\$122.28 billion) in 2011 (Reuters Agency, 2012). Utilization of CO_2 by the chemical industry exists for more than one century as, for instance, the synthesis of urea (50 Mt/y) (Aresta, 1999) (Eq. 11), salicylic acid (Song, 2006) (Eq. 12) and inorganic carbonates (20Mt/y) (Aresta, 1999).

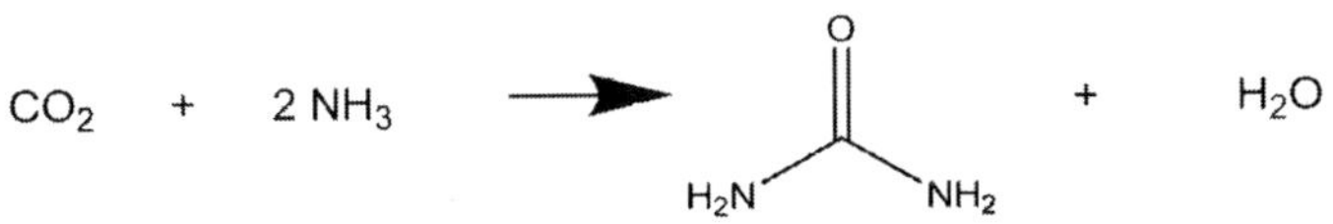

(11)

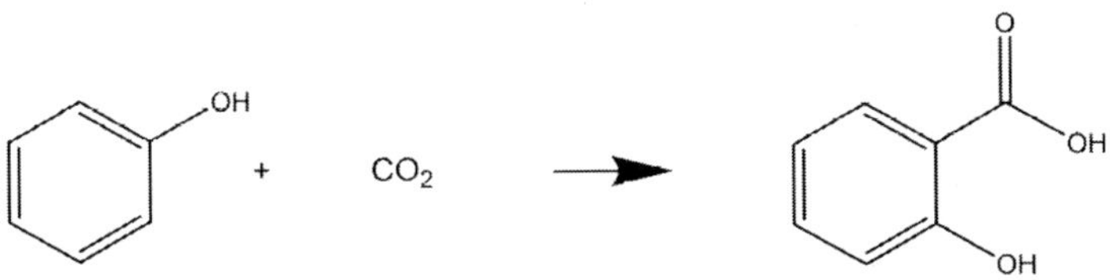

(12)

According to Song (2006), the worldwide production of urea in 2002 was about 110 million metric tonnes, which contains 51.8 million metric tonnes of nitrogen with an estimated value of US\$11.5 billion. This corresponds to about 81 million metric tons of CO_2, and 22 million metric tonnes of carbon.

4.1. Thermodynamic and chemical considerations of CO_2 conversion

Chemical reactions are driven by the difference in Gibbs free energy between the products and reactants at certain conditions. The obstacle for utilizing CO_2 as feedstock to industrial processes is its low energy level - CO_2 is a highly stable molecule. Consequently, a substantial input of energy, effective reaction conditions, and often catalysts, are necessary for its chemical conversion. In other words, many reactions for CO_2 conversion involve positive change in enthalpy, requiring an energy input. There are many large-scale chemical industrial processes that are operated based on endothermic reactions in the chemical industry (e.g., ammonia Haber process).

Song (2006) states properly that it is more energy-demanding if one were to use only CO_2 as a single reactant, but it becomes easier in thermodynamically terms if CO_2 is used as a co-reactant with another substance that has higher Gibbs free energy, such as CH_4, graphite and H_2. Song (2006) illustrates this trend by the change in the reaction heat for reactions with CO_2 as the single reactant (Eq. 13) and with CO_2 as a co-reactant (Eq. 14).

$$CO_2 \longrightarrow CO + \tfrac{1}{2}\,O_2 \qquad \Delta H^0 = +\,293\,\frac{kJ}{mol\,CO_2} \tag{13}$$

$$CO_2 + H_2 \longrightarrow CO(g) + H_2O \qquad \Delta H^0 = +51\,\frac{kJ}{mol\,CO_2} \tag{14}$$

Therefore, energy input is necessary to transform CO_2 into chemicals. Four methods are possible: (i) reaction with high-energy molecules (e.g., ethylene oxide, H_2, unsaturated compounds and organometallic compounds); (ii) reaction with low energy oxidized compounds (e.g., organic carbonates), (iii) shifting chemical equilibrium towards products (via removal of a reaction product) and (iv) supplying physical energy (e.g., light or electricity) (Sakakura et al., 2007). The appropriate selection of chemical reactions makes it possible to obtain a negative Gibbs energy change.

As the carbon of the carbonyl group has an electron deficiency, CO_2 has great affinity for nucleophilic compounds and electron donors, i.e., as an anhydrous carboxylic acid, it promptly reacts with basic compounds. For instance, organometallic compounds, such as Grignard compounds, react promptly with CO_2 even at low temperatures. Reactions with CO_2 can be divided into two groups: (1) formation of a carboxylic group via a nucleophilic attack and (2) oxidative cycle addition yielding a ring of 5 members (Sakakura et al., 2007).

A relevant aspect to be considered is that utilization of CO_2 as feedstock does not necessarily contribute to the mitigation of greenhouse effects, even though CO_2 stands as a green reactant in many cases (Sakakura et al., 2007). Three points are hence relevant:

a. The chemical (or biochemical) fixation of CO_2 does not necessarily imply in reducing CO_2 emissions as its transformation requires energy, both to drive reaction (high temperatures and pressures) and separate products (separation occurs mainly at low pressures and, hence, recycling unreacted CO_2 to the reactor will require recompression at the expense of high energy input);

b. The energy demand of the world is order of magnitude higher than the amount of CO_2 fixed by chemical utilization of CO_2, and

c. In the critical phase of its life cycle, organic chemicals will emit CO_2. Nevertheless, the relevance of CO_2 as raw material stands for being a renewable feedstock, substituting the conventional fossil based routes

Among other utilization, CO_2 is currently used as supercritical solvent, refrigerant fluid, beverage carbonation agent, inert medium (such as fire extinguisher), pressurizing agent, neutralizing agent, gas for greenhouses, "inerting" applications to inhibit unintended chemical reactions, welding (preventing atmospheric oxygen from reacting with molten metal), food processing (suppressing aerobic bacterial activity for preservation in processes like pneumatic conveying or food storage). In any of these applications for inerting, carbon dioxide serves as a cover against atmospheric oxygen and is thus implicitly released into the atmosphere. The carbonation of beverages accounts for around 1.0×10^6 t CO_2/y. Nevertheless, they do not constitute CO_2 sink as it is ultimately released to the atmosphere or remains in a closed loop (Ormerod et al., 1995). Hence, this study does not review such utilizations.

4.2. Supply chain considerations of CO_2 conversion

Bayer (2013) estimates that the chemical industry has over 40,000 final chemicals, produced from approximately 400 intermediate chemicals, derived from ~40 basic chemicals that, in turn, are based on 4 classic feedstocks: petroleum, natural gas, coal and biomass. The Company expands the set of feedstock with the inclusion of CO_2.

For the near and middle term time-period, i.e., next one or two decades, it is reasonable to assume that presently dominant technologies (from an economic standing point) will persist and, consequently, expanding economies´ demand of energy will be met by present technologies. Consequently, GHG emission of chemical processes will expand. This same approach leads to a transition scenario to a low-carbon economy equally dominated by presently installed infrastructure.

Hence, the CO_2 utilization cycle is likely to rely on commercially mature technologies or on technologies presently in large-scale pilot or demonstration plants as *Bridge Technologies*. Therefore, only peripheral technological advances are expected in these technologies such as process intensification, enhanced selectivity and activity of catalysts, and process optimization with increased mass and energy integration.

It is worth noting that 5 chemical commodities, ammonia, methanol, ethylene, propylene and BTX dominate energy consumption and GHG emissions in the chemical industry (IEA, 2013). In the conception of co-processing of fossil feedstock, with biomass and CO_2, conversion routes to produce ammonia, methanol and olefins (e.g., ethylene and propylene) are considered (Figure 2). Although not included in Figure 2, catalytic fast pyrolysis of biomass can lead to the key aromatic compounds, Benzene, Toluene, and Xylene (BTX), with generation of paraxylene from the BTX and subsequent conversion to Purified Terephthalic acid (PTA) and PolyEthylene Terephthalate (PET).

Synthesis gas (SYNGAS), a mixture of hydrogen, carbon monoxide and CO_2, is a versatile intermediate feedstock used in the production of a number of hydrocarbons such as methanol, ammonia, synthetic hydrocarbon liquids, and as a source of pure hydrogen and carbon monoxide. Applications of these products range from petrochemical feedstock to fuels.

According to Rostrup-Nielsen and Christiansen (2011), trends in the use of SYNGAS are dominated by the conversion of inexpensive remote natural gas into liquid fuels ("gas to liquids" or "GTL") and by a possible role in a future "hydrogen economy" mainly associated with the use of fuel cells. Some relevant synthesis to the chemical industry are:

SYNGAS to Ammonia: Ammonia serves as a building block in many pharmaceuticals, fertilizers, ethanolamines, urea and cleaning products, as well as an anti-microbial agent in food processing. 50% of the world´s food production relies on ammonia fertilizers.

SYNGAS to Methanol: The main use for methanol is to produce other chemicals; about 40% is converted to formaldehyde, and further processed into plastics, plywood, paints, explosives and textiles. It is also used in anti-freeze, solvents, and fuels, and can serve as energy carrier.

SYNGAS to Hydrogen: Hydrogen generation is one of the largest energy-consuming steps in the production of the crucial chemical precursors of ammonia and methanol.

SYNGAS to Synthetic Fuels: Liquid hydrocarbons exhibit an excellent volumetric energy density and offer various opportunities for storing electric energy (Kaiser et al., 2013). Kaiser et al. (2013) point generation of SYNGAS by reverse water-gas shift (RWGS) at elevated temperatures as the first step, followed by Fischer-Tropsch (FT) synthesis. If CO is substituted by CO_2, less synthetic fuels are formed, the water-gas shift is repressed, and methane selectivity increases.

4.3. CO_2 to SYNGAS

SYNGAS is a toxic, colorless and odorless mixture. Its efficient commercial production is gaining significant attention worldwide (Raju et al., 2009) as it is a versatile feedstock to produce a variety of fuels and chemicals.

Almost any carbon source ranging from natural gas and oil products to coal and biomass can be used in the SYNGAS production. The lowest cost routes for its production, however, is natural gas (Spath and Dayton, 2003), which is also the cleanest of all fossil fuels. Furthermore, steam methane reforming is a well-established process for the production of SYNGAS and hydrogen (Gangadharan et al., 2012). The H_2/CO ratio varies over a wide range, depending on the primary feedstock and technology employed. Particular SYNGAS ratios are required depending on the chemical product desired, therefore creating flexibility for the chemical industry.

In the twofold context of avoiding emissions and standing as a renewable feedstock, carbon dioxide has been investigated as raw material in SYNGAS production. The new technologies involves CO_2: (i) reforming processes using a hydrocarbon (methane, typically) as reducing agent; (ii) using CO_2 as a co-reactant with hydrogen in the catalytic reverse water gas shift (RWGS); (iii) thermocatalytic (solar assisted) routes; (iv) electro- or photo-catalysis; (v) plasma processes, and (vi) bio-processes, e.g., by hybrid enzyme-nanoparticle systems, bioelectrochemical reduction or using a biomass char and a catalyst such as Ni/Al_2O_3.

CO_2 Reforming of CH_4 (Dry Reforming): The dry (carbon dioxide) reforming of methane has been of interest for a long time, dating back to as early as the 1920s, and was first proposed by Fischer and Tropsch (1928), but it is only in recent years that interest in it has rapidly increased for both environmental and commercial reasons (Zhang et al., 2003). Its name derives from the fact that CO_2 replaces steam in the conventional steam methane reforming process (Hartley & Tam, 2012). This reaction can be represented as

$$CH_4 + CO_2 \longrightarrow 2CO + 2H_2 \qquad \Delta H = +247 \frac{kJ}{mol} \tag{15}$$

According to Hartley and Tam (2012), dry reforming utilizing CO_2 produces synthesis gas with higher purity and lower H_2/CO ratio than either partial oxidation or steam reforming. The produced SYNGAS has an H_2/CO ratio of unity without further post-reformer reactions (Zhang et al., 2003). The interest in this reforming route in recent years (Treacy and Ross, 2004, Shi et al., 2013) is due to two main reasons: (i) it produces SYNGAS with a H_2/CO molar ratio that is suitable for a variety of products including Fischer–Tropsch fuels and (ii) the reaction consumes two types of greenhouse gases, CO_2 and CH_4 (Zhang et al., 2003, Gangadharan et al., 2012). Moreover, SYNGAS production stands as the most promising alternative of CO_2 utilization as it presents flexibility of using installed infrastructure to the manufacture of important chemical commodities.

The biggest limitation to the dry reforming process appears to be the availability of a suitable catalyst. The high temperatures required to reach high conversions, due to the endothermic nature of the process, contribute to carbon deposition (both CO_2 and CH_4 give off carbon deposits), and a catalyst capable of operating at such severely deactivating conditions has not been well developed (Zhang et al., 2003, Shi et al., 2013). According to Shi et al (2013), from the viewpoint of GTL industry, developing a catalyst for CO_2 reforming of CH_4 is a challenge, because the catalyst must exhibit very high production rates to render the GTL methane reformer as small as possible. Nevertheless, progress in the development of suitable catalysts with higher activities and optimized lifetime stabilities have been reported (Bradford and Vannice, 1999, Souza and Schmal, 2003; Zhang et al., 2003, Ginsburg et al., 2005; Kahle et al., 2013; Shi et al., 2013; Zheng et al., 2013; Edwards, 1995; Wurzel et al., 2000; Nagaoka et al., 2001; Li et al., 2004). Nevertheless, there is still no process for the CO_2 reforming currently considered to be commercially feasible. However, a variation of dry reforming has been used industrially (Hartley and Tam, 2012). The CALCOR process (Teuner, 1985, Kurz and Teuner, 1990, Teuner et al., 2001) involves dry reforming of methane, optimized to reduce the hydrogen content of the product gas. Furthermore, hydrogen separation by membrane permeators produces hydrogen gas that combusts a fuel (e.g., methane) producing pure carbon monoxide. The SPARG process (promotion by poisoning) is also a dry reforming reaction process (Gunardson, 1998; O'Connor and Ross, 1998; Rostrup-Nielsen, 2006). The active catalytic sites are blocked by poisoning the feed gas with hydrogen sulfide (H_2S). The adsorption of sulfur at the catalytic sites is favored over carbon growth. The SPARG process is claimed to produce high CO content SYNGAS.

Also combined CO_2 and steam reforming systems have been operational in the industry for a number of years (Gangadharan et al., 2012). By choosing the right proportions between CH_4, water and CO_2 (3/2/1), the combination of steam and dry reforming of methane can generate SYNGAS with a H_2/CO ratio of 2, ideal, for example, for the synthesis of methanol (Rostrup-Nielsen and Christiansen, 2011; Olah et al., 2009). This combination of steam and dry reforming was named bi-reforming.

Bi-reforming could be advantageous in the use of various natural gas sources even these containing substantial amounts of CO_2. Some natural gas as well as biogas sources contain CO_2 concentration up to 50–70%. Bi-reforming can also be used to recycle CO_2 emissions from sources such as flue gases from fossil fuel (coal, petroleum, natural gas, etc.), burning power plants, exhaust of cement factories, among other industries (Olah et al., 2013). This reaction can be represented as

$$\text{Steam reforming: } CH_4 + H_2O \longrightarrow CO + 3H_2 \quad \Delta H = +206.3 \frac{kJ}{molCO_2} \tag{16}$$

$$\text{Dry reforming: } CH_4 + CO_2 \longrightarrow 2CO + 2H_2 \quad \Delta H = +247.3 \frac{kJ}{molCO_2} \tag{17}$$

$$\text{Bi-reforming: } 3CH_4 + 2H_2O + CO_2 \longrightarrow 4CO + 8H_2 \tag{18}$$

Bi-reforming is adaptable for reforming varied natural gas (containing hydrocarbon homologues) and CO_2 sources, e.g., shale gas (Olah et al., 2013):

$$3C_nH_{(2n+2)} + (3n-1)H_2O + CO_2 \longrightarrow (3n+1)CO + (6n+2)H_2 \tag{19}$$

Numerous authors (Ashcroft et al., 1991; O´Connor and Ross, 1998; Wang et al, 200;, Jarungthammachote, 2011) have studied a similar idea, which combines dry reforming with partial oxidation. The idea again being that the combination helps overcome the endothermic requirement of dry reforming with the exothermic nature of partial oxidation, resulting in lower total energy consumption. In addition, it allows altering the H_2/CO ratio by controlling the extent to which each reaction takes place (Hartley and Tam, 2012). The combination of exothermic and endothermic reactions is called autothermal reaction (ATR). The ATR technology requires addition of CO_2 or CO_2-rich gas, in order to adjust the SYNGAS composition to the desired H_2/CO ratio.

$$\text{Dry reforming: } CH_4 + CO_2 \longrightarrow 2CO + 2H_2 \quad \Delta H = +247.3 \frac{kJ}{mol} \tag{20}$$

$$\text{Partial Oxidation of Methane: } CH_4 + 0.5O_2 \longrightarrow CO + 2H_2 \quad \Delta H = -35.6 \frac{kJ}{mol} \tag{21}$$

The combined Dry Reforming and Partial Oxidation is hence:

$$CH_4 + 0.5CO_2 + 0.25O_2 \longrightarrow 1.5CO + 2H_2 \quad \Delta H = +211.7 \frac{kJ}{mol} \tag{22}$$

Integrating steam reforming and partial oxidation with CO_2 reforming could reduce or eliminate carbon formation on reforming catalyst, thus increasing catalyst life and process efficiency. Therefore, the tri-reforming, a synergetic combination of CO_2 reforming, steam reforming, and partial oxidation of methane in a single reactor for effective production of industrially useful SYNGAS (Song, 2006) could solve two important problems encountered in individual processing. Incorporating oxygen in the reaction generates heat *in situ* that could increase energy efficiency; oxygen also reduces or eliminates carbon formation on the reforming catalyst. The tri-reforming can be achieved with natural gas and flue gases using the waste heat in power plants and the heat generated *in situ* from oxidation with the oxygen that is present in flue gas (Zhou et al., 2008; Zangouei et al., 2010; Moon et al., 2004).

The tri-reforming process is presented in Eqs. (23) to (26) (Song and Pan, 2004):

$$\text{Steam reforming: } CH_4 + H_2O \longrightarrow CO + 3H_2 \quad \Delta H = +206.3\frac{kJ}{mol} \tag{23}$$

$$\text{Dry reforming: } CH_4 + CO_2 \longrightarrow 2CO + 2H_2 \quad \Delta H = +247.3\frac{kJ}{mol} \tag{24}$$

Partial Oxidation of Methane:

$$CH_4 + 0.5O_2 \longrightarrow CO_2 + 2H_2 \quad \Delta H = -35.6\frac{kJ}{mol} \tag{25}$$

$$CH_4 + 0.5O_2 \longrightarrow CO + 2H_2O \quad \Delta H = -880\frac{kJ}{mol} \tag{26}$$

Song (2006) reports experimental and computational results to support that tri-reforming produces SYNGAS with desired H_2/CO ratios (1.5–2.0) and eliminates carbon formation in the CO_2 reforming of CH_4. Song (2006) suggests that tri-reforming is especially suited to using CO_2 in concentrated sources without prior CO_2 separation, as in non-conventional (low-quality CO_2-rich) natural gas, and has been demonstrated in pilot scale in Korea.

In general, produced SYNGAS from methane reforming is converted catalytically *in situ* via one of two main routes. The first is to use Fischer-Tropsch synthesis, a process that catalytically converts SYNGAS to hydrocarbons of varying molecular weights. The second is methanol synthesis. The latter has better atomic economy, since the oxygen atom in CO is included in the product and CO_2 can be blended into SYNGAS as a reactant. However, production of methanol is very inefficient in this reaction: only 10-15% one pass conversion typically at 5.0-10.0 MPa and 523-573 K, due to the severe thermodynamic limitations of this exothermal reaction ($CO+2H_2 \rightarrow CH_3OH$) (Shi et al., 2013).

Finally, CO_2 reforming of methane can also be used as a chemical energy storage alternative and an energy transmission system (Richardson and Paripatyadar, 1990, Levitan et al., 1991; Levy et al., 1993). According to Zhang et al. (2003), in this system, solar energy is used to drive the endothermic forward reaction, and the energy thus stored can be transported via pipelines such as SYNGAS and liberated at will by the reverse reaction at any location or time. The highly endothermic reaction could be an option to store solar energy in hot regions (Zhang et al., 2013).

Reverse Water Gas-Shift (RWGS): The reverse water gas shift (RWGS) reaction has been known from over two centuries and is a well-researched and understood process for SYNGAS ratio alteration (Hartley and Tam, 2012). In fact, both the water gas shift (WGS) and the RWGS reactions are mostly used in combination with reforming of hydrocarbons to adjust the H_2/CO ratio, as shown in Eq. (27) (Song, 2006). Depending on the reaction conditions, the equilibrium for the WGS can be pushed in either the forward or the reverse direction. Efforts to explain the RWGS reaction mechanism are reported (Goguet et al., 2006, Meunier et al., 2007, Wang et al., 2013), and two main mechanisms have been proposed: the *redox* mechanism and the *associative formate* mechanism. The reversibility of the WGSR is important in the production of ammonia, methanol, and Fischer-Tropsch synthesis where the ratio of H_2/CO is critical. Many industrial companies exploit the RWGS reaction as a source of the synthetically valuable CO from cheap CO_2. In fact, catalytic RWGS reaction is the main route to produce SYNGAS from CO_2.

$$H_2 + CO_2 \longleftrightarrow CO + H_2O \quad \Delta H = +51\frac{kJ}{mol} \tag{27}$$

RWGS provides a source of hydrogen at the expense of carbon monoxide, which is important for the production of high purity hydrogen. This is a mildly endothermic reaction, as shown in Eq. (27).

Although high temperature reactions are effective for obtaining a high conversion, WGS reaction is an equilibrium-limited reaction that exhibits decreasing conversion with increasing temperature. In order to take advantage of both the thermodynamics and kinetics of the reaction, industrial scale WGS reaction is conducted in multiple adiabatic stages consisting of a high temperature shift (HTS) followed by a low temperature shift (LTS) with intersystem cooling (Byron, 2010). The initial HTS takes advantage of the high reaction rates, but is thermodynamically limited, which results in incomplete conversion of carbon monoxide and a 2-4% carbon monoxide exit composition. To shift the equilibrium towards hydrogen production, a subsequent low temperature shift reactor is employed to produce a carbon monoxide exit composition of less than 1% (Byron, 2010). A catalyst is required under these conditions because of the lower reaction rate at low temperatures. The RWGS reaction uses a variety of catalysts, including palladium, platinum on titania, copper, cobalt with manganese/zinc oxide and rhodium with ceria (Tanaka et al., 2003, Saito and Murata, 2004, Meunier et al., 2007). Many research groups are looking at copper as a catalyst due to its effectiveness and its relatively low cost (Armstrong et al., 2013). However, there has been renewed interest in the WGSR at extreme temperatures, because of recent advances in high-temperature materials for hydrogen separation membranes (Bustamante et al., 2002).

4.4. CO_2 to methanol

Methanol is one of the most important commodity chemicals as it is used as a raw material in several intermediate chemicals and end uses. Methanol is produced industrially from SYNGAS from natural gas or coal mainly containing CO, H_2 and a small amount of CO_2 in presence of a catalyst. Nevertheless, direct CO_2 hydrogenation has also been reported. Other non-conventional routes are electro- or photoprocesses, as well as the use of enzymes. The importance of methanol synthesis is demonstrated by widespread scientific publications of various reaction routes (Razali et al. 2012), and the development of several pilot plants to use waste carbon dioxide for methanol production.

Among the new technologies, in terms of potential for application, the CO_2 catalytic hydrogenation to methanol appears to have the highest degree of commercialization. It may be already commercially interesting when cheap sources of renewable H_2 are available, or to store excess electrical energy, as an alternative to actual systems. It is estimated that this reaction could reach the industrial stage in less than five years. This development would be pushed by experience in pilot or pre-commercial industrial plants, such as the Mitsui Chemicals Inc.'s plant (pilot in Japan capable of producing 100 t of methanol per year, and large unit expected in Singapore) and a plant by Carbon Recycling International (installed at the end of 2010) (Quadrelli et al., 2011). Mitsui´s pilot plant uses CO_2 from an ethylene production plant of Osaka Works Petrochemical Complex (ADEME, 2010). It synthesizes methanol by CO_2 hydrogenation and the simultaneous water gas shift reactions. The process claims 96% selectivity (Hartley and Tam, 2009). Carbon Recycling International is capable of producing 3000 t/y of methanol (ADEME, 2010). This unit has a capacity of about 10 t of methanol from 18 t of CO_2 (Carbon Recycling International, 2009; Van-Dal and Bouallou, 2013), with CO_2 from the Svartsengi geothermal plant and an aluminum production plant. Hydrogen is generated from the electrolysis of water using a renewable source of electricity.

Methanol from SYNGAS: Synthesis gas composed of the proper ratio of hydrogen, carbon monoxide and carbon dioxide is converted to methanol.

Alternatives paths to methanol are via CO from RWGS reacting with hydrogen according to Eq. (28) and via CO_2 being hydrogenated following Eq. (29).

$$CO + 2\,H_2 \longrightarrow CH_3OH \qquad \Delta H = -90.6\frac{kJ}{mol} \tag{28}$$

$$CO_2 + 3\,H_2 \longrightarrow CH_3OH + H_2O \qquad \Delta H = -49.5\frac{kJ}{mol} \tag{29}$$

From Eq. (28), production of methanol involves SYNGAS production as intermediate stage. Hence, two steps are required for the manufacture of methanol: reduction to SYNGAS and reaction to form methanol. There are process variations for implementing the sequence. Before being sent to the methanol production unit, the SYNGAS must thus be subjected to the WGS reaction to enhance its hydrogen content. Alternatively, H_2 from other sources can be added. Recent efforts have been aimed at production of methanol in a one-step process without intermediate formation of SYNGAS. Homogeneous or heterogeneous catalysts are typically preferable. The conventional process occurs at relatively low pressures (5 to 10 MPa) and 210 to 350 °C employing a $Cu/ZnO/Al_2O_3$ catalyst.

Catalytic Hydrogenation Conversion of Carbon Dioxide to Methanol: The most direct and studied route to methanol from CO_2 is the catalytic conversion of CO_2 with hydrogen. Carbon dioxide hydrogenation to methanol is a relatively mature process. The main issue is the cost (and associated carbon footprint) of the H_2 necessary for the reaction. Any available energy source (alternative energies such as solar, wind, geothermal, and atomic energy) can be used for the production of needed hydrogen and chemical conversion of CO_2. The process can use lower operational pressures of 3 MPa at 240 °C. This direct CO_2 hydrogenation exhibits low conversions resulting in high volumes of recycled gas. Literature indicates that methanol is synthesized following a 3:1 hydrogen to carbon dioxide stoichiometry using catalysts of copper oxide, zinc oxide, incorporating either titania, aluminum oxide, chromium oxide and alternatively lanthanum or gallinium (Lachowska and Skrzypek, 2004, Lee et al., 2004, Stoczynski et al., 2004).

An alternative approach to the use of solid catalysts and a gas phase process is to employ the so called low-temperature methanol synthesis (LTMS) (Dixneuf, 2011). LTMS is based on the catalytic hydrogenation of methanol to formic acid (HCOOH) with subsequent etherification to methanol formate (alternative to methyl formate from SYNGAS), followed by hydrogenation of formate to two methanol molecules using Pincer-type ruthenium(II) catalyst (Balaraman et al, 2011; Dixneuf, 2011; Huff and Sanford, 2011). A liquid-phase allows CO_2 and H_2 conversion to methanol of about 95% with very high selectivity in a single pass

(Olah, 2009). Waugh (2012) has published a review on catalytic methanol synthesis which includes the use of carbon dioxide as a feedstock.

Photoreduction of CO_2 to methanol: Photoelectrochemical reduction of carbon dioxide or photocatalysis generally uses semiconductors to promote reaction in the presence of sun light. The semiconductor is used as a catalyst to absorb solar energy and generate electrons and protons needed for the reduction of carbon dioxide. While hydrogenation of carbon dioxide requires high temperature and high pressure conditions, photocatalysis carries out under relatively mild conditions with advantageous energy input – sun light – a continuous and readily available source (Le, 2009).

Considerable research effort has been made on CO_2 activation by visible light photocatalysts due to the natural abundance of sunlight. Nevertheless, the efficient photoreduction of CO_2 with H_2O remains one of the most challenging tasks of environmental catalysis.

CO_2 can be reduced in water vapor or solvent by photocatalysts such as TiO_2 and ZnS. Eq. (30) describes the overall reaction.

$$CO_2 + 2H_2O \xrightarrow{h\nu} CH_3OH + 3/2\ O_2 \tag{30}$$

Due to the high energy requirements, this method is often combined with electrochemical methods via photoelectrocatalysis to drive the reaction (Hu et al., 2013). The catalysts traditionally used are transition metal complexes, TiO_2, ZnO, CdS, and functionalized metal surfaces (Yamashita et al., 1998; Kuwabata et al., 1994). A wide variety of CO_2 photoreduction has been achieved on the surface of TiO_2 under UV irradiation. The yield of photoproducts can be changed substantially under different experimental conditions such as UV wavelength, UV intensity, additives of reaction media and reactor configuration. Other variables, such as CO_2 pressure, moisture and residence time are also important in photoreducing CO_2 (Wu & Lin, 2005).

Electrochemical Production of Methanol from CO_2 and H_2O: The direct reduction of CO_2 to CH_3OH is known to occur at several types of electrocatalysts including oxidized Cu electrodes. The current stage of the technology is still very experimental. The majority of tests have been performed on a laboratory scale with a purpose of either kinetic analysis or proof-of-concept to examine product distribution for different material and condition combinations (Beck et al., 2010). An advantage of electrochemical CO_2 reduction is that unlike many other hydrocarbon processes it can occur at ambient conditions.

The electrochemical reduction of carbon dioxide to methanol is thermodynamically possible, but there seems to be no well-established technique to achieve this reaction with high current efficiencies close to l00%.

Nevertheless, methanol production has been reported with the use of ruthenium: gallium arsenide and RuO_2-TiO_2 mixed cathodes (Le, 2009). Cole and Bocarsly (2010) have reviewed electrochemical reduction processes, including electrochemical CO_2 conversion to methanol. Few studies have investigated the feasibility of this technology, and none has been found to provide an in-depth analysis of its potential industrial implementation.

Applications of Methanol: Methanol has traditionally been used as feed for production of a range of chemicals including acetic acid, formaldehyde and MTBE (Olah, 2009). In recent years, methanol has also been used for other markets such as production of Di-methyl-ether (DME) and olefins by the so-called methanol to olefins process (MTO) or as blendstock for motor fuels. As a liquid fuel, methanol is of interest especially for use in fuel cells (Olah, 2009).

Methanol to Olefins (MTO) explores alternative pathways to produce small olefins, in particular ethylene and propylene. Conventional steam cracker feeds are either natural gas liquids (NGL) or heavy liquids (i.e., naphtha). Ethane cracking, however, is increasing its share as feedstock. A promising alternative route is dehydration of methanol (MTO). Methanol-to-olefins (MTO) was first developed by ExxonMobil (1980s) as part of its methanol-to-gasoline (MTG) process. In the 1990s, UOP and Norsk Hydro built an MTO pilot plant in Norway. Since then, Lurgi has developed its own version of this process, methanol-to-propylene (MTP).

4.5. CO_2 to DMC

Dimethyl carbonate (DMC) is a biodegradable and nontoxic chemical acceptable environmentally as a chemical destination of CO_2. It is exempted from VOC classification and can be used as raw material for producing valuable chemicals, including aromatic polycarbonate, and qualifies as an octane booster component in gasoline and diesel. It is a safer and nontoxic substitute of well-established methylating-carbonylating hazardous chemicals like dimethyl sulfate and phosgene (Souza et al., 2013). Although DMC is presently produced on a relatively small scale, approximately 400 kt/y, its demand has grown strongly in recent times because of its green properties.

Currently, DMC is produced mainly by oxidative carbonylation of methanol (Aoussi et al., 2010). The direct methylation reaction is possible where, according to Ferreira et al. (2013), the most used catalyst is tin, employed as an oxide compound or as an organometallic complex, according to Eq. (31):

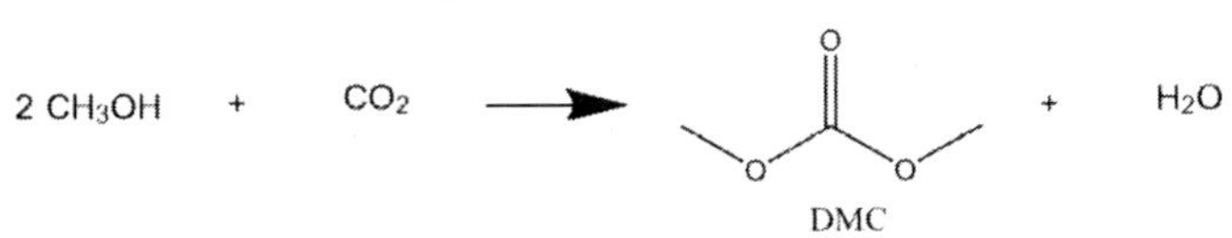

(31)

However, direct methylation presents low yields, inferior to 10%, due to the chemical inertness of CO_2 and to the deactivation of catalysts induced by water formation in the reaction (Aouissi, 2010). For large-scale production of DMC from CO_2, one route seems to be promising: the indirect route (IR) for two-step conversion of CO_2 with ethylene oxide (EO) to ethylene carbonate (EC), which then reacts with excess methanol (MeOH) giving DMC and ethylene glycol (EG) as shown in Eqs. (32) and (33).

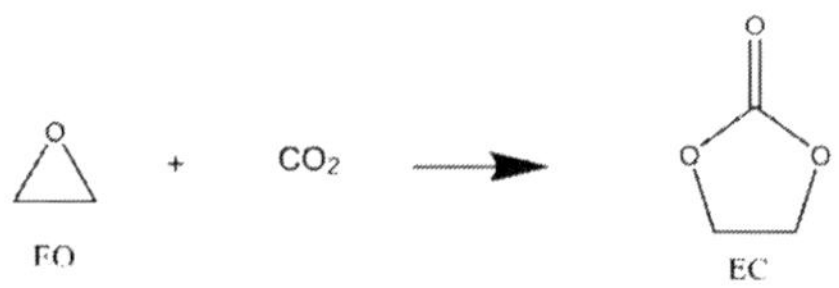

(32)

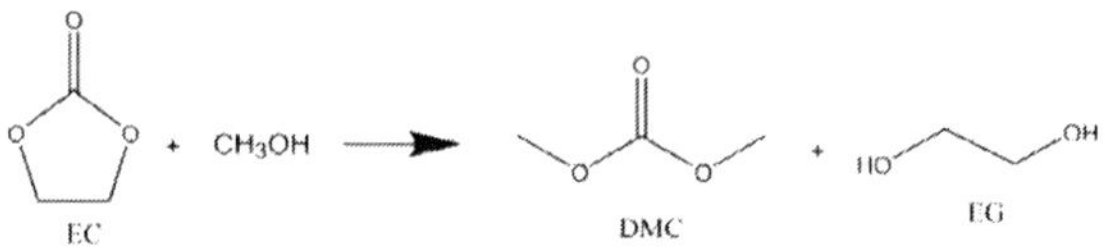

(33)

Souza et al. (2013) evaluated IR's performance from technical, economical and environmental standpoints. Accordingly, the authors proposed a process flowsheet with two serial reactors' system followed by an integrated separation section, with extractive distillation using methyl-iso-butyl ketone (IR-MIBK) and ethylene glycol (IR-EG). For environmental performance assessment, Souza et al. (2013) defined Chemical Sequestration of CO_2 (CSC) as the amount of CO_2 consumed by chemical reaction minus the amount of CO_2 emitted by heat, power and purges. Considering a production of $1.3x10^5$t/y of DMC produced, the authors reported CSC values for IR–MIBK and IR–EG of -15.9kt/y and -8.1kt/y, respectively. The negative values of CSC indicate that both alternatives are net emitters and illustrates that fixation of CO_2 in chemical products does not necessarily imply into net CO_2 reduction, and that the main aspect in CO_2 utilization as feedstock is of substituting fossil carbon source for a renewable alternative. Furthermore, Souza et al. (2013) estimated net present values for IR–MIBK and IR–EG of \$$71.5x10^6$ and \$$106.5x10^6$, respectively, and payback times of 5.5 and 4.5, respectively, concluding for economic feasibility of DMC production. It is worth noting that DMC production from CO_2 is already in use at Asahi Chemical Industry.

4.6. Biochemical conversion of CO_2

Biofixation of CO_2 with microalgae is a promising route of utilization of CO_2, as it exhibits fast growth (e.g., Picardo et al., 2013a) and produces numerous high-added-value bioproducts (Grima et al., 2003). Additionally, they do not contain lignin, a fact that renders microalgae better adapted to biochemical valorization. As advantages of CO_2 bioconversion with relation to energy crops, microalgae grow in variable climates on non-arable land with non-potable water, releasing competition with food crops, and are able to use direct flue gases as their carbon source (Fernández et al., 2012). Alternatively, to inject directly flue gases into microalgae cultures, adequate design and operation of the carbonation culture system unit are also necessary, otherwise almost all of the CO_2 fed to the culture would be released into the atmosphere. In this aspect, photobioreactors are more appropriated arrangements. Furthermore, photobioreactors provide cell requirements such as light, temperature, pH, and mixing (Fernández et al, 2012).

Chisti (2007) concluded that microalgae are the only alternative for the sustainable production of biodiesel. Accordingly, the fact that microalgae biomass is rich in lipids and are, hence, high energy density feedstock for fuels and chemicals is of relevance. Picardo et al (2013a) proposed a screening procedure for microalgae selection to meet production objectives such as SYNGAS for the production of synthetic fuels, since the biomass, or the residual biomass obtained after extraction of bioproducts, can be gasified to yield SYNGAS. An attractive alternative in this route is to employ CO_2 as oxidation agent (Butterman and Castaldi, 2007). Butterman and Castaldi (2007) report that the injection of CO_2 and H_2O in gasification increases char reactivity that results in more efficient use of the feedstock with less residual to be post-processed.

According to Grima et al. (2003), production of microalgal biomass can be carried out in fully contained photobioreactors or in open ponds and channels. Biomass productivity depends on species, operational conditions and the choice of ponds (~20g/m^2.d) or photobioreactor (~50g/m^2.d) geometry. Open-culture systems are almost always located outdoors and rely on natural light for illumination while closed photobioreactors may be located indoors or outdoors, although outdoor location is more common. Grima et al. (2003) list as biomass harvesting operations centrifugation, filtration or gravity sedimentation, which may be preceded by a flocculation step.

Microalgae contain lipids and fatty acids as membrane components, storage products, metabolites and sources of energy. Microalgae have been found to contain proportionally high levels of lipids (for some species this value can reach 50% oil by weight), with a convenient fatty acids profile and an unsaponifiable fraction allowing a biodiesel production with high oxidation stability (Grima et al., 2013). Lipid accumulation is promoted by stress, notably by nitrogen starvation (Picardo et al., 2013b).

Elemental analysis of carbon content of biomass points to ~50% (Picardo et al., 2013a), what leads to conclude that approximately 2t of CO_2 can be converted into 1t of biomass, potentially amenable to 0.2t of lipids. Its massive extension to the energy sectors constitutes a vast potential for large-volume CO_2 utilization. Fernandez et al. (2012) recognize that microalgae are not a storage strategy because the biomass produced cannot be stored for a long time. Its contribution to reducing CO_2 emissions is only possible if biofuels are produced to replace the fossil fuels use, and allowing the production of other commodities, or by-products from flue gases, which allows one to obtain revenues to mitigate the penalty of carbon capture. To illustrate the potential industrial application of microalgae, Figure 19 shows a schematic of microalgae bioconversion of CO_2 and its downstream processing in a biorefinery arrangement producing long-chain fatty acids (PUFA´s, MUFA´s and PUFA´s), biodiesel, green diesel, gasoline, biogas, urea, N_2 and propylene carbonate.

Monteiro et al. (2010) employed Pareto optimization of what was named an "industrial ecosystem" comprised of a biorefinery of microalgal biomass aiming at maximizing sustainability of the productive arrangement. The authors concluded that increasing the weight of environmental objectives against economic performance might make sectors of the proposed original superstructure of amenable processes unattractive. Therefore, the final structure of a biorefinery of microalgae depends on the priorities set for the productive complex.

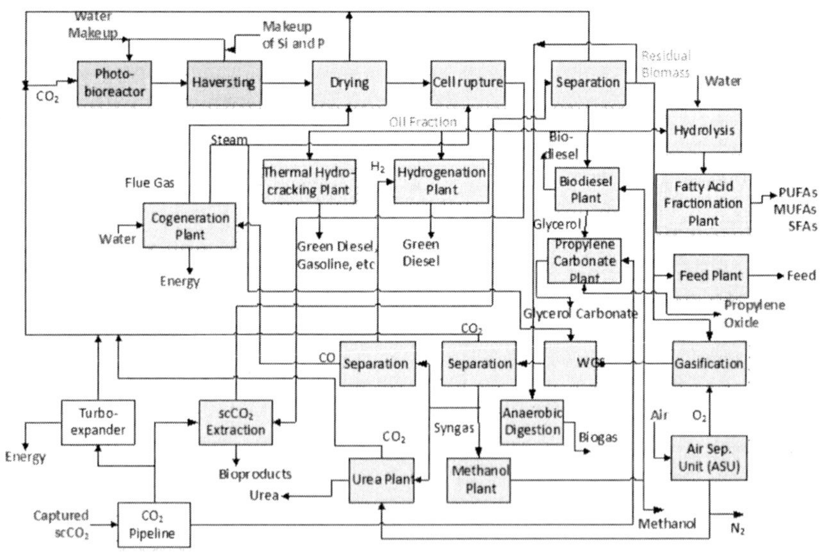

Figure 19. Bioconversion of CO_2 integrated into a biorefinery arrangement. Main external inputs are marked in red.

4.7. Some pilot and commercial scale CO_2 utilization processes

Polycarbonate: Polycarbonate (PC) is a plastic with impact resistance and heat resistance, mainly produced (4t/y) by reacting CO and Cl_2 to form phosgene as an intermediate material. The phosgene process has a number of disadvantages, including the risk of environmental harm (Ushikubo, 2013). Asahi Kasei Corporation succeeded in commercializing the first non-phosgene polycarbonate production using ethylene oxide and CO_2, a by-product of ethylene oxide synthesis. The Asahi Kasei Process has as co-products high-purity monoethylene glycol (MEG). The process employs reactive distillation in the monomer production and gravity-utilized, non-agitation polymerization reactor in the melt polymerization. The monomer process consists of 3 production steps, ethylene carbonate (EC) from CO_2 and EO, dimethyl carbonate (DMC) and MEG from EC and MeOH, and diphenyl carbonate (DPC) and MeOH from DMC and phenol (PhOH). All intermediates are recycled. The by-produced PhOH is recycled to the monomer process (Sinshuke et al., 2010). According to Ushikubo (2013), the new process reduces CO_2 emissions by 0.173kg per one kg of polycarbonate. Five commercial plants using the Asahi Kasei process are operating in Taiwan (150,000 t/y), Korea (2 plants of 65,000 t/y), Russia (65,000 t/y) and Saudi Arabia (260,000 t/y) (Shinsuke et al., 2010).

Monoethylene glycol (MEG): MEG is used as an antifreeze and as a raw material for the production of polyester fibers and resins, mainly PET (Ushikubo, 2013). An expanding market share is foreseen in natural gas industry, where MEG is added into the pipeline or the gas conditioning process, either as hydrate inhibitors or for dehydration purposes to protect downstream pipelines. The pipeline can extend for thousands of kilometers and MEG is injected to inhibit hydrate formation, avoiding plugging (*e.g.,* Statoil´s Snøhvit field, Pettersen, 2011). Concerning the production process, a new technology was developed by Mitsubishi Chemical Corporation with 99% selectivity while the conventional (non-catalytic) process has selectivity around 89% (Kawabe, 2010). The conventional technology produces as co-product DEG and TEG (di- and triethylene glycols, whose demand is expanding at only 2-3% as opposed to MEG expansion (world demand amounts to 17t/y). Mitsubishi technology uses a two-step catalytic synthesis: production of ethylene carbonate (EC) as intermediate followed by EC hydrolysis under almost stoichiometric condition, while the conventional hydrolysis occurs at a higher H_2O/ethylene oxide molar ratio, according to Eqs. (34) and (35).

$$EO + CO_2 \longrightarrow EC \ (\text{carbonation}) \tag{34}$$

$$EO + H_2O \longrightarrow EG + CO_2 \ (\text{hydrolysis}) \tag{35}$$

The product purification is simpler (water removal and MEG purification distillation columns, while the conventional process has 4 distillation columns: water, MEG, DEG and TEG columns). Furthermore, CO_2 remains in closed loop. Ushikubo (2013) reports that several commercial plants operating with the new (and greener) technology save resources and energy and reduces the amount of wastewater and CO_2 production. It is worth noting that MEG is a co-product of the production of DMC in a process where Eq. (4.24) is replaced by Eq. (40).

Polyurethane: Bayer (2013) targets the production of polyurethane, via the utilization of CO_2 as feedstock. In this route, CO_2 is converted to polyols (HO-R-OH) which reacts with isocyanate to yield polyurethane. The conversion of CO_2 starts with its reaction with an epoxide, (propylene oxide) of higher energy content, in a catalytic route. Polyol with 30%CO_2 had 2.64 kg of equivalent CO_2 emissions. A maximum theoretical value of 43% of CO_2 can be incorporated in the polyol (Bayer, 2010). The development started in 1669, ended its laboratory scale in 2009, and is moving to industrial implementation of the named "Dream Production" with a pilot plant in Leverkusen to produce polyol, for testing purposes. In early 2013, the new method was successfully converted from the production of discrete quantities to continuous production, a key intermediate step for the industrial-scale production of CO_2-based polyurethane, which Bayer is targeting for 2015.

4.8. Emerging CO_2 utilization processes

The fixation of CO_2 into chemicals and polymers will not substantially contribute to a reduction in antropogenic GHG emissions given the current energy demand. Nevertheless, using CO_2 as a feedstock meets the requirements of sustainable development. An insight into advanced process concepts focus on chemical sequestration of CO_2 creating manufactured products from captured CO_2 with large potential markets. Integration of capture technologies into energy production schemes or oil and gas refining installations is the idea behind cPSE approach, namely, to abate chemical emissions while producing industrial products.

Formic acid: According to Armstrong et al. (2013), the amount of energy required to utilize carbon dioxide as a feedstock largely depends on the oxidation state of the intended products. The next-highest oxidation state molecules from CO_2 are formic acid (HCOOH) and carbon monoxide (CO). So, carbon dioxide utilization to manufacture formic and other carboxylic acids is a relatively low-energy transformation. Formic acid has numerous applications, including food technology, agriculture, and the leather and rubber industries. Moreover, it has recently been considered as a promising candidate material for hydrogen storage and it is an important chemical with numerous applications. Moreover, formic acid has limited uses for further conversion, except reduction to methanol. The industrial methods used for its production employ CO as a raw material. Maihom et al. (2013) concluded that a first step occurs where CO_2 is hydrogenated to a formate intermediate. In the second step, the formate is further hydrogenated into formic acid. The hydrogenation of CO_2 would complete the chemical loop for hydrogen storage using CO_2. The complementary step is the catalyzed decomposition of formic acid to pure H_2 and reusable CO_2.

Carbon dioxide-based copolymers: The synthesis of organic carbonates has been one of the most widely studied areas of CDU. Typically, CO_2 is inserted into a molecule without the loss of any atoms in either the co-reactant or the gas itself (Armstrong, 2013). Carbonates are formed by the insertion of a CO_2 molecule into a guest co-reactant, typically an epoxide. Poly(propylene carbonate) (PPC), an alternating copolymer of CO_2 and propylene oxide, is one of the emerging low-cost biodegradable plastics. The fast development in catalyst design and performance improvement for PPC has created new chances for the chemical industry. In particular, high molecular weight PPC from rare earth ternary catalyst is becoming an economically viable biodegradable plastic with tens of thousands of tons produced per year, providing a new solution to overcoming the problem of high cost in biodegradable plastics (Qin and Wang, 2010). According to Qin and Wang (2010), with the continuous improvement in catalyst systems, commercialization of CO_2 copolymer is possible. The authors report industrial activities by Empower Materials producing polypropylene carbonate (QPAC®40), polyethylene carbonate (QPAC®25), polybutylene carbonate (QPAC®60), and polycyclohexene oxide(QPAC®130) on a pilot scale.

Electrochemical Reduction of CO_2. Delacourt (2010) studied the electrochemical conversion of CO_2 into SYNGAS. The driver of the proposed route is that renewable energies (e.g., solar and wind) are only alternatives to fossil fuel as they are not available on demand, thus requiring storage. Delacourt (2010) lists as a storage opportunity the conversion to liquid fuels (e.g., methanol), in which SYNGAS is the required intermediate, by converting solar energy into electricity through photovoltaic arrays, and then by using this electricity to produce fuels by electrolysis. Evolved H_2 reacts with CO_2 in a water-gas-shift reactor to make CO (and H_2O). The resulting SYNGAS is converted to methanol. Delacourt (2010) decided for a low-temperature technology (room temperature) although reported that high-temperature electrolysis (800 to 900°C) could be an attractive alternative. Because of the relatively low solubility of CO_2 in water under ambient conditions, gas-diffusion electrodes were applied to operate at higher current densities, and ion-exchange membrane was used as the electrolytic medium to limit gas crossover resulting in a decrease of the current efficiency of the electrochemical cell. Catalysts capable of reduction of CO_2 to CO at low overpotentials were selected.

Light-Driven Technologies. The rubisco enzyme is probably the most abundant enzyme of the biosphere. The fixation of CO_2 and its transfer to organic substrates in the Calvin cycle leads by way of starch to an annual production of 10^{11} t of biomass (Walther et al., 1999). With the development of catalysts able to reproduce the key steps of photosynthesis, water and sunlight would ultimately be the only needed sources for clean energy production. Light driven technologies under development include (a) photoelectrochemical cells where CO_2 present in a moistened gas stream is converted into organic molecules based on the photooxidation of water into oxygen gas O_2, protons H_+, and electrons. The conversion of CO_2 occurs at the photocathode and involves the generated protons, electrons and the "fuel" CO_2 (Kayaert et al., 2013); (b) direct water oxidation - photocatalytic water splitting - to produce H_2 and O_2 over a metal-oxide-based photocatalyst using solar energy (Maeda and Domen, 2013); (c) hydrogen-producing systems consisting of a hydrogen-evolving catalyst linked to a photosensitizer (Badura et al., 2012). Although promising alternatives, biomimetic CO_2 conversions are still in its early stage of technological development.

CO_2 Mineralization for Environmental Remediation: Lim et al. (2013) reviewed the application of carbonation to solidify or stabilize solid combustion residues from municipal solid wastes, paper mill wastes, etc. and contaminated soils, and to manufacture precipitated calcium carbonate. For instance, the red mud - a highly alkaline waste of Bayer's process - can be treated by absorption of CO_2. Machado (2012) analyzed the process of red mud carbonation with the exhausted gases from the alumina production calcinators, by developing a dynamic model representative of the mass and energy balances involved in the process, and chemical reactions occurring in the mud under carbonation. Machado (2012) was able to predict the species behavior, as well as the decrease in mud pH and the rebound phenomenon observed when the CO_2 concentration is reduced. The transient profile of the main process responses indicated a substantial reduction of CO_2 concentration in the output gas, in consequence of tons of CO_2 captured, and a significant reduction in mud pH. Concerning other environmental applications, Lim et al. (2013) report that carbonated products can be utilized as aggregates in the concrete industry and as alkaline fillers in the paper (or recycled paper) industry. Mineral carbonation consist in reacting CO_2 and Ca or Mg-bound compounds such as wollastonite (CaSiO3), olivine (Mg2SiO4), and serpentine ($Mg_3Si2O_5(OH)_4$). As a result, CO_2 is stably stored in final products such as $CaCO_3$ and $MgCO_3$. Last, the accelerated carbonation of solid wastes containing alkaline minerals such as Ca and Mg before their landfill treatment is effective for decreasing the mobility of heavy metals by adjusting pH to below 9.5 at which their solubility is lowest.

4.9. Non-conversion utilization of CO_2

CO_2 utilization that does not involve its chemical conversion is an alternative destination of captured emissions. Among such alternatives the injection of supercritical CO_2 into depleted oil wells to enhance the further recovery of oil is well established. Indeed, this is presently the only commercially viable technology adding value to large volumes to CO_2 in the order of magnitude of emissions from fossil fuel based energy generation. It has been estimated that CO_2 injection can enhance oil recovery from a depleting well by about 10 to 20 % of the original oil in place. Similarly, CO_2 can be used to recover methane from unmined coal seams. It has been estimated that, in the U.S. alone, 89 billion barrels of oil could technically be recovered using CO_2, leading to a storage of 16 Gt of CO_2 in the depleted oil reservoirs (DNV, 2011).

The use of supercritical CO_2 as a solvent in processing chemicals (e.g., flavor extraction) is also well established. New uses of supercritical CO_2 in chemical processing are emerging, and have the added benefit of reducing water usage. Supercritical CO_2 is also being explored as a heat transfer fluid for some geothermal applications. These non-conversion methods of utilization constitute a significant fraction of the total CO_2 emissions (DNV, 2011).

Enhanced Oil Recovery (CO_2-EOR)*:* Through CO_2-EOR, oil producers inject CO_2 into wells to help sustain production in otherwise declining oil fields. The main goal of this technology is to draw more oil to the surface. In 2012, CO_2-EOR accounted for 6% of current U.S. domestic oil production. The limited CO_2 source is the main barrier to reaching higher levels of CO_2-EOR production due to insufficient supplies of affordable CO_2. With the discovery of offshore gas fields with high CO_2 contents in Brazil, there is a great opportunity to implement CO_2-EOR at those fields.

Furthermore, the offshore removal of acid gases poses a choice of onshore processing against offshore processing. Factors like safety and operability may favor onshore processing in comparison with offshore processing. The proper on land disposal of the CO_2 removed from natural gas requires the construction of CO_2 pipelines to transport CO_2 to offshore EOR applications. Another aspect is the high cost of ship hulls as shifting CO_2 removal to onshore facilities releases the weight shipped, which could overload the cost of building the required CO_2 pipelines.

In the option of onshore processing, CO_2 rich natural gas would be available as feedstock to SYNGAS production from CO_2 reforming, besides CO_2 separation and transport back to oil fields for CO_2-EOR. The current estimated cost gap for CGS from power, steel and cement plants is several times larger than the current CO_2 market price, and downward pressure on this market price is likely to increase. Investments in CO_2 reuse technologies need to be assessed as a screening procedure among potential alternatives.

5. CONCLUDING REMARKS

Technologies for utilization of CO_2 amenable to commercial scales are presently a very small fraction of anthropogenic CO_2 emissions, and very endothermic due to the inertness of CO_2, what reduces their abatement potential. Furthermore, chemical and biochemical conversion of CO_2 presents a sequestration potential that is orders of magnitude lower than the CO_2 emissions associated to energy generation from fossil fuels.

Geographical synergies of CO_2 supply (power plant emissions or natural gas processing) should guide in the medium term feasible utilization alternatives. The main synergy is identified in offshore gas processing and EOR, which, due to the economic benefit, process scale and maturity, stands as the most relevant utilization route in the short to medium term. Furthermore, most of the emerging alternatives reviewed are at their early stage of technological development.

However, CO_2 stands as a promising renewable feedstock to the chemical industry, which has been limited to oil, natural gas, coal and, recently, biomass. Such as posed, SYNGAS based conversions to the downstream supply chain is a route for flexibility of raw materials. Gasification of a variety of feedstock can lead to SYNGAS. Furthermore, expanding non-conventional gas supply enforces natural gas reforming in the upstream of the chemical supply chain. CO_2 captured from emissions and natural gas processing may drop into the supply chain via Dry Reform. As SYNGAS derived products, hydrogen, methanol and synthetic fuels (e.g., olefins, naphta, diesel, lubricants and kerosene) from Fischer-Tropsch process are likely to dominate the scenario.

Additionally, methanol (MeOH) is expected to grow in relevance either as hydrogen carrier and as intermediate product such as feedstock to MeO (Methanol to Olefins) process, as well as trans-esterification agent in biodiesel and dimethyl carbonate (DMC) production processes. Nowadays, the interest in DMC has grown significantly because it is considered to be a safe and nontoxic substitute for well-established methylating and carbonylating agents (e.g., phosgene), and has potential as an oxygen-containing fuel additive. There are several technological routes to produce DMC, however, the one route considered promising for large-scale commercialization is the trans-esterification of ethylene carbonate (EC) with methanol. In this indirect route, EC is obtained by a previous reaction of CO_2 with ethylene oxide. The route yields DMC and ethylene glycol (EG) as co-products in equimolar ratio.

The use of CO_2 as a carbon source in the synthesis of chemicals, in contrast to disposal, reduces dependence on fossil fuels, generates profit and is in line with a sustainable chemical industry. However, the actual use of CO_2 corresponds to about 0.4% of the potential CO_2 suitable to be converted to chemicals (Navarro et al., 2013).

Finally, large-scale utilization of CO_2 require energy efficient CO_2 capture technologies and an expansion of CO_2 transportation infrastructure.

ACKNOWLEDGEMENTS

O. Araujo and J.L. Medeiros kindly acknowledge CNPq for scholarships and financial grants; and CAPES for grant no. 113/2008.

REFERENCES

1. ADEME (Agence de l'Environnement et de la Maîtrise de l'Energie), Panorama des voies de valorisation du CO2 http://www2.ademe.fr/servlet/getDoc?cid=96&m=3&id= 72052&p1= 30&ref=12441 (2010).

2. Aouissi, A., Al-Othman, Z.A., Al-Amro, A. Gas-Phase Synthesis of Dimethyl Carbonate from Methanol and Carbon Dioxide over Co1.5PW12O40 Keggin-Type Heteropolyanion. Int. J. Mol. Sci., 11, 1343-1351 (2010).

3. Aresta, M. Perspectives in the use of carbon dioxide. Quím. Nova, 22 (2), (1999).

4. Aresta, M., Aresta, Michele. Carbon Dioxide as Chemical Feedstock, WILEY-VCH Verlag GmbH & Co., (2010).

5. Armstrong, Katy, Dawson, George and Styring, Peter. Recent Advances in Catalysis for Carbon Dioxide Utilization. The Catalyst Review, vol. 26, issue 2, 6-13, (2012).

6. Ashcroft, A.T., Cheetham, A.K., Green, M.L.H., Vernon, P.D.F, Partial oxidation of methane to synthesis gas using carbon dioxide Nature 352, 225 - 226 (18 July 1991).

7. Badura, A., Guschin, D., Esper, B., Kothe, T., Neugebauer, S., Schuhmann, W., Rögner, M. Photo-Induced Electron Transfer Between Photosystem 2 via Cross-linked Redox Hydrogels. Electroanalysis 20(10), 1043–1047 (2012).

8. Bayer. Use of carbon dioxide for the production of plastics. http://www.materialscience. bayer.com/~/media/Bms/Documents /Publications/CO2-Projekte_kurz_EN.ashx. Accessed on 11/27/2013 (2013).

9. Bayer. A Dream Comes True http://solutions.bayertechnology.com/fileadmin/ user_upload/sat_pages/technologyImpulse/articles /A-Dream-Comes-True.pdf. Accessed on 11/27/2013 (2010).

10. Beck, J., R. Johnson, R., Naya, T., Electrochemical Conversion of Carbon Dioxide to Hydrocarbon Fuels, EME 580, Spring (2010).

11. Bradford, M. C. J., Vannice, M. A., CO2 Reforming of CH4, Catalysis Reviews, Vol. 41(1), 1-42, (1999).

12. Bustamante, F., Enick, R. Rothenberger,K., Howard, B., Cugini, A., Ciocco, M. Morreale, B., Kinetic Study of the Reverse Water Gas Shift Reaction in High-Temperature, High Pressure Homogeneous Systems, Fuel Chemistry Division Preprints, 47(2), 663, (2002).

13. Butterman, H.C., Castaldi, M.J. Influence of CO2 Injection on Biomass Gasification. Ind. Eng. Chem. Res., 46, 8875-8886 (2007).

14. Byron, S.R.J., Loganthan, M., Shantha, M.S., A Review of the Water Gas Shift Reaction. International Journal of Chemical Reactor Engineering 8, 1–32, (2010).

15. Chisti, Y. Biodiesel from microalgae. Biotechnol Adv, 25, 294–306 (2007).

16. Cole, E.B., Bocarsly, A.B., Photochemical, Electrochemical, and photoelectrochemical Reduction of Carbon Dioxide, in "Carbon Dioxide as Chemical Feedstock", M. Aresta, Editor, Wiley-VCH Verlag GmbH & Co.: Weinheim. (2010).

17. de Medeiros, J.L., Versiani, B., Araújo, O.Q.F. A model for pipeline transportation of supercritical CO2 for geological storage. The Journal of Pipeline Engineering, 4th Quarter, 253-279 (2008).

18. de Medeiros, J.L., Nakao, A., Grava, W.M., Nascimento, J.F., Araújo, O.Q.F. Simulation of an Offshore Natural Gas Purification Process for CO2 Removal with Gas Liquid Contactors Employing Aqueous Solutions of Ethanolamines. Industrial & Engineering Chemistry Research, 52, 7074-7089 (2013a).

19. de Medeiros, J.L., Barbosa, L.C., Araújo, O.Q.F. An Equilibrium Approach for CO2 and H2S Absorption with Aqueous Solutions of Alkanolamines: Theory and Parameter Estimation. Industrial & Engineering Chemistry Research, 52, 9203-9226 (2013b).

20. Delacourt, C. Electrochemical reduction of carbon dioxide and water to syngas (CO + H2) at room temperature. Available at http://charles.delacourt.free.fr/Postdoc-manuscript_Charles-Delacourt.pdf (2010).

21. DNV. Carbon Dioxide Utilization Electrochemical Conversion of CO2 – Opportunities and Challenges. Available online at http://www.dnv.com/binaries/DNV-position_paper_CO2_Utilization_tcm4-445820.pdf (2011).

22. Edwards, J.H., Potential sources of CO2 and the options for its large-scale utilisation now and in the future. Catalysis Today, 23, 59-66, (1995).

23. Ferreira, H.B.P., Vale, D.L., Mota, C.J.A., Miranda, J.L. Experimental Design for CO2 Conversion into Dimethylcarbonate (DMC) using Bu2SnO at Subcritical Conditions. Brazilian Journal of Petroleum and Gas, 6(3), 93-104, (2012).

24. Fernández, F.G.A., González-López, C.V., Sevilla, J.M.F., Grima, E.M. Conversion of CO2 into Biomass by Microalgae: How Realistic a Contribution May it be to Significant CO2 Removal? Appl Microbiol Biotechnol, 96, 577–586 (2012).

25. Gangadharan, P., et al., Evaluation of the economic and environmental impact of combining dry reforming with steam reforming of methane. Chem. Eng. Res. Des. (2012), http://dx.doi.org/10.1016/j.cherd.2012.04.008

26. Genesis. Equation of State Prediction of Carbon Dioxide Properties. Project Kingsnorth Carbon Capture & Storage Project. CP-GNS-FAS-DRP-0001. http://www. decc.gov.uk/assets/decc/11/ccs/chapter6/6.23-equation-of-state-prediction-of-carbon-dioxide-properties.pdf, (2011).

27. Ginsburg, J. M.; Pina, J.; El Solh, T.; de Lasa, H. I.Coke formation over a nickel catalyst under methane dry reforming conditions: Thermodynamic and kinetic models Ind. Eng. Chem. Res., 44 (14) 4846– 4854, (2005).

28. Grima, E.M., Belarbia, E.-H., Acién Fernández, F.G., Medina, A.R., Chisti, Y. Recovery of microalgal Biomass and Metabolites: Process Options and Economics. Biotechnology Advances, 20, 491–515 (2003).

29. Gunardson, H. Industrial Gases in Petrochemical Processing; Marcel Dekker: New York, (1998).

30. Hartley, M., Tam, I., Non-Sequestration Utilization Options for Carbon Dioxide (CO2), Nexant´s ChemSystems PERP09/10S10, (2012).

31. Ho, M.T., Wiley, D. E. Economic Evaluation of Membrane Systems for Large Scale Capture and Storage of CO2 Mixtures. Report, UNESCO Centre for Membrane Science, The University of New South Wales (2005).

32. Hu, B., Guild, C., Suib, S.L., Thermal, electrochemical, and photochemical conversion of CO2 to fuels and value-added products, Journal of CO2 Utilization, 1, 18–27, (2013).

33. IEA. Technology Roadmap - Energy and GHG Reductions in the Chemical Industry via Catalytic Processes. Technology Perspectives. Available online at: http://iea.org/media/freepublications/technologyroadmaps/TechnologyRoadmapCatalyticProcessesAnnexes.pdf, (2013).

34. Jarungthammachote, S., Combined Partial Oxidation and Carbon Dioxide Reforming Process: A Thermodynamic Study, American Journal of Applied Sciences 8 (1): 9-14, (2011).

35. Jiang, Z.; Liao, X.; Zhao, Y., Comparative study of the dry reforming of methane on fluidized aerogel and xerogel Ni/Al2O3 catalysts. Applied Petrochemical Research, p. 1-9, (2013).

36. Kahle, L. C. S., Roussière, T., Maier, L., Delgado, K.H., Wasserschaff, G., Schunk, S.A., Deutschmann, O., Methane Dry Reforming at High Temperature and Elevated Pressure: Impact of Gas-Phase Reactions, Industrial & Engineering Chemistry Research, 52 (34), (2013).

37. Kaiser, P., Unde, R.B., Kern, C., Jess, A. Production of Liquid Hydrocarbons with CO2 as Carbon Source based on Reverse Water-Gas Shift and Fischer-Tropsch Synthesis. Chemie Ingenieur Technik. Special Issue: Reaktionstechnik. 85(4), 489–499, (2013).

38. Kawabe, K. Development of Highly Selective Process for Mono-Ethylene Glycol Production from Ethylene Oxide via Ethylene Carbonate Using Phosphonium Salt Catalyst. Catal Surv Asia 14, 111–115, (2010).

39. Kayaert, S., Martens, J., Masschaele, K. Photo-Electrochemical Cell. United States Patent Application 20130026029, (2013).

40. King, D. The Future of Industrial Biorefineries. World Economic Forum, (2010).

41. Kuwabata, S., Nishida, K., Tsuda, R., Inoue, H., Yoneyama, H., Photochemical Reduction of Carbon Dioxide to Methanol Using ZnS Microcrystallite as a Photocatalyst in the Presence of Methanol Dehydrogenase, J. Electrochem. Soc. 141(6), 1498, (1994).

42. Kuwabata, S., Tsuda, R., Yoneyama, H., Electrochemical Conversion of Carbon Dioxide to Methanol with the Assistance of Formate Dehydrogenase and Methanol Dehydrogenase as Biocatalysts. J. Am. Chem. SoC., 116, 5431-5443, (1994).

43. Kurz, G. , Teuner, S., Calcor process for carbon monoxide production, Erdoel & Kohle, Erdgas, Petrochemie, 43(5), 171-172, (1990).

44. Lachowska, M., Skrzypek, J., Ga, Mn, and Mg Promoted Copper/Zinc/Zirconia-Catalysts for Hydrogenation of Carbon Dioxide

to Methanol, Carbon Dioxide Utilization for Global Sustainability, Elsevier, Amsterdam, (2004).

45. Le, M.T.H., Electrochemical Reduction of CO2 to Methanol, MSc Thesis. Graduate Faculty of the Louisiana State University and Agricultural and Mechanical College, (2011).

46. Lee, D., Lee, J.-Y., Lee, J.S., Effects of Palladium Particle Size in Hydrogenation of Carbon Dioxide to Methanol over Pd/ZnO Catalysts, Carbon Dioxide Utilization for Global Sustainability, Elsevier, Amsterdam, (2004).

47. Levitan, R., Levy, M., Rosin, H., Rubin, R., Closed-loop operation of a solar chemical heat pipe at the Weizmann Institute solar furnace. Solar Energy Materials vol. 24 issue 1-4, 464-477, (1991).

48. Levy, M., Levitan, R., Rosin, H., Rubin, R., Solar energy storage via a closed-loop chemical heat pipe. Solar Energy, vol. 50, issue 2, 179-189, (1993).

49. Li, H.. Thermodynamic Properties of CO2 Mixtures and Their Applications in Advanced Power Cycles with CO2 Capture Processes. Energy Processes Department of Chemical Engineering and Technology Royal Institute of Technology, Stockholm, Sweden. TRITA-CHE Report 2008:58, (2008).

50. Li, H., Yan, J.. Evaluating Cubic Equations of State for Calculation of Vapor–Liquid Equilibrium of CO2 and CO2-Mixtures for CO2 Capture and Storage Processes. Chem. Eng. and Technology/Energy Process, Royal Institute of Technology, (2009).

51. Lim, M., Han, G.C., Ahn, J.W., You, K.S. Environmental Remediation and Conversion of Carbon Dioxide (CO2) into Useful Green Products by Accelerated Carbonation Technology. Int. J. Environ. Res. Public Health, 7, 203-228, (2010).

52. Li, M.W., Xu, G.H., Tian, Y.L., Chen, L., Fu, H.F. Carbon dioxide reforming of methane using DC corona discharge plasma reaction. J. Physical Chem. A., 108: 1687-1693, (2004).

53. Machado, R.B.P. Avaliação Técnica de Processo de Carbonatação de Lama Vermelha com Gás Exausto de Calcinadores de Alumina. MSc Thesis. Graduate Program in Chemical and Biochemical Process Technology, School of Chemistry, Federal University of Rio de Janeiro, (2012).

54. Maeda, K. Domen, K. 7.22 – Photochemical Water Splitting Using Nanostructured Metal Oxides. Reference Module in Chemistry, Molecular Sciences and Chemical Engineering Comprehensive

Inorganic Chemistry II (Second Edition). From Elements to Applications, 587–614. Vol. 7: Surface Inorganic Chemistry and Heterogeneous Catalysis, (2013).

55. Maiohm, T., Wannakao, S., Boekfa, B., Limtrakul, J. Production of Formic Acid via Hydrogenation of CO2 over a Copper-Alkoxide-Functionalized MOF: A Mechanistic StudyJ. Phys. Chem. C, 117, 17650–17658, (2013).

56. McCoy, S.T. The economics of CO2 transport by pipeline and storage in saline aquifers and oil reservoirs. PhD Thesis, Carnegie-Mellon University, Pittsburgh, USA, (2008).

57. Meunier, F.C., Tibiletti, D., Goguet, A., Shekhtman, S., Hardacre, C., Burch, R., On the complexity of the water-gas shift reaction mechanism over a Pt/CeO2 catalyst: Effect of the temperature on the reactivity of formate surface species studied by operando DRIFT during isotopic transient at chemical steady-state, Catalys Today, 126, 143, (2007).

58. Moon, D.J., Ryu, J.W., Kang, D.M., Lee, B.J., Ahn, B.S., CO2 Reforming by CH4 over Ni-YSZ Modified Catalysts in Carbon Dioxide Utilization for Global Sustainability, Elsevier, Amsterdam, (2004).

59. Monteiro, J.G.M.S., Silva, P.A., Araújo, O.Q.F., de Medeiros, J.L. Pareto Optimization of an Industrial Ecosystem: sustainability maximization. Brazilian Journal of Chemical Engineering, 27, 429-440, (2010).

60. Nakao, A., Macedo, A.P.F., Versiani, B.M., Araújo, O.Q.F., and de Medeiros, J.L., Modeling of Flowcharts of Permeation through Membranes for Removal of CO2 of Natural Gas, 10th International Symposium of Process System Engineering, PSE-2009, Computer Aided Chemical Engineering, 27, 1875-1880, ISBN 978-0-444-53472-9, Elsevier, (2009)

61. Navarro, R., Pawelec, B., Alvarez-Galván, M. C., Guil-Lopez, R. Al-Sayari, S., Fierro, J. L. G. Renewable SYNGAS Production via Dry Reforming of Methane, in M. de Falco et al., CO2: A Valuable Source of Carbon Green Energy and Technology, pp 45-66, Spring-Verlag London, (2013).

62. Nagaoka, K., Seshan, K., Aika, K., Lercher, J.A., Carbon deposition during carbon dioxide reforming of methane-comparison between Pt/Al2O3 and Pt/ZrO2. J. Catalysis, 197: 34-42, (2001).

63. O'Connor A.M., Ross J.R.H. The Effect of O2 Addition on the Carbon Dioxide Reforming of Methane over Pt/ZrO2. Catalysts. Catal. Today, 46 (2-3), 203–210, (1998).

64. Oi, L.E. CO2 removal by absorption: challenges in modelling. Mathematical and Computer Modelling of Dynamical Systems, 16(6), 511–533 (2010).

65. Olah, G.A., Goeppert, A., Prakash, G.K.S., Beyond Oil and Gas: The Methanol Economy, 2nd ed., Wiley VCH, Weinheim, Germany, (2009).

66. Olah, G.A., Goeppert, A., Prakash, Czaun, M., G.K.S., Bi-reforming of Methane from Any Source with Steam and Carbon Dioxide Exclusively to Metgas (CO–2H2) for Methanol and Hydrocarbon Synthesis, J. Am. Chem. Soc., 135 (2), 648–650, (2013).

67. Ormerod, W., Riemer, P., Smith, A. Carbon Dioxide Utilisation. IEA Greenhouse Gas R&D Programme, http://www.ieagreen.org.uk/sr4p.htm, (1995).

68. Pettersen, J. Snøhvit field development. Available online at http://www.ipt.ntnu.no/~jsg/ undervisning/prosessering/gjester /Lysark Pettersen2011.pdf, (2011).

69. Petrobras. Visão Geral da Petrobras. Available online at: http://www.investidorpetrobras.com.br/pt/apresentacoes/visao-geral-da-petrobras.htm

70. Pettersen, J., Snøhvit field development, TEP4520, Statoil, (2011).

71. Picardo, M.C., de Medeiros, J.L., Monteiro, J.G.M., Chaloub, R.M., Giordano, M., Ofélia, Q.F. A methodology for screening of microalgae as a decision making tool for energy and green chemical process applications. Clean Technologies and Environmental Policy, 15, 275-291, (2013a).

72. Picardo, M.C., de Medeiros, J.L., Ofélia, Q.F., Chaloub, R.M. Effects of CO2 Enrichment and Nutrients Supply Intermittency on Batch Cultures of Isochrysis galbana. Bioresource Technology, 143, 242-250, (2013b).

73. Qin, Y., Wang, X. Carbon dioxide-based copolymers: Environmental benefits of PPC, an industrially viable catalyst. Biotechnol. J., 5, 1164–1180, (2010).

74. Raju, A.S.K., Park, C.S., Norbeck, J.M., Synthesis gas production using steam hydrogasification and steam reforming. Fuel Process. Technol. 90, 330–336, (2009).

75. Richardson, J.T., Paripatyadar, S.A. , Carbon dioxide reforming of methane with supported rhodium Appl. Catal. 61, 293-309, (1990).

76. Rostrup-Nielsen, J., 40 years in catalysis, Catalysis Today, 111, 4-11, (2006).

77. Rostrup-Nielsen, J., Christiansen, L.J. Concepts in syngas Manufacture. Catalytic Series, V. 10. World Scientific, 392 pp, 2011. ISBN: 978-1-84816-567-0, (2011).

78. Saito, M., Murata, K., Development of high performance Cu/ZnO-based catalysts for methanol synthesis and the water-gas shift reaction, Catalysis Surveys from Asia, 8, 285-294, (2004).

79. Sakakura, T., Choi, J., Yasuda, H. Transformation of carbon dioxide. Chem Rev, 107, 2365-2387, (2007).

80. Shinsuke, F., Isaburo, F., Masahiro, T., Kazuhiro, O., Hiroshi, H., Muneaki, A., Kazumi, H., Kyosuke, K. A Novel Non-Phosgene Process for Polycarbonate Production from CO2: Green and Sustainable Chemistry in Practice. Catalysis surveys from Asia, 14(3-4), 146-163, (2010).

81. Song, C., Pan, W., Tri-Reforming of Methane: A Novel Concept for Synthesis of Industrially Useful Synthesis Gas with Desired H2/CO Ratios Using CO2 in Flue Gas of Power Plants without CO2 Separation, Prepr. Pap.-Am. Chem. Soc., Div. Fuel Chem., 49 (1), 128, (2004).

82. Song, C. Global Challenges and Strategies for Control, Conversion and Utilization of CO2 for Sustainable Development Involving Energy, Catalysis, Adsorption and Chemical Processing. Catalysis Today, 115, 2–32, (2006).

83. Souza, M.M.V.M., Schmal, M., Methane conversion to synthesis gas by partial oxidation and CO2 reforming over supported platinum catalysts, Catalysis Letters, 91(1–2), 11-17, (2003).

84. Souza, L.F.S., Ferreira, P.R.R., de Medeiros, J.L.M., Alves, R.M.B., Araújo, O.Q.F. Production of DMC from CO2 via Indirect Route: Technical–Economical–Environmental Assessment and Analysis Sustainable Chem. Eng., dx.

85. Span, R., Wagner, W. A New Equation of State for Carbon Dioxide Covering the Fluid Region from the Triple-Point Temperature to 1100 K at Pressures up to 800 MPa. J. Phys. Chem. Ref. Data, 25(6), (1996).

86. Spath, P.L., Dayton, D.C., Preliminary Screening – Technical and Economic Assessment of Synthesis Gas to Fuels and Chemicals with Emphasis on the Potential for Biomass-Derived SYNGAS. National Renewable Energy Lab Golden Co., (2003).

87. Stoczynski, J., Grabowski, R., Koslowska, A., Lachowska, M., Skrzypek, J., Effect of Additives and a Preparation Method on Catalytic Activity of Cu/ZnO/ZrO2 System in the Carbon Dioxide Hydrogenation to

Methanol, Carbon Dioxide Utilization for Global Sustainability, Elsevier, Amsterdam, (2004).

88. Tanaka, Y., Utaka, T., Kikuchi, R., Sasaki, K., Eguchi, K., CO removal from reformed fuel over Cu/ZnO/Al2O3 catalysts prepared by impregnation and coprecipitation methods, Applied Catalysis A: General, 238, 11–18, (2003).

89. Teuner, S., Make carbon monoxide from carbon dioxide, Hydrocarbon Processing, International Edition, 64(5), 106-7, (1985).

90. Teuner, S. et al., The Calcor Standard and Calcor Economy Processes, in Oil and Gas European Magazine, 44-46, (2001).

91. Treacy, D., Ross, J.R.H. The Potential of the CO2 reforming of CH4 as a method of CO2 mitigation. A thermodynamic study. Prepr. Pap.-Am. Chem. Soc., Div. Fuel Chem., 49 (1), 126–127, (2004).

92. Trusler, J. P. M. Equation of State for Solid Phase I of Carbon Dioxide Valid for Temperatures up to 800 K and Pressures up to 12 GPa. J. Phys. Chem. Ref. Data, 40 (4), 043105-1-043105-19,

93. Van-Dal, E.S., Bouallou, C. Design and simulation of a methanol production plant from CO2 hydrogenation. Journal of Cleaner Production, 57, 15 October 2013, 38–45, http://dx.doi.org/10.1016/j.jclepro.2013.06.008

94. Ushikubo, T. Green Chemistry in Japan. Chemistry International, 35(4), July-August 2013. Available online at www.iupac.org/publications/ci, (2013).

95. Wagener, D.H.V., Rochelle, G.T. Alternative Stripper Configurations for CO2 Capture by Aqueous Amines. AIChE Spring Meeting, (2010).

96. Wagener, D.H.V., Rochelle, G.T. Stripper configurations for CO2 capture by aqueous monoethanolamine. Chemical Engineering Research and Design, 89, 1639–1646, (2011).

97. Walther, D., Ruben, M., Rau, S. Carbon Dioxide and Metal Centres: from Reactions Inspired by Nature to Reactions in Compressed Carbon Dioxide as Solvent. Coordination Chemistry Reviews 182, 67–100, (1999).

98. Wang, W., Stagg-Williams, S.M., Noronha, F.B., Mattos, L.V.,Passos, F.B., Partial Oxidation and Combined Reforming of Methane on Ce-promoted Catalysts, Prepr. Pap.-Am. Chem. Soc., Div. Fuel Chem. 49 (1), 132-133, (2004).

99. Wu, J.C.S., Lin, H.-M., Photo reduction of CO2 to methanol via TiO2 Photocatalyst, International Journal of Photoenergy, 7, 115-119, (2005).

100. Wurzel, T., Malcus, S., Mleczko, L., Reaction engineering investigations of CO2 reforming in a fluidized-bed reactor. Chem. Eng. Sci., 55, 3955-3966 (2000).

101. Yamashita, H., Fujii, Y., Ichihashi, Y., Zhang, S. G., Ikeue, K., Park, D. R.,Koyano, K., Tatsumi, T., Anpo, M., Selective formation of CH3OH in the photocatalytic reduction of CO2 with H2O on titanium oxides highly dispersed within zeolites and mesoporous molecular sieves, Catalysis Today 45, 221-227, (1998).

102. Zangouei, M., Moghaddam, A.Z., Razeghi, A., Omidkhah, M.R., Reforming and Partial Oxidation of CH4 over Ni/Al2O3 Catalysts in Fixed-bed Reactor International Journal of Chemical Reactor Engineering, 8. Available at: http://works.bepress.com/ mohammadreza _omidkhah/2, (2010).

103. Zhang, X., Lee, C. S.-M., Mingos, D.M.P., Hayward, D. O., Carbon dioxide reforming of methane with Pt catalysts using microwave dielectric heating. Catalysis Letters, 88(3–4), 129-139, (2003).

104. Zhou, H., Cao, Y., Zhao, H., Liu, H., Wei-Ping, P., Investigation of H2O and CO2 Reforming and Partial Oxidation of Methane : Catalytic Effects of Coal Char and Coal Ash, Energy & Fuels, 22, 2341-2345, (2008).

Chapter 4

Carbon Sequestration in Tropical Forest Stands: Its Control by Plant, Soil and Climatic Factors

K. A. J. M. Kuruppuarachchi[1], Gamini Seneviratne[2], Buddhika D. Madurapperuma[3]

[1]*Department of Botany, The Open University of Sri Lanka, Nawala, Sri Lanka*

[2]*Microbial Biotechnology Unit, Institute of Fundamental Studies, Sri Lanka*

[3]*Department of Environmental Science and Management, Humboldt State University, Arcata, CA, USA*

ABSTRACT

This study investigated the relationships amongst floristic, soil and climatic parameters and their control on carbon sequestration (CS) in two selected forest stands of Sri Lanka. Representative sampling sites were selected from the dry zone (Sigiriya forest sanctuary) and the wet zone (Udawattakele forest reserve) of Sri Lanka. Litter and soil samples were collected from each sampling site randomly in monthly intervals to cover an annual cycle. Plant biomass carbon stocks were calculated using standard biomass equations. Soil carbon stocks were determined by chemical oxidation and loss on ignition (LOI) methods. Principle Factor Analysis and multiple regression were used to quantify the relationships among the plants, soil and climatic variables. Plant biomass carbon stocks of the forests were governed by labile and stable C fractions, soil moisture, and plant diversity. The soil fulvic fraction acts as a focal point of interacting the variables such as soil N, free litter fraction (FLF) and humic fraction. During dry period in the dry zone forest, CS was governed by maximum relative humidity through an atmosphere-floor litter-soil continuum. Air temperature and FLF play a vital role in determining soil N. In addition, MacIntosh distance (U) diversity index showed a significant positive relationship with soil N. The dry zone forests are seen to be more climatic sensitive and vulnerable than the wet zone forests in Sri Lanka due to influence of more climatic parameters that govern the soil organic carbon fractions.

Keywords: Tropical Forests, Carbon Sequestration, Soil Organic Carbon

1. INTRODUCTION

The Kyoto protocol mainly summarized the effective ways of reducing greenhouse gases such as CO_2, which can ideally be controlled through forest plantation supporting carbon sequestration (CS). Therefore, it is imperative to trace controlling factors of CS in the forest ecosystems, because the factors can then be manipulated in order to improve C storage in forests. In particular, forests in tropics should be managed to maximize C retention in aboveground and belowground biomass and to conserve soil C (Beedlow et al., 2004) . The multiple associations between traits and ecosystem processes can help to identify predictable trait-service clusters. These clusters depend on traits of plants and soil organisms that underlie nutrient cycling (Bello et al., 2010) . Such combined approaches are a crucial step in ecosystem service monitoring and in balancing the delivery of multiple and sometimes conflicting services in ecosystem management.

Studies on CS confirmed that tropical forestry is the most cost-effective greenhouse gas mitigation strategy (Ramachandran et al., 2007) . Sri Lanka is recognized as vulnerable to climatic change under the small island category (UNFCCC, 1997; IPCC, 2001) with huge community types due to diversity of rainfall and the elevation (Legg & Jewell, 1995) . Nearly 80% of the land mass covers the dry zone and the rest are wet and intermediate zones (Panabokke, 1996) . The southwestern region in the wet zone is mainly covered by the wet evergreen forest. The dry zone has tropical dry mixed evergreen forests (Bandaratillake, 1996) . The forest cover in Sri Lanka is approximately 2.1 million hectares, representing 31% - 32% of the total land mass (NPPD, 2002) .

Plant functional traits control a variety of forest processes, including soil C storage (De Deyn et al., 2008) . The plant traits regulate net soil C storage by controlling C assimilation, its transfer and storage in the belowground biomass, and its release from soil through respiration, fire, and leaching. However, understanding of these processes is still incomplete. There is a huge knowledge gap for understanding the control of floristic diversity on ecosystem C storage in tropical forests (Clark et al., 2001; IPCC, 2006; Rossi et al., 2009) . Therefore, more research is needed to evaluate forest C pools and their control by floristic, climatic, and edaphic factors. An attempt is made here to recognize the factors, which control and/or contribute to biomass CS in the dry and the wet zone forests. This article studies the relationships among the plant, soil and climatic parameters and their control on CS of selected dry and wet forests of Sri Lanka.

2. MATERIALS AND METHOD

2.1. Study Area

Representative sampling sites were selected from a dry zone forest (Sigiriya sanctuary) and a wet zone forest (Udawattakele) of Sri Lanka (5°54'N - 9°52'N; 79°39'E - 81°53'E). The Dry Zone Forest (DZF), Sigiriya sanctuary, surrounds the ancient rock fortress of Sigiriya, an archeologically important world heritage site at 150 m altitude. In the 1990s, it was declared a sanctuary with 5099 hectares and belongs to Ceylonese monsoon forest (Green, 1990) . Proximity to Sigiriya wildlife office of Kimbissa was selected as the sampling site (Figure 1).

The Wet Zone Forest (WZF) Udawattakelle national heritage, Sri Lanka lies as a small hill range arising from the north of Kandy Lake on the periphery of the ancient "Temple of Tooth Relic" Kandy-The Kings Palace (7°19'N - 80°39'E). Udawattakelle forest reserve (111 ha) was declared as a sanctuary in 1938. The sanctuary has been designated as a national Man and Biosphere (MAB) reserve (Green, 1990) . Altitudes of the area vary from 500 to 800 m. The forest is an isolated cluster with surrounding land settlements. Udawattakelle is a wet semi-evergreen forest in the central hilly region with high species richness (Pushpakumara & Hitinayake, 2001) and about 460 plant species including 135 tree species and 11 liana species, of which 9 species are endemic to Sri Lanka (Wedathanthri & Hitinayake, 1999; Senarathna, 2005) . Vegetation comprises dense forest, mostly plantation and secondary formation. A plaque in front of "Senanayake Buddhists hermitage" was selected to lay out permanent sampling plots with a flat terrain.

2.2. Climate

Sri Lanka is divided into three climatic zones based on annual rainfall and its distribution: Dry Zone (1250 - 1525 mm) in parts of the southeast and northwest; Intermediate Zone (1525 - 2280 mm); and Wet Zone (2280 - 5100 mm) mostly in the southwestern parts (Premaratne & Premalal, 2005) . Climatic data for the present study were collected from the nearest meteorological stations (dry zone: Palwehera; wet zone: Katugasthota). Monthly average climatic data for ten years at the dry zone revealed 27.7°C as average air temperature, 1380 mm as

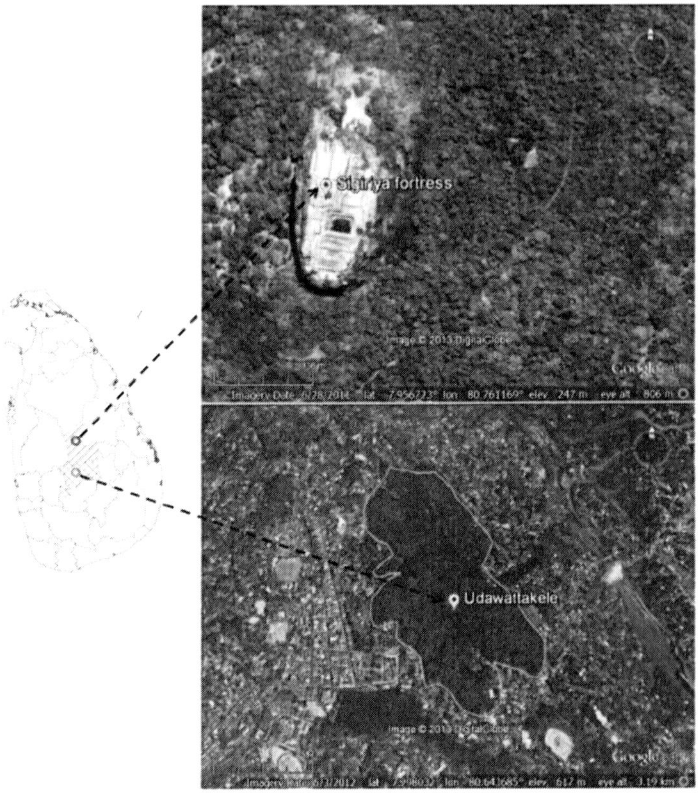

Figure 1. Map of study sites depicting Sigiriya forest sanctuary and Udawattakele forest reserve in Sri Lanka.

annual rainfall, and 79% as monthly relative humidity. Rainfall pattern of the study period (January-December 2007) slightly deviated from the normal pattern, i.e. inter monsoonal rainfall during March and April was lower in the study period. Northeastern monsoonal rainfall was higher than the normal rainfall. Generally, the dry zone experiences a five-month dry spell from May to end of September. For the Wet zone area, Udawattakelle, climatic conditions include Southwest monsoons from mid-May to July/August and Northeast monsoons October/November to late December. Annual rainfall is 1850 mm without a prominent dry spell and temperature is around 24˚C throughout the year (Cramer, 1993; Naoum & Tsanis, 2003; Petkovsek & Mikos, 2005) .

2.3. Design of Experimental Plots

Four sampling plots, 2 m × 2 m, were demarcated randomly in the dry zone forest and the wet zone forest. Around a central square of 1 m × 1 m of each plot, a trench was dug to a depth of 30 cm and a width of 15 cm (Figure 2). A thick polythene sheet was fixed vertically to cut off the fine root growth into the 1 m × 1 m square, and the trench was filled with soil. A Nylon mesh sized 2 mm (2 m × 2 m) was fixed 20 cm above the plot as a litter-trap. Then, the plot was fenced by thick polythene sheets (black and white) to protect from wild animals

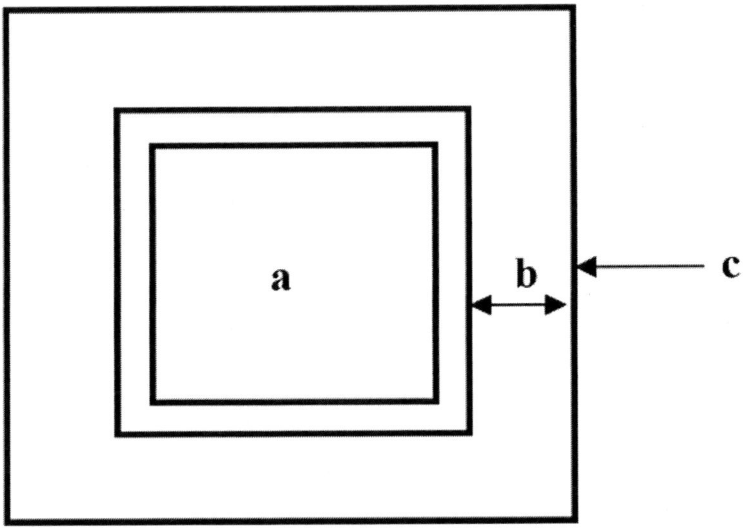

Figure 2. Permanent sampling plot (the layout). Position-a: located in the 1 m × 1 m square of the plot (no root growth or litterfall). Position-b: located outside the 1 × 1 m square (root growth occurs, but no litterfall). Position-c: located outside the 2 × 2 m plot (root growth as well as litterfall occur, i.e. forest floor).

and to stop litter movement. A floor litter layer of a 30 cm × 30 cm in each plot was collected. The remainder of the litter of the plot was removed. The collected litter was oven-dried (65°C for 72 hrs.), and dry weight was recorded and stored.

2.4. Soil and Litter Sampling

Soil and litter samples were collected at monthly intervals to cover a complete annual cycle. Litter accumulation in the litter trap (2 m × 2 m) was sorted into fine (leaves, flowers etc.) and coarse (woody parts) litter. Initial fresh weight was recorded and then the samples were oven-dried (65°C for 72 hrs.), weighed and stored. In addition, a litter sample was collected from 30 cm × 30 cm sharp-edged steel square outside the plot, oven-dried, weighed and stored. Three soil samples were collected using an auger (cross section area 19.6 cm^2), one from the central square of the 1 m × 1 m (position-a) and one from outside the 1 m × 1 m square within the 2 m × 2 m plot (position-b) to a depth of 25 cm. Another soil core sample was taken to a depth of 25 cm outside the permanent plot (position-c).

2.5. Preparation of Soil Samples

Soil samples were air-dried and sieved (<2 mm). Root fractions (fine roots <2 mm, and coarse roots >2 mm) were hand-picked, oven-dried, weighed and stored for later analyses. Very fine root fraction was separated by the floating method (Anderson & Ingram, 1998) . The fine root fraction was oven dried at 60˚C for constant weight (Hodge et al., 2000) , and the root biomass was recorded as a percentage of total soil weight. Litter samples were first cleaned with a brush and separated into fine (leaves, flowers and buds) and coarse (twigs, plant materials such as seeds and seed coats) litter. Initial fresh weight was recorded. Then, they were oven dried at 65˚C to a constant weight and dry weights recorded (Hairiah et al., 2006) . Dry samples were ground and stored for further chemical analyses.

2.6. Soil Analyses

Soil moisture was determined by oven drying the samples at 105˚C to a constant weight. Soil pH was measured using a glass calomel electrode fixed to a pH meter (Schofield and Taylor, 1955) with a soil to water ratio of 1:2. The wet oxidation method without external heating was used to determine C content colorimetrically at 600 nm by using a UV-Visible spectrophotometer (Walkley, 1934; Walkley & Black, 1947; Anderson & Ingram, 1998) . Total N in the soil was determined by using the Micro-Kjeldahl procedure (Bremner & Tabatabai, 1972) . The digestion was performed by heating the sample with concentrated H_2SO_4 at 420˚C for 3 hrs. Ammonium-N content in the digest was determined by using Autoanalyzer (Tecator 1030 KjeltecAutoanalyzer) (Bremner & Tabatabai, 1972) . Fractionation of soil C was done by the weight loss on ignition (LOI) method. In this procedure, soil was subjected to sequential thermal oxidation from 100˚C to 550˚C using a muffle furnace (Gallenkamp-box furnace F SL-340-0160). Weight loss steps of temperatures between 150˚C - 200˚C, 200˚C - 400˚C and 400˚C - 550˚C corresponded to oxidation of free soil litter fraction (labile), fulvic fraction (intermediate) and humic fraction (stable) respectively (Evans et al., 2001; Ratnayake et al., 2007) .

2.7. Litter Analyses

Dry and wet weights of litterfall (coarse and fine litter separately) as well as floor litter layer were recorded in each sampling time and litter density in grams per square meter was calculated (Hairiah et al., 2006) . Annual litterfall was then calculated for the study sites. Litter moisture was measured through oven drying the litter (65°C) till a constant weight was reached, and weight difference between dry and wet litter was recorded. Thereby, litter moisture was calculated.

2.8. Aboveground and Belowground Biomass and C Stock

Two internationally accepted biomass regression equations (Brown et al., 1989; Anderson & Ingram, 1998) were applied to calculate aboveground biomass stock using DBH. Thus, an equation developed for dry forests with annual rainfall < 1500 mm; $B = \exp(-1.996 + 2.32 \times \ln D)$; $R^2 = 0.89$ (Brown et al., 1989) was used, where B and D are biomass and DBH, respectively. The equation used for calculation of aboveground biomass was $B = \exp(-2.134 + 2.530 \times \ln(D))$ for wet zone forest (Anderson & Ingram, 1998) .

According to IPCC guidelines for national greenhouse gas inventories (IPCC, 2006) , aboveground C stock of tropical forests is calculated as Biomass ×0.49 (Hughes et al., 2000) . Aboveground to belowground biomass ratio, and hence its C ratio, for tropical rain forests (annual rainfall >2000 mm) and moist deciduous forests (annual rainfall between 1000 - 2000 mm) was found to be 0.2 (Mokany et al., 2006) . Therefore, belowground biomass or C content is calculated as aboveground biomass or C ×0.2.

2.9. Climatic, Soil, and Plant Relationships

Plant, soil and climatic data of the two forests were pooled and analyzed for detecting grouping of variables responsible for plant CS using Principal Factor Analysis (PFA). The main applications of factor analytical techniques are to reduce the number of variables and to detect structure in the relationships among variables (i.e. to classify variables) (De Genus & Denys, 2004) . Therefore, factor analysis is applied as a data reduction and structure detection method. In data analysis, first, Principal Factor Analysis (PFA) was performed and factor plots were constructed to identify highly correlated variables clustering on the plot (Vereecken & Herbst, 2004) . Varimax rotation was applied to rotate the factor structure. Then, the correlated variables were used in multiple regression analysis using MaxR for selection of variables to enter the equation. In addition, T-test was used to compare means and simple regression analysis was performed to establish univariate relationships found in data analysis. Then diagrammatical models were developed to represent relationships of the above factors.

The data pooling is justifiable as plant CS is governed basically by physico-chemical parameters, but not by forest types etc. (Hessen et al., 2004; Toledo et al., 2010) . For example, the same forest type (e.g. montane forests) existing in different regions have different plant C stocks, mainly due to different physico-chemical parameters (Keith et al., 2009) . The pooling also helped to have a sufficient data-set for the analysis.

3. Results

Our results showed that the soil moisture, NH_4^+ -N, soil organic carbon (SOC) and total N contents were higher in the wet zone forest than the dry zone forest (Table 1). In the dry zone forest NO_3^--N content and bulk density were higher than the wet zone forest. In the wet zone forest, aboveground biomass (201 t/ha) and belowground biomass (48 t/ha) C stocks were significantly high compared to the dry zone forest (60 t/ha and 17 t/ha, respectively). There was a higher amount of fine root biomass and floor litter C stock in the dry zone forest than in the wet zone forest. Total ecosystem C stock of the wet zone forest (ca. 361 t/ha) was significantly higher ($p \leq 0.05$) than the dry zone forest (ca. 176 t/ha).

The soil C fractions in the dry zone forest and the wet zone forest were dominated by fulvic fraction followed

Table 1. Initial soil properties and carbon stocks of the dry zone forest and the wet zone forest.

Parameter	Dry Zone Forest	Wet Zone Forest
Soil properties		
Moisture (%)	10.3 ± 4.5	18.2 ± 1.4
pH	6.19 ± 0.38	5.78 ± 0.26
NH_4^+ -N (µg/g)	0.029 ± 0.007	0.173 ± 0.034
NO_3^- -N (µg/g)	0.182 ± 0.013	0.049 ± 0.008
SOC (%)	1.48 ± 0.25	1.58 ± 0.15
Total N (%)	0.097 ± 0.023	0.138 ± 0.08
Bulk density (g/cm^3)	1.4 ± 0.28	1.27 ± 0.15
Carbon stock (t/ha)		
Above ground biomass C stock	60.2 (34.2%)b	201.0 (55.6%)a
Below ground biomass C stock	16.9 (9.6%)b	48.0 (13.3%)a
Total biomass C stock	77 (43.8%)b	249 (68.9%)a
Fine root biomass C stock	2.7 (1.53%)a	2.4 (0.66%)a
Annual aboveground biomass increment	4.47 (2.54%)a	4.85 (1.34%)a
Litter fall C stock	4.72 (2.68%)a	5.73 (1.58%)b
Floor litter C stock	2.36 (1.34%)a	2.01 (0.55%)b
Total soil C stock	94 (53.3%)b	108 (29.9%)a
Total ecosystem C stock	176b	361a

Mean ± SE. Means in the same column followed by the same letter are not significantly different at 5% probability level, according to T-test. Prediction of total biomass of forest stands calculated using biomass regression equation. i.e., $B = \exp(-1.996 + 2.32 \times \ln(D))$ for dry forest (Brown et al., 1989) $B = \exp(-2.134 + 2.530 \times \ln(D))$ for moist forest (Anderson & Ingram, 1998). (B = Biomass Stock, D = DBH). Values within parentheses are magnitudes of component C stocks as a percentage of total ecosystem C stock.

by humic fraction then free soil litter. Fulvic fraction represented ca. 50% of total C, ca. 30% humic fraction and ca. 20% free soil litter (Figure 3). From January to August, there was a declining trend of free soil litter resulting in a simultaneous increase of fulvic fraction. Compared to the wet zone forest, the dry zone forest consists of higher amount of fulvic fraction. There was an accumulation of free soil litter fraction with a decline of fulvic fraction. There was a considerable decrease of humic fraction in the wet zone forest during January-February and then humic fraction increased in the short dry season (April-May). However, in the dry zone forest more humic fraction accumulates in wet season (August-September).

Pooled data of the two forests were analyzed using Principal Factor analysis (PFA) with varimax rotation. A single cluster was separated in the factor plot, which included factors influencing plant biomass C stocks (Figure 4). The variables included in the same cluster were biomass C, free litter fraction, fulvic fraction, humic fraction, soil moisture, organic C, total litterfall, soil N, alpha index, and rainfall. Multiple regression analysis of the variables yielded the following relationship.

$$\text{Plant biomass C stock} = -3238 + 6.81\text{FLF} - 8.54\text{HF} + 2215\text{SM} + 13463\text{MD}; R^2 = 0.94$$

where FLF, HF, SM and MD are free litter fraction, humic fraction, soil moisture, and McIntosh distance (MD), respectively. In addition, a positive correlation for soil nitrogen and McIntosh distance (MD) was resulted in second order regression model (Figure 5). Finally, we developed two models using the most influential factors, which contribute to CS in the dry zone forest and the wet zone forest (Figure 6 and Figure 7).

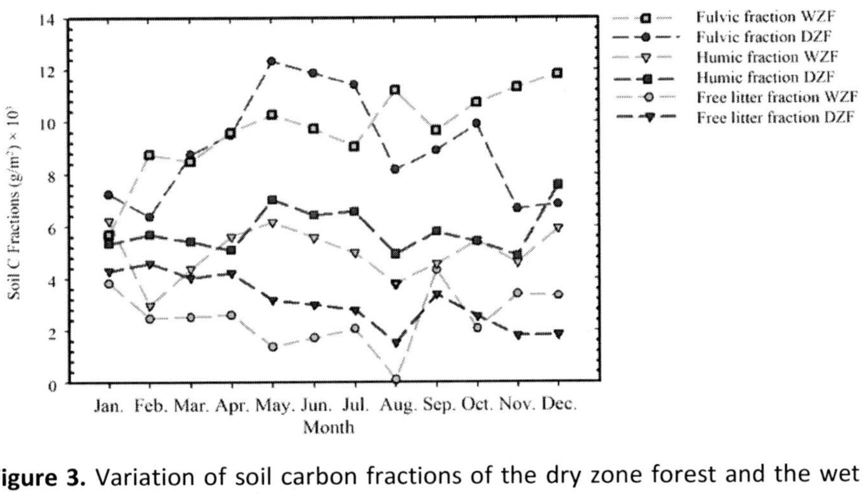

Figure 3. Variation of soil carbon fractions of the dry zone forest and the wet zone forest within an annual cycle.

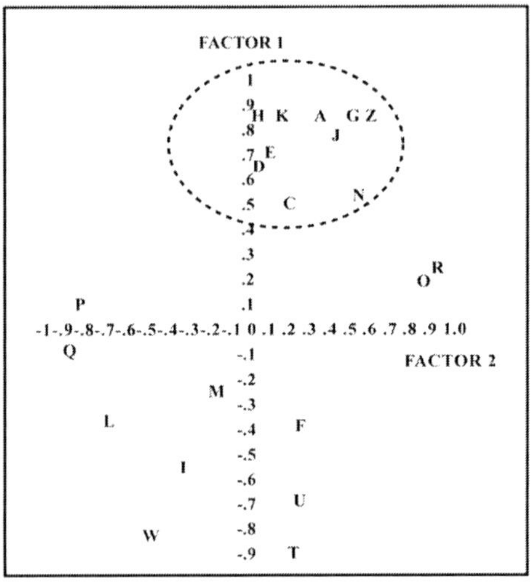

Figure 4. Factor plot of variables responsible for plant biomass carbon sequestration of the dry and wet zone forests (pooled data). Highly correlated variables clustered on the plot are encircled with a dotted circle. A = biomass carbon, B = biomass stock, C = free litter fractions, D = fulvic fractions, E = humic fractions, F = fine root biomass, G = soil moisture, H = soil organic carbon, I = total floor litter, J = total litterfall, K = soil nitrogen, L = soil pH, M = C/N ratio, N = alpha index, O = Shannon J, P = berger-parker dominance, Q = simpons diversity, R = hill number (Species richness), S = mackintosh distance (U), T = mackintosh evenness (E), U = total tree density, V = total basal area, W = minimum temperature, X = maximum temperature, Y = average temperature, Z = rainfall and alliance variables.

4. DISCUSSION

Soil moisture is fundamental to plant growth, because water exchange is an important complement to the forest C cycle (Chapin et al., 2009) . Depending on the availability, labile C constrains microbial action that determines nutrient cycling (Ekberg et al., 2007; Kuzyakov et al., 2007) , which in turn limits plant growth. On the other hand, aboveground floristic composition affects the accumulation of labile C fractions (Bardgett et al., 2005) . Clark et al. (2003) reported that species composition and element inputs fundamentally alter the relationship between climate drivers and ecosystem C dynamics. This could be due to acceleration of soil microbial action with the availability of soil moisture (Lupwayi et al., 2010) .

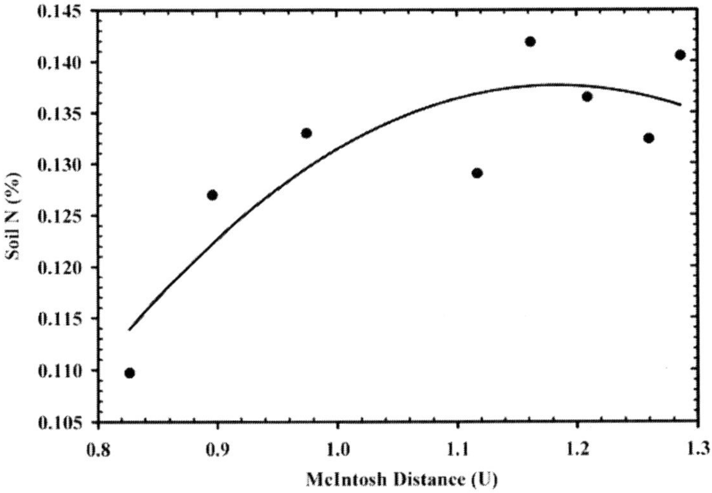

Figure 5. Relationship between Mackintosh distance (U) and soil nitrogen of the dry and wet zone forests (pooled data). Equation for the model is Y = 0.14/(1 + 3841e – 11.7x) and R^2 = 0.82.

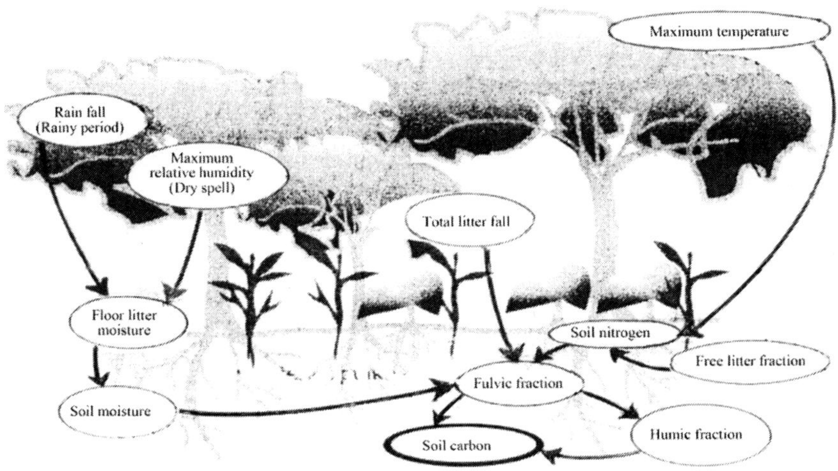

Figure 6. Diagrammatic representation of the network of factors controlling soil carbon sequestration in the dry zone forest. Arrows indicate influences or flows of matter.

The most abundant plants in the dry zone of Sri Lanka belong to Fabaceae, which can fix nitrogen using symbiotic association. This could be a reason for high NO_3^--N content in the dry zone forest soil. The reason for higher fine root growth of dry zone forest is for up-taking water efficiently and the survival successional stager's process (in five months dry spell). The distribution of aboveground and belowground biomass is largely influenced by the plant species and growth forms within spatial gradients in soil moisture and other edaphic conditions (Wang et al., 2008) .

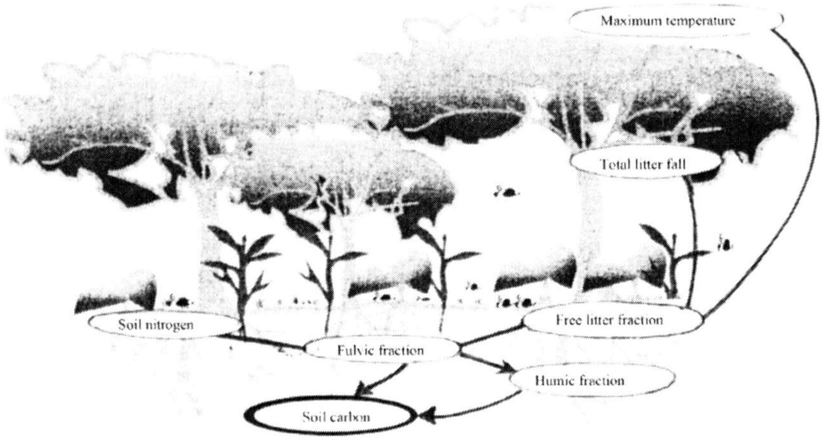

Figure 7. Diagrammatic representation of the network of factors controlling soil carbon sequestration in the wet zone forest. Lines represent interactions whereas arrows indicate influences or flows of matter.

Leaf and root litter are major nutrient suppliers to natural forest ecosystems. Wet zone plants have broad leaves and therefore they contributed more leaf litter in the soil. Litter decomposition in the wet zone is high due to high moisture content, which aids in active microbial growth. This leads to an increase of soil C stock in the wet zone forest. In contrast, the dry zone forest showed a high accumulation of floor litter C stock, which resulted in defoliation due to environmental stress and due to less microbial decomposition in five months dry spell (Table 1). However, soil C stock in the dry zone forest was less compared to the wet zone forest resulting in a reduced rate of decomposition of organic matter, which produces less CO_2. Comparatively higher plant biomass C, total soil C, and ecosystem C were recorded at the wet zone forest due to high soil moisture and labile organic matter fraction (Table 1).

There was an increased production of fine roots in the dry zone forest compared to the wet zone forest attributed to a simultaneous leaf flush on the canopy with a green-up (Kuruppuarachchi et al., 2013) . Fine root growth in the dry zone forest seemingly was fuelled by N supply from the organic matter through microbial growth and action utilizing labile soil C and the water supply through the atmosphere-floor litter-soil continuum (Figure 6). The fine root growth in the dry zone forest indicates that the plants have developed a survival mechanism in response to dryness to absorb water as efficiently as possible.

McIntosh distance (U) determines plant species richness. When the index is high, the species richness is also high. Thus, high values of the index contribute to high plant biomass C stocks. However, the biomass C accumulation is soil nutrient-limited, particularly N (Cleveland & Townsend, 2006) . Apparently, this has been overcome by increased soil N accumulation, possibly through biological nitrogen fixation under high species richness and abundance (i.e. McIntosh index) (Figure 5). In addition, McIntosh distance (U) index showed the significant positive relationship with soil N. Generally, soil N is limited in tropical forests (Cleveland & Towsend, 2006) . High species richness and spatial distribution (i.e. high McIntosh index) would reduce the N limitation.

Janzen et al. (1997) and Weil and Magdoff (2004) reported that stable C fractions establish a relatively larger pool of C than the labile fractions in the soils as shown by our results (Figure 3). This is attributable to the decomposition and transformation of the free soil litter to the fulvic fraction (Zalamea et al., 2007; Gulde et al., 2008) . This could be due to fresh C supply from floor litter layer that dissolved into organic C, which helps rapidly decompose humic fractions by active microbial growth after the rains (Qualls, 2005; Johnson et al., 2007; Müller et al., 2009) .

Organic matter fractions sequestered in the two forests showed that all three fractions were significantly higher in the wet zone forest than the dry zone forest (Table 1). Thus, the wet zone forest soil had a larger C storage than that of the dry zone forest soil. It may be due to absence of moisture limitation soils in the wet zone forests as dry zone forest. Previous studies reported that soil C storage in tropical forests ranges between 90 - 200 t C/ha (Brown & Lugo, 1984; Kimble et al., 1990; Amundson, 2001; Lewis et al., 2009) . Therefore, it is apparent that the soil C sequestration capacity of the two forests stands lies in the lower range of the tropics. The SOC results suggest that the dry zone forest needs excess water to maintain stable carbon fraction hence CS and wet forest prefer water stress for soil CS.

Results of biotic/abiotic relationships towards CS indicates that plant biomass C stock is governed by labile (FLF) and stable (HF) C fractions, soil moisture and plant species diversity (MD) of the selected forests. The size of stable C pool i.e. humic fraction governs plant growth, because a larger pool in the soil restricts the availability of C sources to microbes, as reflected by negative relationship in the above equation. Soil moisture is the fundamental need for plant growth (Chapin et al., 2009) . Labile fraction (FLF) of forest soil drives nutrient cycling, which is important for microbial action (Ekberg et al., 2007) . A large stable C pool would restrict the supply of the labile C source to microbes, as reflected by the negative relationship with biomass C accumulation. The soil fulvic fraction was found to act as a focal point of interacting with or influencing the other variables such as soil N, free litter fraction, humic fraction etc. Thus, these variables determine soil CS of the wet zone forest and the dry zone forest (Figure 6 & Figure 7). Our study revealed that air temperature and labile fraction (FLF) play a vital role in determining soil N, as also shown by Singh and Kashyap (2007) and Rice (2006) , respectively.

The dry zone forest CS model showed that maximum relative humidity and rainfall during dry and wet periods, respectively, governed soil moisture through an atmosphere-floor litter-soil continuum (Figure 6). Soil moisture directly influenced the fulvic fraction in the forest soil. The model developed for the dry zone forest indicated that the control of CS is more complicated than that of the wet zone forest (Figure 6 and Figure 7).

It is projected that climate change will affect the dry zone more (the northeast and the east) with a decline in rainfall and an increase in temperature, evapotranspiration and soil moisture deficit (De Silva, 2006; De Silva et al., 2007; Eriyagama et al., 2010) . Similarly, it has been predicted that there would be a northward shift of tropical wet forest into areas currently occupied by tropical dry forest (Somaratne & Danapala, 1996) . Thus, it is clear that more interactions with climatic parameters would arise in the dry zone forests in the future climatic change scenarios. As such, the dry zone forests are seen to be more climatic sensitive and vulnerable than the wet zone forests.

5. CONCLUSION

This study showed that the humic fraction is more stable and a mirror of the free litter and the fulvic fractions, which contributes ca. 30% of the total soil C. With availability of soil moisture, free litter fraction transformed into fulvic fraction resulted in ca. 20% and ca. 50% of the total soil C respectively. The study suggests that storing SOC in deeper soil layers would be an effective management strategy for CS, which transforms free litter fraction to fulvic and humic fractions, while reducing CO_2 emission though microbial activities. Fine root biomass accumulation in the dry zone forest aids in efficiency of water absorption of plants especially in the dry spell. Consequently, leaf flushing and canopy green-up were observed in the dry spell of the dry zone forest to facilitate root growth. The dry zone forest showed that maximum relative humidity is the most influential factor to govern CS, through an atmosphere-floor litter-soil continuum. Carbon sequestration process in the dry zone would be more vulnerable to climatic change in future due to less rainfall and hence moisture scarcity in the soil.

ACKNOWLEDGEMENTS

This research study was carried out at the Microbial Biotechnology Unit (MBU), Institute of Fundamental Studies (IFS), Kandy, Sri Lanka. We would like to thank the relevant authorities of the IFS for providing us excellent working facilities and we are very grateful to Ms. Kumuduni Karunaratne and Mr. K.K. Karunadasa, IFS Technical Officers, for providing invaluable support in the laboratory as well as in the field.

REFERENCES

1. Amundson, R. (2001). The Carbon Budget in Soils. Annual Review of Earth and Planetary Sciences, 29, 535-562.
2. Anderson, J. M., & Ingram, J. S. I. (1998). Tropical Soil Biology and Fertility: A Handbook of Methods (2nd ed.). CABI International, Wallingford, UK.
3. Bandaratillake, H. M. (1996). Eucalyptus Plantations in Sri Lanka. In M. Kashio, & K. White (Eds.), Environmental, Social, Economic, and Policy Issues (pp. 193-212). Bangkok: FAO Regional Office for Asia and the Pacific, RAP Publication 2.
4. Bardgett, R. D., Bowman, W. D., Kaufmann, R., & Schmidt, S. K. (2005). A Temporal Approach to Linking Aboveground and Belowground Ecology. Trends in Ecology & Evolution, 20, 635-641.
5. Beedlow, P. A., Tingey, D. T., Phillips, D. L., Hogsett, W. E., & Olszyk, D. M. (2004). Rising Atmospheric CO2 and Carbon Sequestration in Forests. Frontiers in Ecology and the Environment, 2, 315-322.
6. de Bello, F., Lavorel, S., Díaz, S., Harrington, R., Cornelissen, J. H. C., Bardgett, R. D., Berg, M. P., Cipriotti, P., Feld, C. K., Hering, D., da Silva, P. M., Potts, S. G., Sandin, L., Sousa, J. P., Storkey, J., Wardle, D. A., & Harrison, P. A. (2010). Towards an Assessment of Multiple Ecosystem Processes and Services via Functional Traits. Biodiversity and Conservation, 19, 2873-2893.
7. Bremner, J. M., & Tabatabai, M. A. (1972). Use of an Ammonia Electrode for Determination of Ammonium in Kjeldahl Analysis of Soils. Communications in Soil Science and Plant Analysis, 3, 159-165. Brown, S. A., Gillespie, J. R., & Lugo, A. E. (1989). Biomass Estimation Methods for Tropical Forests with Applications to Forest Inventory Data. Forest Science, 35, 881-902.
8. Brown, S., & Lugo, A. E. (1984). Biomass of Tropical Forest: A New Estimate Based on Forest Volume. Science, 223, 1290-1293.
9. Chapin, F. S., McFarland, J., McGuire, A. D., Euskirchen, E. S., Ruess, R. W., & Kielland, K. (2009). The Changing Global Carbon Cycle: Linking

Plant-Soil Carbon Dynamics to Global Consequences. Journal of Ecology, 97, 840-850.

10. Clark, D. A., Brown, S., Kicklighter, D. W., Chambers, J. Q., Thomlinson, J. R., Ni, J., & Holland, E. A. (2001). Net Primary Production in Tropical Forests, an Evaluation and Synthesis of Existing Field Data. Ecological Applications, 11, 371-384.
11. Clark, D. A., Piper, S. C., Keeling, C. D., & Clark, D. B. (2003). Tropical Rain Forest Tree Growth and Atmospheric Carbon Dynamics Linked to Inter Annual Temperature Variation during 1984-2000. Proceedings of the National Academy of Sciences of the United States of America, 100, 5852-5857.
12. Cleveland, C. C., & Townsend, A. R. (2006). Nutrient Additions to a Tropical Rain Forest Drive Substantial Soil Carbon Dioxide Losses to the Atmosphere. Proceedings of the National Academy of Sciences of the United States of America, 103, 10317-10321.
13. Cramer, L. H. (1993). A Forest Arboretum in the Dry Zone. Kandy: Institute of Fundamental Studies.
14. De Deyn, G. B., Cornelissen, J. H. C., & Bardgett, R. D. (2008). Plant Functional Traits and Soil Carbon Sequestration in Contrasting Biomes. Ecology Letters, 11, 516-531.
15. De Genus, F., & Denys, D. (2004). Pitfalls in Factor Analytic Techniques. The American Journal of Psychiatry, 161, 579-580.
16. De Silva, C. S. (2006). Impacts of Climate Change of Water Resources in Sri Lanka. 32nd WEDC International Conference, Colombo, 13-17 November 2006, 289-295.
17. De Silva, C. S., Weatherhead, E. K., Knox, W. J., & Rodriguez-Diaz, J. A. (2007). Predicting the Impacts of Climate Change—A Case Study of Paddy Irrigation Water Requirements in Sri Lanka. Agricultural Water Management, 93, 19-29.
18. Ekberg, A., Buchmann, N., & Gleixner, G. (2007). Rhizospheric Influence on Soil Respiration and Decomposition in a Temperate Norway Spruce Stand. Soil Biology and Biochemistry, 39, 2103-2110.
19. Eriyagama, N., Smakhtin, V., Chandrapala, L., & Fernando, K. (2010). Impacts of Climate Change on Water Resources and Agriculture in Sri Lanka: A Review and Preliminary Vulnerability Mapping. International Water Management Institute, IWMI Research Report No. 135.
20. 21. Evans, J. L., Fernandez, I. J., Rustad, L. E., & Norton, S. A. (2001). Method for Evaluating Carbon Fractions in Forest Soils: A Review. Maine Agricultural and Forestry Experiment Station Technical Bulletin, 178, 1-42.
21. Green, M. J. B. (1990). IUCN Directory of South Asian Protected Areas. Cambridge, UK: IUCN.

22. Gulde, S., Chung, H., Amelung, W., Chang, C., & Six, J. (2008). Soil Carbon Saturation Controls Labile and Stable Carbon Pool Dynamics. Soil Science Society of America Journal, 72, 605-612.
23. Hairiah, K., Sulistyani, H., Suprayogo, D., Widianto, L., Purnomosidh, P., Widodo, R. H., & Van Noordwijk, M. (2006). Litter Layer Residence Time in Forest and Coffee Agroforestry Systems in Sumberjaya, West Lampung. Forest Ecology and Management, 224, 45-55.
24. Hessen, D. O., Agren, G. I., Anderson, T. R., Elser, J. J., & de Ruiter, P. C. (2004). Carbon Sequestration in Ecosystems: The Role of Stoichiometry. Ecology, 85, 1179-1192.
25. Hodge, A., Stewart, J., Robinson, D., & Fitter, A. H. (2000). Competition between Roots and Soil Micro-Organisms for Nutrients from Nitrogen-Rich Patches of Varying Complexity. Journal of Ecology, 88, 150-164.
26. Hughes, R. F., Kauffman, J. B., & Jaramillo-Luque, V. J. (2000). Ecosystem-Scale Impacts of Deforestation and Land Use in a Humic Tropical Region of Mexico. Ecological Applications, 10, 515-527.
27. IPCC (2001). Climate Change 2001: Impacts, Adaptation, and Vulnerability, Summary for Policymakers. A Report of Working Group II of the Intergovernmental Panel on Climate Change. Geneva: IPCC.
28. IPCC (2006). IPCC Guidelines for National Greenhouse Gas Inventories. H. S. Eggleston, L. Buendia, K. Miwa, T. Ngara, & K. Tanabe (Eds.). Hayama: Institute for Global Environmental Strategies, IPCC.
29. 30. Janzen, H. H., Campbell, C. A., Gregorich, E. G., & Ellerf, B. H. (1997). Soil Carbon Dynamics in Canadian Agroecosystems. In R. Lal, J. M. Kimble, R. F. Follett, & B. A. Stewart (Eds.), Soil Process and the Carbon Cycle (Advance in Soil Science) (pp. 57-80). Boca Raton, FL: CRC Press.
30. Johnson, J. M. F., Barbour, N. W., & Weyers, S. L. (2007). Chemical Composition of Crop Biomass Impactsits Decomposition. Soil Science Society of America Journal, 71, 155-162.
31. Keith, H., Mackey, B. G., & Lindenmayer, D. B. (2009). Re-Evaluation of Forest Biomass Carbon Stocks and Lessons from the World's Most Carbon-Dense Forests. Proceedings of the National Academy of Sciences of the United States of America, 106, 11635-11640.
32. Kimble, J., Cook, T., & Eswaran, H. (1990). Organic Matter in Soils of the Tropics. Proceeding of the Symposium: Characterization and Role of Organic Matter in Different Soils. 14th International Congress of Soil Science, Kyoto, 12-18 August 1990, 250-258.
33. Kuruppuarachchi, K. A. J. M., Seneviratne, G., & Madurapperuma, B. D. (2013). Drought Induced Fine Root Growth and Canopy Green-Up of Tropical Dry Zone Vegetations in Sri Lanka. Journal of Tropical Forestry and Environment, 3, 17-23.
34. Kuzyakov, Y., Hill, P. W., & Jones, D. L. (2007). Root Exudates Components Change Litter Decomposition in a Simulated Rhizosphere

Depending on Temperature. Plant Soil, 290, 293-305. http://dx.doi.org/10.1007/s11104-006-9162-8
35. Legg, C., & Jewell, N. (1995) A 1:50,000 Forest Map of Sri Lanka: The Basis for a National Forest Geographic Information System. Battaramulla: The Forester, Forestry Information Service, Sri Lanka Forest Department.
36. Lewis, S. L., Lopez-Gonzalez, G., Sonké, B., Affum-Baffoe, K., Baker, T. R., Ojo, L. O., Phillips, O. L., Reitsma, J. M., et al. (2009). Increasing Carbon Storage in Intact African Tropical Forest. Nature, 457, 1003-1006.
37. Lupwayi, N. Z., Grant, C. A., Soon, Y. K., Clayton, G. W., Bittman, S., Malhi, S. S., & Zebarth, B. J. (2010). Soil Microbial Community Response to Controlled-Release Urea Fertilizer under Zero Tillage and Conventional Tillage. Applied Soil Ecology, 45, 254-261.
38. Mokany, K., Raison, J. R., & Prokushkin, A. S. (2006). Critical Analysis of Root: Shoot Ratios in Terrestrial Biomes. Global Change Biology, 12, 84-96.
39. Müller, M., Alewell, C., & Hagedorn, F. (2009). Effective Retention of Litter-Derived Dissolved Organic Carbon in Organic Layers. Soil Biology & Biochemistry, 41, 1066-1074.
40. Naoum, S., & Tsanis, I. K. (2003). Temporal and Spatial Variation of Annual Rainfall on the Island of Crete, Greece. Hydrological Processes, 17, 1899-1922.
41. NPPD (National Physical Planning Department) (2002). National Physical Planning Policy, Final Document. Battaramulla: National Physical Planning Department, Ministry of Western Region, Government of Sri Lanka.
42. Panabokke, C. R. (1996). Soils and Agro-Ecological Environments of Sri Lanka. Colombo: Natural Resources, Energy and Science Authority of Sri Lanka.
43. Petkovsek, G., & Mikos, M. (2005). Estimating the R Factor from Daily Rainfall Data in the Sub-Mediterranean Climate of Southwest Slovenia. Hydrological Sciences Journal, 49, 20-23.
44. Premaratne, S., & Premalal, G. G. C. (2005). Country Pasture/Forage Resource Profiles, Sri Lanka. http://www.fao.org/ag/agp/agpc/doc/Counprof/srilanka/srilanka.htm
45. Pushpakumara, D. K. N. G. P., & Hitinayake, H. M. G. S. B. (2001). Invasive Tree Species in Udawattakele Forest Reserve. Sri Lankan Biodiversity Review, 1, 33-63.
46. Qualls, R. G. (2005). Biodegradability of Fractions of Dissolved Organic Carbon Leached from Decomposing Leaf Litter. Environmental Science & Technology, 39, 1616-1622.
47. Ramachandran, A., Jayakumar, S., Haroon, R. M., Bhaskaran, A., & Arockiasamy, D. I. (2007). Carbon Sequestration: Estimation of Carbon

Stock in Natural Forests Using Geospatial Technology in the Eastern Ghats of Tamil Nadu, India. Current Science, 92, 323-331.

48. Ratnayake, R. R., Seneviratne, G., & Kulasooriya, S. A. (2007). A Modified Method of Weight Loss on Ignition to Evaluate Soil Organic Matter Fractions. International Journal of Soil Science, 2, 69-77.
49. Rice, C. W. (2006). Organic Matter and Nutrient Dynamics. Encyclopedia of Soil Science: Second Edition. Boca Raton, FL: Taylor and Francis Group.
50. Rossi, J., Govaerts, A., de Vos, B., Verbist, B., Vervoort, A., Poesen, J., Muys, B., & Deckers, J. (2009). Spatial Structures of Soil Organic Carbon in Tropical Forests: A Case Study of Southeastern Tanzania. Catena, 77, 19-27.
51. Senarathna, P. M. (2005). Sri Lankawe Wananthara (1st ed.). Nugegoda: Sarasavi Publishers.
52. Singh, J. S., & Kashyap, A. K. (2007). Variations in Soil N-Mineralization and Nitrification in Seasonally Dry Tropical Forest and Savanna Ecosystems in Vindhyan Region, India. Tropical Ecology, 48, 27-35.
53. Somaratne, S., & Dhanapala, A. H. (1996). Potential Impacts of Global Climate Change on Forest Distribution in Sri Lanka. Water, Air, and Soil Pollution, 92, 129-135.
54. Toledo, M., Poorter, L., Claros, M. P., Alarcón, A., Balcázar, J., Leano, C., Licona, J. C., Llanque, O., Vroomans, V., Zuidema, P., & Bongers, F. (2010). Climate Is a Stronger Driver of Tree and Forest Growth Rates than Soil and Disturbance. Journal of Ecology, 99, 254-264.
55. Vereecken, H., & Herbst, M. (2004). Statistical Regression. In Y. Pachepsky, & W. J. Rawls (Eds.), Developments in Soil Science Series (pp. 3-19). Amsterdam: Elsevier BV.
56. UNFCCC (1997). Kyoto Protocol. Geneva: UNFCCC. http://unfccc.int/kyoto_protocol/items/2830.php
57. Walkley, A. (1947). A Critical Examination of a Rapid Method for Determining Organic Carbon in Soils: Effect of Variations in Digestion Conditions and of Inorganic Soil Constituents. Soil Science, 63, 251-264.
58. Walkley, A., & Black, I. A. (1934). An Examination of the Degtjareff Method for Determining Soil Organic Matter and a Proposed Modification of the Chromic Acid Titration Method. Soil Science, 37, 29-38.
59. Wang, C., Long, R., Wang, Q., Jing, Z., Du, Y., & Cao, G. (2008). Effects of Soil Resources on Species Composition, Plant Diversity, and Plant Biomass in an Alpine Meadow Qinghai-Tibetan Plateau. Israel Journal of Ecology and Evolution, 54, 205-222.
60. Wedathanthri, H. P., & Hitinayake, H. M. G. S. B. (1999). Invasive Behavior of Myroxylonbalsamum at Udawattakele Forest Reserve. 5th Annual Forestry and Environment Symposium: Challengers in Natural

Resource Conservation in the 21st Century, Department of Forestry and Environmental Science, University of Sri Jayewardenepura, Nugegoda, 10-11 December 1999, 14.

61. Weil, R. R., & Magdoff, F. (2004). Soil Organic Matter in Sustainable Agriculture. Department of Natural Resource Sciences and Landscape Architecture. Boca Raton, FL: CRC Press.

62. Zalamea, M., González, G., Ping, C.-L., & Michaelson, G. (2007). Soil Organic Matter Dynamics under Decaying Wood in a Subtropical Wet Forest: Effect of Tree Species and Decay Stage. Plant Soil, 296, 173-185.

Chapter 5

Carbon Dioxide Storage and Sequestration in Unconventional Shale Reservoirs

Mohammad O. Eshkalak, Umut Aybar

Petroleum and Geosystems Engineering Department, The University of Texas at Austin, Austin, TX, USA

ABSTRACT

Carbon Dioxide (CO_2) storage and sequestration in unconventional shale resources has been attracting interest since last couple of years due to the very unique characteristics of such formations have made them a feasible option for this object. Shale formations are found all around the world and the conventional assets are easily accessible, and also the huge move of operators toward developing unconventional reservoirs during past years leaves many of such formations ready for sequestering CO_2. Today, the use of long horizontal wells that are drilled on a pad has the lowest amount of environmental footprint in which for storage and sequestration purpose also provides much more underground pore spaces available for CO_2. In this paper we study the state of the art of the technology of CO_2 storage and sequestration and provide different and fresh look for its complex phenomena from a mathematical modeling point of view. Moreover, we hope this study provides valuable insights into the use of depleted shale gas reservoirs for carbon sequestration, which as a result, a cleaner atmosphere will be achieved for the life of our next generations. Also, we present that the depleted shale gas reservoirs are very adequate for this purpose as they already have much of the infrastructure required to perform CO_2 injection available in sites.

Keywords: Carbon Dioxide Sequestration, Shale Reservoirs, Modeling and Simulation, Clean Environment

1. INTRODUCTION

Geologic formations and abundant reservoirs have attracted much attention within the engineering community for the purpose of reducing carbon dioxide of the atmosphere through storing and sequestration process in recent years. Worldwide energy demand is supported by the significant role of conventional and unconventional resources of oil and gas. This huge dependency has resulted in critical environmental issues such as increased CO_2 emission known as the greenhouse gas, and for the purpose of having a clean atmosphere for next generations, CO_2 storage and sequestration has gotten a common procedure for this purpose. During the last decade, energy equations are changed due to the advances obtained in drilling and fracturing of unconventional reservoirs both in the US and worldwide. The main simple difference between conventional and unconventional resources is the fact that unconventional are scattered all around the world but conventional exist just in some parts of the world. This strongly means that since unconventional are everywhere on the map then the CO_2 sequestration in such reservoirs are very reasonable and economically feasible [1] [2] .

Shale gas is mostly methane that is trapped within the shale formations. Shale is located thousands of feet below earth surface. Gas-productive shale formations in the continental US are of thermogenic or biogenic origin and are found in Paleozoic and Mesozoic rocks. Shale is a fine-grained sedimentary rock that, when deposited as mud, can collect organic matter. When the organic matter decays over time, petroleum and natural gas products are formed within the rock's pores. Shales typically hold dry gas, but some formations produce liquid products as well. Conventional gas reservoirs form from the migration of natural gas from an organic-rich source into permeable reservoir rock. Unconventional gas-rich shales, however, generally function as both the source and reservoir for natural gas. Shales have low permeability, which means that trapped gas cannot move easily within the rock. Because of this, a technique called hydraulic fracturing is employed to produce the natural gas. This unique characteristic benefits the sequestration of CO_2. Hydraulic fracturing cracks the shale rock through the injection of water, sand, and chemicals at high pressure.

Underground storage and sequestration of carbon dioxide involves the process of injecting highly pressurized CO_2 into the formations considering the capability of the rock to permanently keep it from leaking toward the earth surface. There are five applicable options for safely sequestering the carbon dioxide into the geologic formations. Saline formations, basalts, un-mineable coal seams, conventional oil and gas reservoirs and depleted unconventional shale resources are technically studied during the course of this paper. As of now, considering the advances gained through last couple of years in carbon management, scientists have proposed the following statements: 1) estimating

CO_2 storage capacity +/– 30 percent in geologic formations; 2) ensuring 99 percent storage permanence; 3) improving efficiency of storage operations; and 4) developing Best Practices Manuals [3] [4] . These technologies will lead to future CO_2 management for coal-based electric power generating facilities and other industrial CO_2 emitters by enabling the storage and utilization of CO_2 in all storage types.

In the U.S., Nuttall et al. (2005) estimated the CO_2 sequestration capacity in the organic-rich Devonian black shales in the Big Sandy gas field of Eastern Kentucky to be about 6.8 Gt. To illustrate the importance of shale formations for CO_2 storage in Europe, the Dutch resource of shale gas own an estimation of between 48,000 and 230,000 Bcm. If one assumes that technology can be developed to store 1 m^3 of CO_2 for 1 m^3 of CH_4 produced, and that only one percent of the resource can be recovered, a substantial 480 - 2300 Bcm of CO_2, corresponding to 0.9 to 4.35 Gton CO_2, could be stored. In this paper we further analyze the developments, advances and issues related to methods that are in practice and authors will provide alternative technologies that are given a fresh look toward the classical research.

2. SHALE GAS RESOURCES IN THE U.S. AND WORLDWIDE

In 2000 shale gas provided only 1% of U.S. natural gas production; by 2010 it is over 20% and the U.S. government's Energy Information Administration predicts that by 2035, 48.9% of the United States' natural gas supply will come from shale gas. These trends obviously demonstrate that we can expect that shale gas will greatly expand worldwide energy supply. Further, China is estimated to have the world's largest shale gas reserves. Some researches show that shale gas production in the USA and Canada could help Europe to be less dependent to Russian and Persian Gulf countries. Figure 1 shows that unconventional resources (Shale Gas, Tight Gas, Coalbed Methane) containing hydrocarbon are almost all over the world. The most common form of unconventional natural gas is shale gas. In this study, the main focus will be on storage and sequestration in shale gas.

3. SHALE RESERVOIRS MODELING AND SIMULATION

Reservoir simulation and modeling of unconventional resources have been given much more attention over the past years. Many numerical and analytical models have been developed and extensive reservoir studies have been conducted. Commercial reservoir simulators are also improved to handle and capture fluid flow behavior

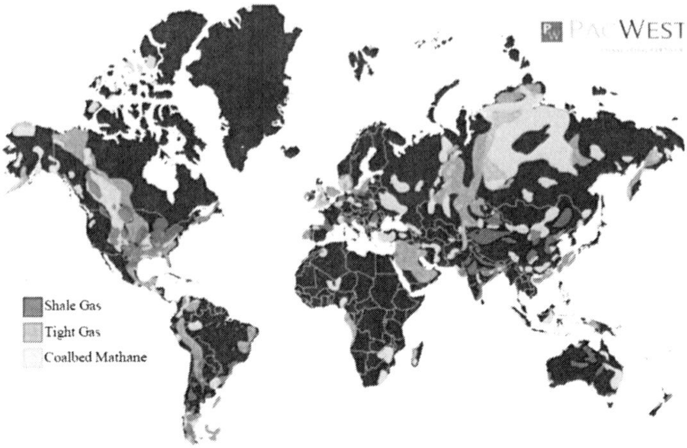

Figure 1. Worldwide shale gas resources.

and natural gas production from unconventional assets, such as shale. However, developing an unconventional reservoir model that accounts for all pressure dependent phenomena and integrates all physics incorporated in gas flow for tight formations is still a challenging target for the petroleum industry. Among analytical and semi- analytical methods, works done by [5] [6] have provided comprehensive progress in the modeling of shale gas reservoirs. Many authors, such as [7] [8] , have used numerical simulation techniques in order to model different aspects of unconventional shale reservoirs.

Analytical reservoir models are widely employed because of their relative simplicity compared to numerical approaches. The aim of analytical solutions is to provide a simple solution which covers the fundamental physics of the phenomenon. In order to accomplish these goals, the analytical solution must have some simplifying assumptions. Having constant and/or homogeneous rock and fluid properties (density, compressibility, permeability, and viscosity) are the common assumptions for analytical solutions. The purpose of these assumptions is to linearize the partial differential equations arising from modeling of fluid flow in a reservoir; in our case, the diffusion equation. There are two methods in the literature customarily preferred for analytical fluid flow solutions: one is proposed by [9] [10] , which has solution in time domain; while the other techniques are introduced by [11] and use Laplace transformation for solutions. Following these two solution techniques, researchers have proposed analytical solutions for hydraulically fractured unconventional reservoirs [12] [13] . Analytical reservoir models are applicable for quite accurate, simple and robust simulations to evaluate production performances of hydraulically fractured unconventional reservoirs.

Eshkalak et al. investigated the geo-mechanical properties of Marcellus shale. They generated a common source for securing these rock mechanical properties, geomechanical well logs, and studied various characteristics, such as minimum horizontal stress, young, bulk shear modulus, as well as poison's ratio that play an important role in defining the stress profiles of an unconventional reservoir. Moreover, having an access to rock geomechanical properties will enhance the understanding of the parameters, such as conductivity and pressure dependency of permeability [14] - [17] .

In Figure 2, the natural gas production for past years and projections for following years are given. The shale gas is the only form of natural gas whose production amount is increasing. All other types of natural gas are either decreasing or remaining constant. It can be easily said that in the future the USA and world natural gas production is the shale gas, the Y axis is trillion cubic feet [1] .

Generally, two-phase fluid flow of water and gas in a dual-permeability model is considered in constructing the geologic model of shale gas reservoirs. The dual-permeability model considers the intercommunication be tween the inter-granular void spaces in contrast to the dual-porosity model. Also, this model considers flow in two domains including the matrix and fracture. This model allows the transfer of both gas and water between the matrix and fracture domains, gas velocity in the matrix and fracture domains is calculated with the Equations (1) and (2):

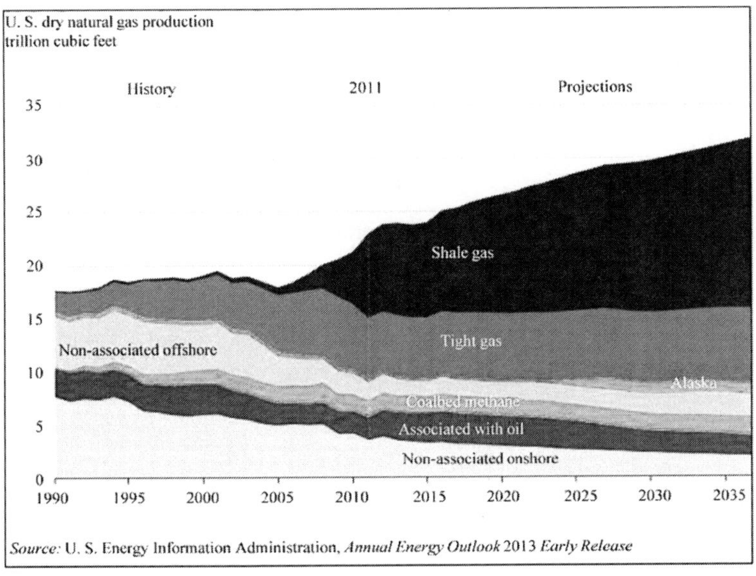

Figure 2. Natural gas production outlook.

$$v_g^m = -\left(\frac{K_g^m}{\mu_g} \nabla P_g + \frac{D_g^m}{C_g^m} \nabla C_g^m \right) \tag{1}$$

$$v_g^f = -\left(\frac{K_g^f}{\mu_g} \nabla P_g + \frac{D_g^f}{C_g^f} \nabla C_g^f \right) \tag{2}$$

where v_g is gas velocity, K_g is gas permeability, D_g is gas diffusivity, P_g is gas pressure, C_g is gas concentration, and μ_g is gas viscosity. Subscripts m and f represent matrix and fracture domains. Velocity of the water flowing in matrix and fracture are determined with Equations (3) and (4), respectively:

$$v_w^m = -\left(\frac{K_w^m}{\mu_w} \nabla P_w^m \right) \tag{3}$$

$$v_w^f = -\left(\frac{K_w^f}{\mu_w} \nabla P_w^f \right) \tag{4}$$

where, v_w is water velocity, K_w is water permeability, P_w is water pressure, and μ_w is water viscosity.

3.1. Flow in Matrix

The equations of gas transport thus are simplified for matrix domain as shown in Equation (5):

$$\frac{\partial}{\partial t}\left(\frac{C_g^m P_g^m}{Z} \right) = \nabla \left(\frac{P_g^m}{Z} \frac{K_g^m}{\mu_g} \nabla P_g^m + D_g^m \nabla \frac{P_g^m}{Z} + \frac{D_g^m}{C_g^m} \frac{P_g^m}{Z} \nabla C_g^m \right) - \frac{RT}{M}\left(q_w^{mf} + q_w^m \right) \tag{5}$$

where Z is the gas compressibility factor, R is the gas constant, T is temperature, M is gas molecular weight, and q_g is gas mass flow rate per unit matrix-block volume. Subscript m and f represent the exchange between matrix and fracture. For the water phase, the same equation is shown in Equation (6):

$$\frac{\partial}{\partial t}\left(\frac{\varnothing_m s_w^m}{B_w}\right)=\nabla\left(\frac{K_w^m}{B_w \mu_w}\nabla P_w^m\right)-\frac{RT}{M}\left(q_w^{mf}+q_w^m\right) \quad (6)$$

where $\varnothing_m$ is matrix porosity, s_w is water saturation, and B_w is water compressibility factor.

3.2. Flow in Fracture

After some manipulation and simplifications, the gas flow governing equation in fracture becomes as the following Equation (7):

$$\frac{\partial}{\partial t}\left(\frac{C_g^f P_g^f}{Z}\right)=\nabla\left(\frac{P_g^f}{Z}\frac{K_g^f}{\mu_g}\nabla P_g^f+D_g^f\nabla\frac{P_g^f}{Z}+\frac{D_g^f}{C_g^m}\frac{P_g^f}{Z}\nabla C_g^f\right)+\frac{RT}{M}\left(q_w^{mf}-q_w^m\right) \quad (7)$$

For the water phase, Equation (8) represents the related formula.

$$\frac{\partial}{\partial t}\left(\frac{\varnothing_f s_w^f}{B_w}\right)=\nabla\left(\frac{K_w^f}{B_w \mu_w}\nabla P_w^f\right)+\frac{RT}{M}\left(q_w^{mf}-q_w^m\right) \quad (8)$$

Equations (9) to (12) represent the auxiliary relations used in the solution method.

$$C_g^m=\varnothing_f s_w^m \quad (9)$$

$$C_g^f=\varnothing_f s_w^f \quad (10)$$

$$s_g^m+s_w^m=1 \quad (11)$$

$$s_g^f+s_w^f=1 \quad (12)$$

4. What We Have Learnt from CO_2 Sequestration in Saline Aquifers

CO_2 emission reduction becomes one of the main concerns of the world; hence deep saline aquifers can be used as CO_2 storage [18] to decrease CO_2 emission. Aquifers are the favorable storage places comparing to depleted reservoirs and coal seams, mainly because of safety issues [19] [20] . Poorly-sealed abounded wells are reducing the safety of storage capability of depleted reservoirs, while water-bearing permeable rock nature of saline aquifers is increasing the safety factor [21] .

To perform a successful CO_2 sequestration project in saline aquifers, monitoring and verification of storing process is a must. Underground distribution of CO_2 needs to be monitored during the life of project. It is essential to identify when the minimum volume and saturation of CO_2 has been reached in the storage reservoir.

Lithology, geomechanics and geology of the storage reservoir is main elements which drive CO_2 detection capability within the reservoir. Having said that, it must also be noted that all of these reservoir characteristics have some level of uncertainties. There is no doubt about that these uncertainties may have an impact on CO_2 distribution detection in the reservoir. Therefore, these uncertainty elements also need to be taken into account for the CO_2 sequestration and CO_2 distribution detection planning stages [22] .

CO_2 detectability also relies on the phase in which CO_2 is injected in the storage reservoir. CO_2 is customarily injected as a liquid form to be able to transform to a supercritical phase. Supercritical state has both a liquid and a gas characteristic; hence supercritical fluid expands like a gas, while keeping its density as a liquid level. CO_2 density is adequate to fill the pores at depths lower than 2600 ft.

State of CO_2, and the environment into which it is sequestered, have a significant bearing on the detectability of CO_2. Typically, CO_2 is injected as a liquid which transforms to a supercritical fluid [23] . In this state, it has properties of both a liquid and a gas so that it expands like a gas, but with a density of a liquid. At depths below 800 m, CO_2 density is high enough to allow efficient filling of pore space. There is also a reduction in the buoyancy difference between the CO_2 and other pore fluids.

Miscibility is another important characteristic which varies with CO_2 sequestration conditions. As it is a known fact that CO_2 and natural gas are miscible, therefore CO_2 is able to relocate water in the pores. However, since our interest here is CO_2 sequestration in saline aquifers, multiphase relationship of CO_2 and the aquifer dictates the pore space volume which can be filled by CO_2 [23] [24] .

5. CO_2 SEQUESTRATION CAPACITY OF SHALE

Coming from the extremely tight and low permeability nature of shale reservoirs, to produce from or to inject any fluid into shale reservoirs is tough. However, technological improvements in hydraulic fracturing technique made these reservoirs producible. Hydraulic fracturing basically opens artificial flow channels for reservoir fluid to flow with cracking the formation via pumping of highly pressurized fluid into the formation. Once these shale reservoirs depleted, existing fractures in the reservoir could be storage for CO_2 sequestration. Another advantage of shale reservoirs is that they do not require capital expense cost such as new well drilling which is the case for saline aquifers [25] .

Viability of CO_2 storage in shale reservoirs can be predicted by reservoir models, since it is critical to identify CO_2 storage capacity of the reservoir. There are two major criteria while performing feasibility study for CO_2 sequestration in shale formations. The first criterion is the diffusivity which gives an idea about formation deliverability. The first criterion in this regard is gas adsorption/desorption characteristics of shale formations. It is an ongoing study and many researchers claim that CO_2 may adsorb the shale formation more favorable than methane.

Storage capacity of shale formations for CO_2 storage is promising. It is also claimed that CO_2 injection time in shale reservoirs would be much faster than the methane production from these reservoirs. In addition, it is estimated that around 14.5 billion tons of CO_2 can be sequestrated into shale formations in the next two decades. The estimated capacity is almost half of the U.S. CO_2 emissions from power plants within the next 20 years, which is equal to 20% of total expected CO_2 emission of the U.S.

Based on numerical investigation, it concluded that the maximum CO_2 storage capacity for eastern U.S [24] shale gas is 1.12 million metric tons per square kilometer, and the sorbed CO_2 storage capacity is estimated to be 0.72 million metric tons per square kilometer. The total CO_2 storage capacity of organic-rich shales at supercritical conditions as a function of pore pressure was measured by [26] by considering the pore compressibility and sorption effects. The results state that kerogen, the organic part of the shale, acts as a molecular sieve and accounts for the gas sorption on shales. The sorption capacity of shales is affected by its TOC content, clay minerals and micropore structure. There is a large body of literature investigating the sorption capacity [27] .

6. CO_2 TRAPPING MECHANISMS

Trapping mechanisms are important in CO_2 sequestration feasibility studies. Four trapping mechanisms are present to our knowledge which are given in Figure 3, and can be named as: mineral trapping, solubility trapping, residual CO_2 trapping, and structural/stratigraphic trapping. Trapping time is considered up to ten thousand years same as the nuclear storage projects. To give an idea about 10,000 years consideration time, it has been 11,000 years up to today since the last ice age [28] .

The most common type of trapping mechanisms is geological structures for CO_2 storage in depleted reservoirs. Deep saline aquifers also have structural trapping mechanism coming from their depositional environment. When upward migration of CO_2 is blocked by impermeable rock layer, CO_2 is trapped due to buoyancy effect.

Moving from geological structures for trapping, solubility type of trapping takes place when CO_2 dissolved in formation water [28] . Reservoir pore pressure and temperature, along with the formation water salinity are the key elements for solubility trapping. CO_2 solubility rises with increasing pressure, on the other hand decreases with increasing temperature and salinity. With this mechanism CO_2 is trapped as a liquid phase sinking under the gas phase due to gravitational forces. Solubility type of trapping is secure and favorable for CO_2 storage [29] .

All in all, CO_2 sequestration feasibility studies are not depending on one unique trapping mechanism. It is concluded that different geophysical characterizations will be necessary throughout the life of the project. For instance, earlier and later stages of the sequestration project might be sensitive to different characteristics, hence it is suggested that time lapse surveys must be performed through the life of the project to confirm sequestrated CO_2 is trapped securely [17] [30] .

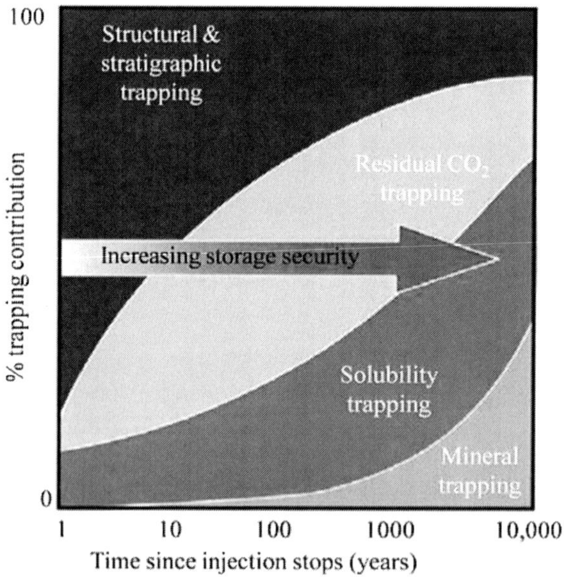

Figure 3. Schematic representation of the security of CO_2 trapping mechanisms over time.

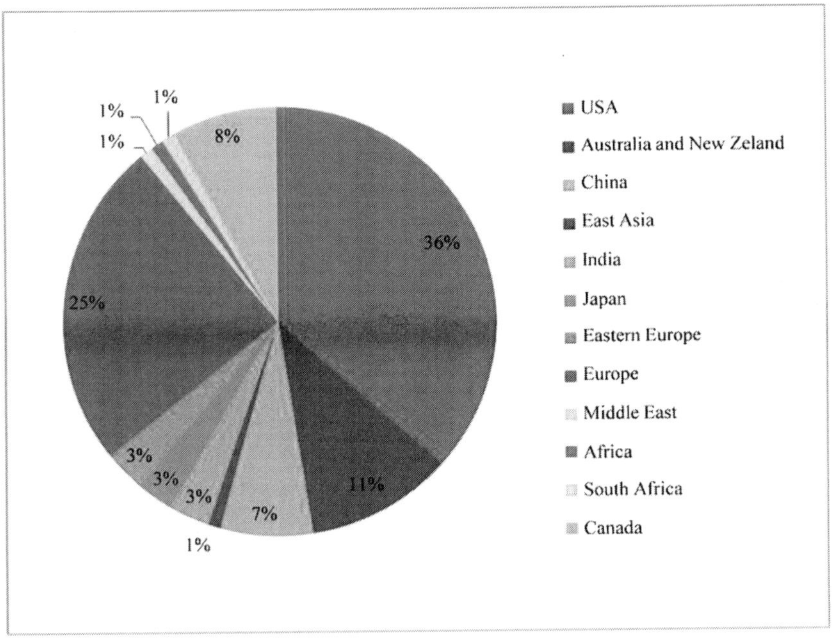

Figure 4. Countries with active CCS program.

7. ACTIVE CARBON CAPTURE AND SEQUESTRATION (CCS) PROJECTS WORLDWIDE

Carbon Capture and Sequestration projects are mainly performed in the U.S. and Europe, pie chart distribution of the world CCS projects is given in Figure 4. Canada, Australia and New Zealand, and China are also other major players of CCS projects [31] - [33] .

8. CONCLUSIONS

Unconventional shale reservoirs are main hydrocarbon resources of today's world, they are developed for oil and gas production. However, depleted unconventional shale reservoirs can be good candidates for CO_2 sequestration and storage operations. Extremely low permeability nature of shale reservoir might seem like their lack, but on the other hand decent amount of CO_2 can absorb on shale fracture surfaces. Existing natural and hydraulic fracture networks in shale reservoirs made these reservoirs attractive to permanent CO_2 storage projects. That being said, when it comes to reservoir modeling for CO_2 storage, many characteristic factors need to be taken into account such as buoyancy, heterogeneity of shale reservoirs, and the existence of formation water, because they will directly affect the storage capacity of the particular reservoir.

In conclusion, our extended literature review shows that shale reservoirs are good candidates for CO_2 storage with the capacity of 5 to 10 kg/t per formation. These results verify the feasibility of CO_2 sequestration in shale reservoirs. It can be stated that the long term feasibility of CO_2 storage in shale reservoirs needs to be extended. However, the information available to our knowledge manifests that shale reservoirs can be good storage candidates for permanent CO_2 storage, therefore studies need to be focused on to make these projects practical. It is concluded that unconventional shale reservoirs have many favorable characteristics for CO_2 storage, and it is expected that these reservoirs will become very attractive for CO_2 sequestration projects all around the world in the very near future.

REFERENCES

1. National Energy Technology Laboratory (2010) Carbon Sequestration Atlas of the United States and Canada. 3rd Edition, the U.S. Department of Energy, Washington DC, p162. 2.
2. Li, Z.W., Dong, M.Z., Li, S.L. and Huang, S. (2006) CO_2 Sequestration in Depleted Oil and Gas Reservoirs—Caprock Characterization and Storage Capacity. Energy Conversion and Management, 47, 1372-1382.
3. Rice University, News and Media Relations (2011) Shale Gas and U. S. National Security. Rice University News and Media Relations Team, Huston.
4. Brown, M.L. (2009) Analytical Trilinear Pressure Transient Model for Multiply Fractured Horizontal Wells in Tight Shale Reservors. M.Sc. Thesis, Colorado School of Mines, Golden.
5. Aybar, U., Eshkalak, M.O., Sepehrnoori, K. and Patzek, T.W. (2014) The Effect of Natural Fracture's Closure on Long-Term Gas Production from Unconventional Resources. Journal of Natural Gas Science and Engineering, 21, 1205-1231.
6. Esmaili, S., Kalantari-Dahaghi, A. and Mohaghegh, S.D. (2012) Forecasting, Sensitivity and Economic Analysis of Hydrocarbon Production from Shale Plays Using Artificial Intelligence & Data Mining. SPE Canadian Unconventional Resources Conference, Calgary, 30 October-1 November 2012, SPE162700.
7. Kalantari-Dahaghi, A. and Mohaghegh, S.D. (2011) Numerical Simulation and Multiple Realizations for Sensitivity Study of Shale Gas Reservoirs. SPE Production and Operations Symposium, Oklahoma City, 27-29 March 2011, SPE 141058.

8. Carslaw, H.S. and Jaeger, J.C. (1959) Conduction of Heat in Solids. 2nd Edition, Clarendon Press, Oxford.
9. Van Everdingen, A.F. and Hurst, W. (1949) The Application of the Laplace Transformation to Flow Problems in Reservoirs. Journal of Petroleum Technology, 1, 305-324. http://dx.doi.org/10.2118/949305-G
10. Aybar, U., Eshkalak, M.O., Sepehrnoori, K. and Patzek, T.W. (2014) Long Term Effect of Natural Fractures Closure on Gas Production from Unconventional Reservoirs. SPE Eastern Regional Meeting, Charleston, 21-23 October 2014, SPE 171010.
11. Ozkan, E., Raghavan, R.S. and Apaydin, O.G. (2010) Modeling of Fluid Transfer from Shale Matrix to Fracture Net- work. SPE Annual Technical Conference and Exhibition, Florence, 19-22 September 2010, SPE 134830.
12. Torcuk, M.A., Kurtoglu, B., Fakcharoenphol, P. and Kazemi, H. (2013) Theory and Application of Pressure and Rate Transient Analysis in Unconventional Reservoirs. SPE Annual Technical Conference and Exhibition, New Orleans, 30 September-2 October 2013, SPE 166147.
13. Eshkalak, M.O., Mohaghegh, S.D. and Esmaili, S. (2014) Geomechanical Properties of Unconventional Shale Reservoirs. Journal of Petroleum Engineering, 2014, Article ID: 961641.
14. Eshkalak, M.O., Mohaghegh, S.D. and Esmaili, S. (2013) Synthetic, Geomechanical Logs for Marcellus Shale. Digital Energy Conference and Exhibition, The Woodlands, 5-7 March 2013, SPE 163690.
15. OmidvarEshkalak, M. (2013) Synthetic Geomechanical Logs and Distributions for Marcellus Shale. MSc Thesis, West Virginia University, Morgantown.
16. Michael, K., Arnot, M., Cook, P., Ennis-King, J., Funnell, R., Kaldi, J., Kirste, D. and Paterson, L. (2009) CO_2 Storage in Saline Aquifers I—Current State of Scientific Knowledge. Energy Procedia, 1, 3197-3204.
17. Aybar, U., Yu, W., Eshkalak, M., Sepehrnoori, K. and Patzek, T. (2015) Evaluation of Production Losses from Unconventional Shale Reservoirs. Journal of Natural Gas Science and Engineering, 23, 509-516.
18. JafarGandomi, A. and Curtis, A. (2011) Detectability of Petrophysical Properties of Subsurface CO_2-Saturated Aquifer Reservoirs Using Surface Geophysical Methods. The Leading Edge, 30, 1112-1121.
19. Eshkalak, M.O., Aybar, U. and Sepehrnoori, K. (2014) An Integrated Reservoir Model for Unconventional Resources, Coupling Pressure Dependent Phenomena. Eastern Regional Meeting, Charleston, 21-23 October 2014, SPE 171008.

20. White, D. and Johnson, J. (2009) Integrated Geophysical and Geochemical Research Programs of the IEA GHG Weyburn-Midale CO_2 Monitoring and Storage Project. Energy Procedia, 1, 2349-2356.
21. Gale, J. and Freund, P. (2001) Coal-Bed Methane Enhancement with CO_2 Sequestration Worldwide Potential. Environmental Geosciences, 8, 210-217. http://dx.doi.org/10.1046/j.1526-0984.2001.008003210.x
22. Benson, S., Hoversten, G.M., Gasperikova, E. and Haines, M. (2004) Monitoring Protocols and Life-Cycle Costs for Geologic Storage of Carbon Dioxide. Proceedings of the 7th International Conference on Greenhouse Gas Control Technologies, Vancouver, 5-9 September 2004, 1259-1264.
23. Benson, S. and Cole, D.R. (2008) CO_2 Sequestration in Deep Sedimentary Formations. Elements, 4, 325-331.
24. Eshkalak, M.O., Aybar, U. and Sepehrnoori, K. (2014) An Economic Evaluation on the Re-Fracturing Treatment of the US Shale Gas Resources. Eastern Regional Meeting, Charleston, 21-23 October 2014, SPE 171009.
25. Bear, J. (1972) Dynamics of Fluids in Porus Media. Elsevier, Amsterdam.
26. Godec, M., Koperna, G., Petrusak, R. and Oudinot, A. (2013) Potential for Enhanced Gas Recovery and CO_2 Storage in the Marcellus Shale in the Eastern United States. International Journal of Coal Geology, 118, 95-104.
27. Kang, S.M., Fathi, E., Ambrose, R., Akkutlu, I. and Sigal, R. (2011) Carbon Dioxide Storage Capacity of Organic-Rich Shales. SPE Journal, 16, 842-855.
28. Eshkalak, M.O., Al-Shalabi, E.W., Sanaei, A., Aybar, U. and Sepehrnoori, K. (2014) Simulation Study on the CO_2- Driven Enhanced Gas Recovery with Sequestration versus the Re-Fracturing Treatment of Horizontal Wells in the U.S. Unconventional Shale Reservoirs. Journal of Natural Gas Science and Engineering, 21, 1015-1024.
29. Eshkalak, M.O., Al-Shalabi, E.W., Sanaei, A., Aybar, U. and Sepehrnoori, K. (2014) Enhanced Gas Recovery by CO_2 Sequestration versus Re-Fracturing Treatment in Unconventional Shale Gas Reservoirs. Abu Dhabi International Petroleum and Exhibition and Conference, Abu Dhabi, 10-13 November 2014, SPE 172083.
30. Aybar, U. (2014) Investigation of Analytical Models Incorporating Geomechanical Effects on Production Performance of Hydraulically and Naturally Fractured Unconventional Reservoirs. MSc Thesis, The University of Texas at Austin, Austin.
31. Gensterblum, Y., Busch, A. and Krooss, B.M. (2014) Molecular Concept and Experimental Evidence of Competitive Adsorption of H_2O, CO_2 and CH_4 on Organic Material. Fuel, 115, 581-588.

32. Polson, D., Curtis, A. and Vivalda, C. (2012) The Evolving Perception of Risk during Reservoir Evaluation Projects for Geological Storage of CO_2. International Journal of Greenhouse Gas Control, 9, 10-23.
33. Eke, P.E., Naylor, M., Haszeldine, S. and Curtis, A. (2011) CO_2-Brine Surface Dissolution and Injection: CO_2 Storage Enhancement. SPE Projects, Facilities and Construction, 6, 41-53.

Chapter 6

Land Management Effects on Carbon Sequestration and Soil Properties in Reclaimed Farmland of Eastern Ohio, USA

David A. N. Ussiri*, Rattan Lal

Carbon Management and Sequestration Center, The Ohio State University, Columbus, USA.

ABSTRACT

Reclaimed mined soils (RMSs) could restore soil quality and ecosystem productivity while sequestering C and off-setting some of C emissions associated with coal utilization. The study was conducted to evaluate the effects of tillage and pasture management on soil physical properties, soil organic carbon (SOC) and microbial biomass carbon (MBC) in RMSs managed for agricultural use in eastern Ohio. Soil bulk density (ρ_b) of the top 50 cm ranged from 1.11 to 1.93 $Mg \cdot m^{-3}$. The ρ_b of the RMSs was significantly more than that of the undisturbed soils. Water stable aggregates (WSA) and mean weight diameter (MWD) of the 0 - 10 cm soil layer were significantly lower under reclaimed conventional tillage (RCT) than reclaimed no tillage (RNT) and reclaimed pasture (RP), probably due to tillage-induced disturbance. The SOC pool of the top 50 cm layer was 64.2, 66.5, 75.4, 86.1 and 101.1 $Mg \cdot C \cdot ha^{-1}$ for undisturbed pasture (Und P), RNT, RCT, RP and undisturbed hardwood forest (Und HWF), respectively (LSD = 7.7 $Mg \cdot ha^{-1}$). The RMSs under pasture accumulated SOC at higher rates than RMSs under cropland. Reclaimed pasture land use increased SOC pool by 14% or 0.5 $Mg \cdot ha^{-1} \cdot yr^{-1}$ and 30% or 0.9 $Mg \cdot ha^{-1} \cdot yr^{-1}$ relative to RNT and RCT land uses, respectively. Our data indicated that RMSs under forest and pastures had higher SOC sequestration rates than RMSs under arable land use, probably due to disturbances associated with farm operations. The MBC of the RMSs were generally lower than those of the undisturbed sites. The disturbances associated with mining and reclamation reduced the MBC by 39%, 53% and 21% under RCT, RNT and RP compared to the undisturbed forest and pasture sites. However, the amount of mineralizable C was not significantly different among land disturbances or land uses.

Keywords: Carbon Sequestration, Surface Mining, Soil Organic Carbon, Microbial Biomass Carbon, Reclaimed Minesoils

1. INTRODUCTION

Surface mining for coal results in drastic landscape disturbance and soil degradation. It strongly changes the antecedent soil profile, along with alterations in physical, chemical and biological properties and processes of soils [1]. Mining and related activities also lead to severe loss of soil organic carbon (SOC) because of topsoil loss, mechanical mixing of soil horizons during removal and storage of topsoil, increased mineralization, soil erosion and leaching from exposed topsoil. Reclamation is the process of restoring the mined land to a useful state [2,3]. Reclamation can improve soil quality, enhance mined soil productivity and increase SOC concentration [4]. Adverse physical (high bulk density, poor soil structure, poor aggregation and low water infiltration rate), chemical (salinity/acidity, low SOC content, low soil fertility, and elemental imbalances) and biological conditions (reduced microorganisms and soil biotic activity) often limit vegetation establishment and restoration of reclaimed minesoils (RMSs) [1,5-8].

The US Surface Mining Control and Reclamation Act of 1977 (SMCRA) mandates that surface mined areas be restored to level similar to that of the pre-mining state. This entails the restoration of landscape topography and the application of topsoil to create a soil environment capable of supporting plant growth. The top 1.5 m of the RMSs should be constructed from medium-textured materials (silt loams, loams or light silt clay loams) for generating high water storage capacity for crops [9]. For the surface mined cropland, the SMCRA requires the reclaimed land to meet or exceed crop yield it had prior to mining and that the yields are comparable to unmined farmland in the surrounding area under equivalent level of management. Unlike reclaimed grassland or forest land where management interventions are minimal, reclaimed cropland soils are intensively managed, with periodic tillage or major seedbed preparation for the purpose of establishing annual, biennial or short-lived perennial herbaceous plants. In cropland, annual plants are established for the exclusive purpose of direct harvest as a cash crop or livestock feed [10]. Tillage has been identified as a major land management factor impacting SOC dynamics and C cycling in agricultural soils [11,12]. Soils under continuous tillage (CT) have lower SOC pools than those under no-tillage (NT) mainly due to mechanical breakdown of soil aggregates and loss of aggregate-protected SOC [13,14]. However, the impacts of management interventions on agricultural land on C sequestration and ecological recovery of the RMSs under cropland are not fully understood.

Several studies have assessed the potential of RMSs for SOC sequestration, and have documented that RMSs could be an important sink for atmospheric CO_2 and thus offset some of the CO_2 emitted as a result of mining and utilization of coal for power generation [8,15-21]. These assessments have documented rapid SOC accumulation with sequestration rates generally higher than in agricultural lands [15,17,20,22,23]. For example, Akala and Lal [15] reported C sequestration rates of 2 to 3 $Mg{\cdot}ha^{-1}{\cdot}yr^{-1}$ in the first 20 yrs after reclamation, decreasing to 0.4 to 0.7 $Mg{\cdot}ha^{-1}{\cdot}yr^{-1}$ by 20 to 30 yrs after reclamation. Shukla et al. [8] reported an average C sequestration rate of 0.75 $Mg{\cdot}C{\cdot}ha^{-1}{\cdot}yr^{-1}$ for the fertilized hay fields in Ohio during the first 20 years after reclamation. Other studies (i.e., [19,20,24] reported sequestration rates as high as 3 to 4 $Mg{\cdot}C{\cdot}ha^{-1}{\cdot}yr^{-1}$ in reclaimed grass and for- est land uses. The restoration of RMSs to grasslands for pastures and hay is the most widespread land use in Eastern United States due to current reclamation regulations which require grading the land to approximate original contour, overall creating lands with gentle slopes that are more readily traversed and amenable to management and high potential for economic returns [2, 10].Very few studies have been conducted to evaluate SOC sequestration on intensively managed RMSs under agriculture land use. Given different land management practices and plant biomass inputs in agricultural land use, it is unclear whether RMSs under arable land use can sustain similar SOC sequestration rates as those under pastoral and silvicultural land uses.

Soil microbiota are the basis of all terrestrial ecosystems recovery from major disturbances, playing significant role in soil formation, soil organic matter (SOM) transformation, nutrients cycling, N fixation and soil quality improvement. Significant degradation of the microbial community can occur after drastic soil disturbance associated with mining and reclamation, both in total microbial biomass and species composition. Studies of RMSs ecosystems indicate that microbial communities may take up to 20 years or more to recover in terms of biomass and diversity similar to that under an undisturbed soil [23,25-27]. While gross measures of microbial biomass C (MBC) reveals little information about functional groups and roles that specific microbes play in soil processes, it is reasonable to assume that MBC comparable to that observed in the undisturbed native soils nearby is a good indicator of a restorative soil system. Understanding the effects of post-reclamation land use, management and its influence on SOC storage and MBC is important to assessing the ecosystem function and the level of reclamation. Therefore, the objectives of this study were to: 1) assess the effects of tillage on soil physical properties and C sequestration in RMSs under arable land use, 2) to evaluate the impact of tillage on microbial biomass in RMSs under arable land use.

2. MATERIALS AND METHODS

2.1. Site Description and Soil Sampling

The experimental site is located in Tuscarawas County near Ragersville, (40˚27'47''N, 81˚38'13''W) Eastern Ohio. Soils of the experimental site belong to Coshocton-Bethesda-Guernsey soil association, well to moderately well drained soils formed from siltstone, shale, and sandstone [28]. The region has temperate continental climate, with average annual temperature of 9.7˚C and precipitation of 983.2 mm, of which 55% is received between April and September. The predominant land use of the region is crop and pasture production.

Prior to mining the study site was under pasture and corn (Zea mays L.) production for gentler slope (2% - 4% slope) and forest for steeper slopes (>5% slope). The site used for this study was mined and reclaimed under SMCRA cropland guidelines in 1984. Mining process involved clearing secondary forest portion, followed by scraping and storing topsoil and subsoil from forest and agricultural sites, and removing overburden to access coal seams. After extracting coal, the site was reclaimed back to original topography by grading the mine spoil, application of stored subsoil and topsoil over the graded overburden to create suitable cropland. The site was then seeded with mixtures of grass species. In 1995, the reclaimed sites were returned to the land owner (farmer). The farmer maintained one portion of the reclaimed site (10 ha) for intensively managed hay production for dairy cows since 1995; reclaimed pasture (RP), while the remaining reclaimed site was converted to cropland in 1995. Part of the cropland have been under conventional tillage and corn (RCT; about 8 ha) while no-till (RNT;

about 10 ha) and corn was practiced on the other part of the field for the past 10 yrs. Both RCT and RNT were under continuous corn for the past 10 yrs and received 5 $Mg{\cdot}ha^{-1}$ of manure every 3 yrs. In addition, corn received 112 $kg{\cdot}ha^{-1}$ N, 25 $kg{\cdot}ha^{-1}$ P and 46 $kg{\cdot}ha^{-1}$ K as NPK fertilizer applied at planting. Nearby remnant unmined and unmanaged pastures (Und P) and hardwood forest (>70 yrs old; Und HWF) soils were sampled as control.

2.2. Soil Sampling and Analysis

The Soil samples were collected between June and July 2007 within each land use from summit, shoulder, and foot-slope landscape positions. Bulk and undisturbed soil core samples were collected from the 0 - 10, 10 - 20, 20 - 30 and 30 - 50 cm depths at each landscape position. Undisturbed core samples were collected using 6 cm diameter and 6 cm long soil cores and used for bulk density determination. Soil bulk density (ρ_b) was computed as the weight to volume ratio of oven-dried (105°C) soil corrected for roots and gravel [29].

Bulk soil samples were air dried and 50 g of aggregates between 5 and 8 mm size were sieved from a part of sample for aggregate analysis. Aggregates were separated using wet-sieving apparatus [30] as described by Nimmo and Perkins [31]. The aggregate analysis was conducted by using a nest of five sieves (5, 2, 1, 0.5 and 0.25 mm) oscillated in water for 30 min. following gentle wetting. Water stable aggregates (WSA) retained on each sieve were backwashed from the sieve with deionized water, oven dried at 60°C for 72 h and weighed. The mean weight diameter (MWD) was computed as the sum of the weighted average diameter of all size classes, where weighting factors were the proportions of the oven-dry (60°C) mass of each size class to the total sample weight [31]. A 2 g subsample of each aggregate-size fraction was finely ground to pass through a 0.25-mm sieve for total organic carbon (TOC) and total nitrogen (TN) determinations.

The remainder of the bulk soil was ground and sieved through a 2-mm sieve. A sub sample (10 g) of each sample was finely ground with a ball mill and passed through a 0.25-mm sieve for total organic carbon (TOC) and total nitrogen (TN) determinations. Concentrations of C and N in composite samples and aggregate fractions were determined by the dry combustion (900°C) method using elemental CN analyzer (VarioMax, Elementar GmbH, Hanau, Germany).

The SOC pool for a specific depth range was calculated using equation:

$$SOC\left(Mg \cdot ha^{-1}\right) = \frac{SOC(\%)}{100} \times \rho_b \times d \times \frac{10^4 m^z}{ha}$$

where ρ_b is bulk density ($Mg \cdot m^{-3}$), of the soil layer, SOC is SOC concentration, and d is soil depth in m.

2.3. Microbial Biomass and Mineralizable Carbon Determination

Fresh soil samples from the top 0 - 10 and 10 - 20 cm depths for each land use were used for this assay. Freshly sampled soils were stored in a cooler, transported to the laboratory, and sieved through a 2-mm sieve. Sieved samples were stored in cold room (4°C) and the microbial biomass carbon (MBC) and mineralizable C (MIN-C) assay were conducted within 1 week of soil sampling. The MBC was quantified by the chloroform fumigation incubation method [32,33] and CO_2 evolved was quantified by gas chromatography (GC). Briefly, duplicate 50 g (oven dry equivalent) of sieved soil samples were placed in glass jars (500 mL), wetted to 50% of soil saturation moisture content (SMC) and incubated for 72 hr at 25˚C in the dark. After 72 hr, one set was transferred into 100 mL beakers, placed in the desiccator lined with moist paper towels to prevent desiccation of soil samples during fumigation. A beaker containing 50 mL of ethanol-free chloroform and anti-bumping granules was placed together with soil samples in the desiccator. The desiccator was evacuated until chloroform bubbled vigorously. Air was let back into desiccator to allow the distribution of chloroform throughout the soil. This was repeated four times, and then the desiccator was evacuated until the chloroform bubbled vigorously for 2 min. The desiccator valve was closed, and placed in the dark for 24 hr. The unfumigated set (control) was kept at 25˚C in the dark during this process.

After 24 hr fumigation, chloroform beaker and paper towels were removed under fume hood, desiccator evacuated several times, each time allowing air to pass into desiccator until soil samples were free of chloroform. After removal of chloroform, soils samples were transferred to mason jars, and inoculated with 1 g of unfumigated soil and mixed thoroughly with spatula.

Both fumigated-inoculated and unfumigated (control) samples were adjusted to 50% of water holding capacity and incubated in closed air tight mason jars in the incubator at 25˚C in the dark for 10 days. Head space air was sampled and analyzed for CO_2 concentration at 1, 2, 4, 7, and 10 d intervals by 5 mL syringe. The concentration of CO_2 evolved in each jar was quantified using GC (Shimadzu, GC 14A, Kyoto, Japan) equipped with thermal conductivity detector (TCD).

After each sampling, samples were aerated for 30 minutes in the lab and soil moisture content was adjusted to 50% SMC. The MBC was calculated as the C concentration difference between fumigated-inoculated and control using a correction factor of $K_c = 0.41$ [33]. After 10 days the CO_2 production monitoring from chloroform-fumigated samples was terminated, but the monitoring from the unfumigated samples continued for the extra period of 90 days for the determination of potentially mineralizable C. Head space air samples were collected at 7 d interval and analyzed for CO_2 concentration using GC. Cumulative CO_2 produced was converted to C based on mass balance and gas laws.

2.4. Statistical Analyses

Analysis of variance was conducted using proc GLM available in SAS (SAS Institute Inc., Cary, NC, 2001). Soil properties were analyzed separately for each depth interval and the entire soil profile (0 - 50 cm depth) to compare differences among land use treatments. When statistically significant effect was detected, Fisher's protected LSD test was performed for mean separations. Statistical significance was determined at the $p < 0.05$.

3. RESULTS AND DISCUSSION

3.1. Soil Bulk Density and Aggregate Stability

Soil ρ_b of the RMS ranged from 1.26 to 1.93 $Mg{\cdot}m^{-3}$, and it increased with increase in soil depth for all management practices (Figure 1). Similar results have been reported in other studies conducted on RMSs [8,17,22,34,35]. The lower ρ_b of the 0 - 10 cm soil layer is consistent with the high biological activity, more root development, greater SOC concentration which enhances soil aggregation and biological activity and freeze/thaw cycles occurring on or near the soil surface. Mining and reclamation increased ρ_b of the soils significantly at all depths compared to the undisturbed land uses ($p < 0.05$; Figure 1). Such a trend may be attributed to mixing of soil layers during scraping and storage of topsoil, and soil compaction caused by heavy equipment involved in mining and reclamation processes. Among the reclaimed sites, ρ_b of the top 50 cm was in the order: RP (1.63 $Mg{\cdot}m^3$) < RCT (1.69 $Mg{\cdot}m^3$) < RNT (1.73 $Mg{\cdot}m^3$) (LSD = 0. 20 $Mg{\cdot}m^3$; $p < 0.05$). Increase in ρ_b under RNT than RCT is consistent with data from agricultural soils which have shown that in the first decade of cultivation NT soils have higher ρ_b than CT soils [36,37], suggesting that under RMSs, tillage management may have impact similar to that observed on unmined agricultural soils during the first 10 yrs.

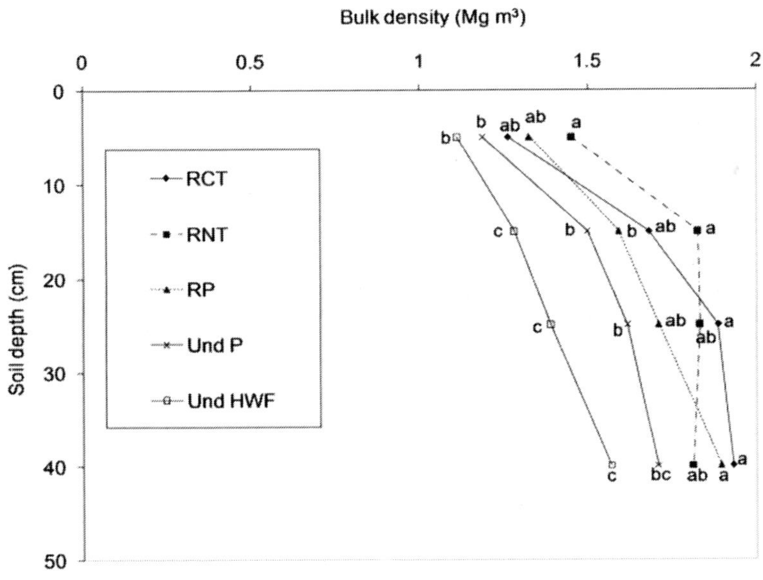

Figure 1. Land use effects on soil bulk density of the reclaimed cropland in Eastern Ohio. For each depth, same lower case numbers (or none) within the same column and sand same depth interval are not statistically significant at p = 0.05 level according to least significant difference (LSD).

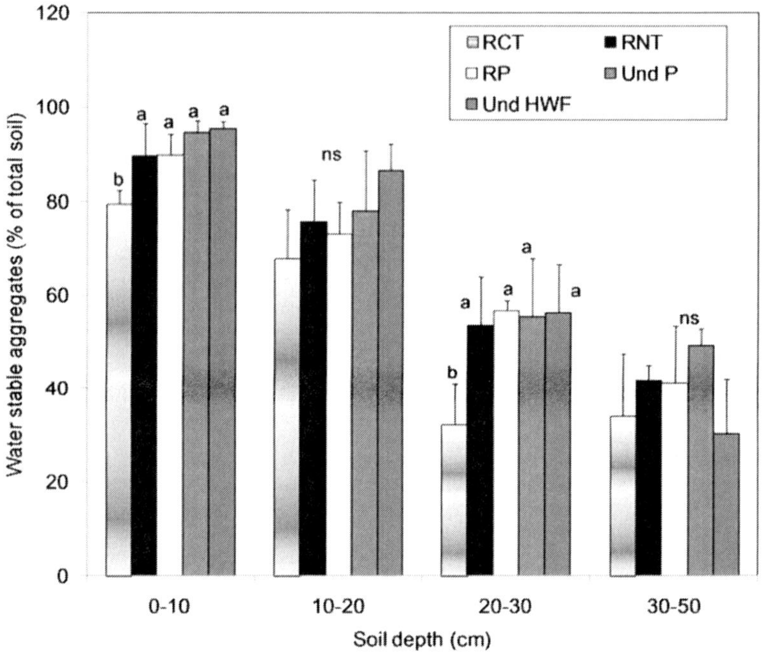

Figure 2. Land use effects on water-stable aggregates of the reclaimed cropland in Eastern Ohio. Error bars represent standard errors of the mean. For each depth group, same lower case numbers (or none) within the same column and sand same depth interval are not statistically significant at p = 0.05 level according to least significant difference (LSD).

The percentage of water-stable aggregates (WSA) > 0.25 mm ranged from 30% to 95%, and decreased with increase in soil depth (Figure 2). The MWD ranged from 0.27 to 4.77 mm (Figure 3). For all sites, MWD was the greatest in the 0 - 10 cm soil layer and decreased with increase in soil depth (Figure 3). The larger WSA and associated MWD in the surface soils may be due to accumulation of fresh and active organic matter (OM) relatively rich in monosaccharides and polysaccharides which enhance aggregate formation and its stability [38,39]. Mean WSA of the top 50 cm soil layer was 53, 65, 65, 67 and 69% in the RCT, RNT RP Und HWF and Und P, respectively (LSD = 10%). On average, the MWD of the top 50 cm was 1.49 ± 0.35, 1.75 ± 0.32, 2.07 ± 0.42, 2.14 ± 0.23, 2.15 ± 0.29 mm for RCT, RNT, Und HWF, RP and Und P, respectively (LSD = 0.45 mm).

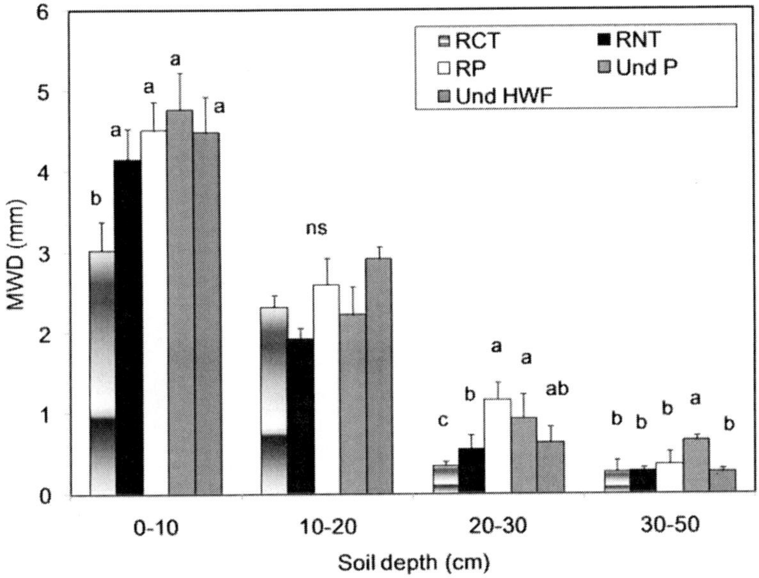

Figure 3. Land use effects on mean weighted diameter of soils from the reclaimed cropland of Eastern Ohio. Error bars represent standard errors of the mean. For each depth group, same lower case numbers (or none) within the same column and sand same depth interval are not statistically significant at p = 0.05 level according to least significant difference (LSD).

Among the reclaimed sites, WSA and MWD were significantly lower for the RCT in the 0 - 10 cm and 20 - 30 cm depths, than those under RNT and RP treatments (Figures 2, 4). Tillage did not influence WSA and MWD in other depth layers. Decrease in WSA and MWD, following cultivation of RMSs suggested a reduced soil structure development in the RMS disturbed by cultivation compared to RNT and RP land uses. In addition, there were no significant differences among RP and undisturbed pasture, suggesting a complete recovery of aggregate stability under pasture land use.

3.2. Soil Organic Carbon and Nitrogen Pools

The SOC and N concentrations were strongly stratified with soil depth (Table 1), and were the highest under the Und HWF on all depth layers compared to those from RMSs sites (Table 1). Among the RMSs sites, SOC concentration was higher under pasture in all depths than RCT and RNT treatments, except in 10 - 20 cm depth where SOC concentration was more under RCT land use ($p < 0.05$). Lower SOC concentration under RCT and RNT than RP treatments could be attributed to accelerated decomposition of SOM caused by soil disturbance associated with these agricultural practices. Total N concentration was significantly more under hardwood forest in the 0 - 10 cm layer, but no significant differences in N concentrations were observed among land uses in sub-soil layer (Table 1).

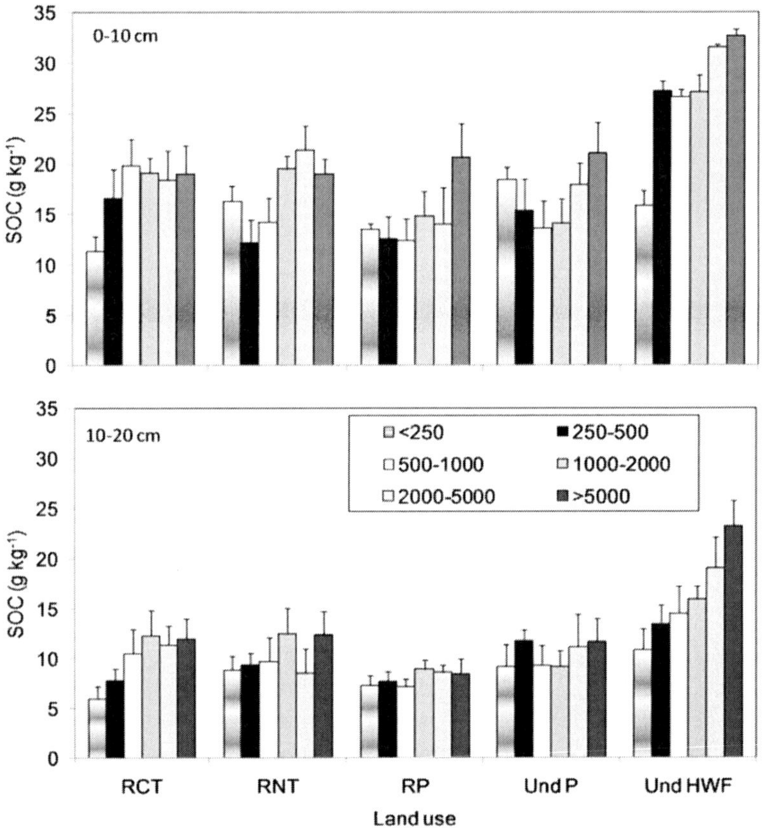

Figure 4. Effects of land use on soil organic carbon concentration of macroaggregates of the reclaimed cropland of Eastern Ohio. Error bars represent standard errors of the mean.

The N pools of the top 0 - 50 cm depth were the lowest under Und P (7.4 ± 1.1) than Und HWF (9.1 ± 1.2 $Mg \cdot N \cdot ha^{-1}$), and on the RMSs land uses RP (9.6 ± 0.2)RCT (9.0 ± 0.6) and RNT (7.8 ± 2.3 $Mg \cdot N \cdot ha^{-1}$; Table 2). Among the RMSs sites, the N pool was significantly lower under RNT than RCT and RP treatments. There were no significant differences among RNT and Und P land uses. In the 0 - 10 cm layer, the N pools were significantly greater under the RP than Und P site. However, there were no significant differences in N pools among RCT, RP, RNT or Und HWF in the top 0 - 10 cm depth. Only few changes in soil N pool were observed among land uses at lower depths (Table 1).

Table 1. Land use effects on soil organic carbon, nitrogen, microbial biomass and mineralizable carbon of the reclaimed cropland of Eastern Ohio.

Land use	SOC	TN	C:N ratio	MBC	Min C
	........($g\ kg^{-1}$ soil)........			........($g\ C\ kg^{-1}$ soil)........	
0 - 10					
RCT	23.6 c	2.5 ab	9.37	0.36 b	4.10 c
RNT	24.3 c	2.0 c	9.33	0.29 c	4.82 a
RP	26.3 b	2.7 a	9.41	0.41b	4.46 b
Und P	20.2 d	2.2c	9.31	0.53a	4.46 b
Und HWF	34.5 a	2.8 a	12.32	0.53 a	5.12 a
10 - 20					
RCT	12.0 b	1.3	8.76	0.09 b	2.18
RNT	7.1 d	1.1	8.49	0.05 b	2.19
RP	10.1 c	1.2	8.83	0.20 a	2.71
Und P	11.9 b	1.2	9.58	0.17 a	2.14
Und HWF	19.7 a	1.7	11.21	0.21 a	2.58
20 - 30					
RCT	5.2 c	0.7	7.45		
RNT	3.6 d	0.7	7.71		
RP	6.7 c	0.7	9.36		
Und P	6.8 b	0.7	8.84		
Und HWF	11.4 a	1.1	10.33		
30 - 50					
RCT	4.0 b	0.6	6.98		
RNT	3.3 b	0.6	6.48		
RP	6.3 a	0.7	9.67		
Und P	3.4 b	0.5	6.67		
Und HWF	7.0 a	0.8	9.36		
0 - 20					
RCT	17.8	1.90		0.22	3.14
RNT	15.7	1.57		0.17	3.50
RP	18.2	1.98		0.29	3.30
Und P	16.1	1.70		0.36	4.09
Und HWF	27.1	2.23		0.37	3.85

The SOC pool of the top 50 cm soil depth was the largest under the Und HWF (101.1 ± 3.0 Mg·C·ha^{-1}) and lowest under the Und P (64.2 ± 0.8 Mg·ha^{-1}; Table 1). Among the RMSs sites, SOC pool of the top 50 cm depth was significantly more under RP (86.1 ± 2.4 Mg·ha^{-1}) than RCT (75.4 ± 2.5 Mg·ha^{-1}) and RNT treatments (66.5 Mg·ha^{-1}; LSD = 7.7 Mg·ha^{-1}, Table 2). Lower SOC pools under

RNT and RCT than RP may be due to increase in the rate of SOM mineralization exacerbated by frequent disturbances caused by the farming operations (i.e. tilling and/or planting). Overall, the SOC pool decreased with increase in soil depth in all land uses.

Same lower case numbers (or none) within the same column and sand same depth interval are not statistically significant at p = 0.05 level according to least significant difference (LSD).

Table 2. Land use effects on soil organic carbon (SOC) and nitrogen pools of the reclaimed cropland in Eastern Ohio. Numbers in brackets are standard deviations.

Soil depth	Land use				
	RCT	RP	RNT	Und P	Und HWF
cm			SOC, Mg ha^{-1}		
0 - 10	29.8 (1.4)b	34.9 (1.0)ab	35.2 (5.0)ab	23.9 (1.2)c	38.2 (2.3)a
10 - 20	20.1 (1.8)b	16.1 (0.3)c	12.9 (1.5)d	17.9 (1.3)bc	25.1 (1.3)a
20 - 30	9.7 (0.9)c	11.4 (0.8)b	6.6 (1.3)d	11.0 (0.1)bc	15.7 (0.5)a
30 - 50	15.7(0.3)b	23.8 (1.8)a	11.7 (2.4)c	11.4 (1.1)c	22.1 (2.4)a
Total	75.4 (2.5)c	86.1 (2.4)b	66.5 (7.4)d	64.2 (0.8)d	101.1 (3.0)a
			Total N, Mg ha^{-1}		
0 - 10	3.2 (0.4)ab	3.6 (0.2)a	3.0 (0.5)ab	2.6 (0.1)b	3.1 (0.3)ab
10 - 20	2.3 (0.1)	1.8 (0.1)	2.0 (0.9)	1.9 (0.5)	2.1 (0.1)
20 - 30	1.3 (0.2)	1.3 (0.2)	1.2 (0.6)	1.2 (0.2)	1.5 (0.2)
30 - 50	2.3 (0.3)	2.9 (0.2)	1.7 (0.5)	1.8 (0.6)	2.4 (0.6)
Total	9.0 (0.6)a	9.6 (0.2)a	7.8 (1.1)b	7.4 (1.1)b	9.1 (1.2)a

Same lower case numbers (or none) within the same column and sand same depth interval are not statistically significant at p = 0.05 level according to least significant difference (LSD).

However, there was a greater stratification of SOC pool in the RNT (53% of the SOC pool in the top 10 cm) than in other land uses. Similar stratification of SOC and N pools have also been reported under unmined agricultural soils [40,41]. These trends indicate that C and N stratification in soil under RNT is probably caused by accumulation of crop residues on the surface, and the lack of stratification under RCT may be due to incorporation of crop residues into the plow layer by tillage.

Similar to the bulk soil, the SOC concentrations in each aggregate-size class was also more in the top 10 cm layer, and generally decreased with increase in soil depth (Figure 4). With the exception of the >250 μm fraction in the 0 - 10 cm layer, the SOC concentration was generally more under Und HWF than in all other land uses for all aggregate size fractions. In the 0 - 10 cm depth, RCT contained more SOC concentrations in 0.25 - 0.5 mm and 0.5 - 1 mm aggregate sizes than those from RNT and RP treatments. In general, the SOC concentration was similar in the 1 - 2 mm fractions among RCT and RNT treatments but lower under RP treatment. In the 2 - 5 mm aggregate fraction, the RNT land use contained more SOC concentration than RCT and RP treatments. For aggregates >5 mm, the SOC concentration was more under RP than RCT and RNT treatments. In the 10 - 20 cm depth, the SOC concentration was the highest under Und HWF and the lowest under RP in all aggregate size fractions. Only minor differences in SOC concentration in aggregates were observed among other fractions and land uses. In addition, only little variations were observed in SOC concentration in the lower soil depths.

The RMSs could potentially act as a sink for atmospheric CO_2 [17,19,20,22,24]. Data from this study indicated that the RMSs under pasture accumulated SOC at higher rates than those under arable land use. Further, RMSs under pastoral land use increased SOC pool by 14% or 0.5 $Mg\cdot ha^{-1}\cdot yr^{-1}$ and 30% or 0.9 $Mg\cdot ha^{-1}\cdot yr^{-1}$ relative to RNT and RCT, respectively. This trend of higher SOC pool under managed pasture relative to arable land suggests that land use plays major role in SOC accumulation in RMSs. These data also suggest that even with similar nutrient input, RMSs under cropland do not maintain high SOC sequestration rates as observed under silviculture and pastoral land uses. Such a trend may be attributed to reduced aggregation in cultivated soils (Figure 3), thereby limiting aggregate-protected SOC in cropland soils.

3.3. Microbial Biomass and Mineralizable Carbon

Concentrations of MBC ranged from 0.29 to 0.53 $g\cdot C\cdot kg^{-1}$ in the 0 - 10 cm layer, and 0.05 to 0.21 $g\cdot C\cdot kg^{-1}$ in the 10 - 20 cm layer (Table 1). Among the reclaimed sites, the MBC was more under RP than RCT and RNT for both 0 - 10 and 10 - 20 cm soil depths. Overall, the MBC was more under the undisturbed soils than under the RMSs (Table 1). Microbial biomass C is a living component of SOC, a reservoir of nutrients and an important determinant of nutrient cycling in soils. The size, composition, activity, and ratio of MBC relative to SOC have been evaluated as indicators of state of the soil ecosystem with regard to degradation and recovery from disturbance [23,42]. These data suggest that disturbances associated with mining and reclamation were detrimental to microbial populations, resulting in reduction of MBC by 39%, 53% and 21% under RCT, RNT and RP compared to the undisturbed forest and pastoral sites for the first 22 years after mining. Similar results indicating 56% decline in MBC 20 years after reclamation and general decline in microbial community have been reported by Mummey et al. [26,27]. Soil microorganisms are sensitive to disturbances [43], and alteration of microbial community in terms of total biomass and species composition after mining and reclamation disturbances have also been reported [23,25,44]. In cultivated RMSs, tillage practices (NT and CT) had no significant influence on MBC in the 0 - 10 or 10 - 20 cm layers (Table 1). These results are contrary to those reported by Jacinthe and Lal [44] who observed significant increase in MBC in the 0 - 5 cm soil layer under NT compared to CT in reclaimed cropland of southwestern Indiana. Lack of response in this study may be attributed to application of cow manure at these sites which tends to favor microbial community restoration. The MBC was correlated with SOC and total N (Figure 5), suggesting that microorganisms in RMSs rely on soil organic matter (SOC and total N) as their main source of Cand energy.

The MIN-C concentration ranged from 4.1 to 5.5 g·C·kg^{-1} and 2.2 to 2.6 g·C·kg^{-1} for 0 - 10 and 10 - 20 cm depths, respectively (Table 1). The MIN-C accounted for up to 25% of the SOC concentration in the top 0 - 20 cm layer. The MIN-C concentration was significantly correlated with soil organic C and total N (Figure 5). The MIN-C concentration was nearly twice as much in the 0 - 10 cm layer compared to 10 - 20 cm layer. MN-C concentration was generally greater under the undisturbed sites (Und HWF and Und P) than the disturbed sites. The concentration of MIN-C did not differ significantly among the three land managements for the RMSs sites. Lower MIN-C under RMSs sites may be attributed to accelerated mineralization of SOC from the stored topsoil, which reduces mineralizable fraction of SOM.

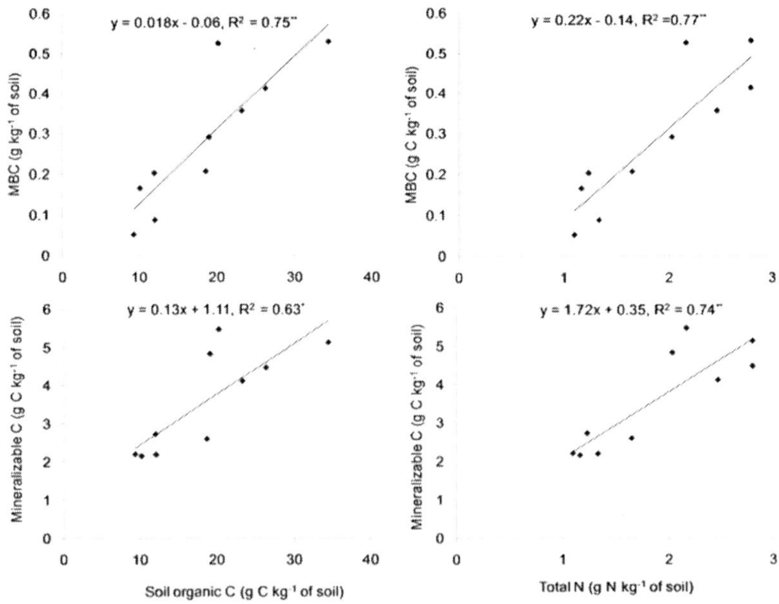

Figure 5. Relationships between soil microbial biomass carbon and soil organic carbon and total nitrogen in top 0 to 20 cm soil layer in reclaimed cropland of eastern Ohio.

4. CONCLUSION

Agricultural use is one of common post-reclamation land use in eastern Ohio. However, C dynamics and sequestration potential of RMSs under crop land is not well understood. This study evaluated tillage and pasture management on soil physical properties and SOC dynamics in 22 yrs old RMSs managed for agricultural use eastern Ohio. Results indicated that RMSs under no tillage and pasture had significantly greater water-stable aggregates and mean-weighted diameter in the top 10 cm soil layer than those under continuous tillage, suggesting improved soil structure under these land uses. The RMSs under pasture sequestered SOC at 30% higher rate or 0.9 $Mg \cdot ha^{-1} \cdot yr^{-1}$ than RMSs under conventional tillage. In addition, RMSs under pasture sequestered SOC at 14% higher rate or 0.5 $Mg \cdot ha^{-1} \cdot yr^{-1}$ than those under no tillage. Overall, the disturbances associated with mining and reclamation reduced the microbial biomass carbon by 21% to 39% compared to undisturbed pasture and forest land uses. The data from this study suggested that agricultural land use could be less efficient for microbial restoration and SOC sequestration compared to forest and grassland under RMSs. However, more data are needed to evaluate the impact of agriculture on ecosystem restoration and SOC sequestration in RMSs.

5. ACKNOWLEDGEMENTS

This research was supported by funds from Department of Energy (DOE) and Ohio Coal Development Office (OCDO), Project D-07-03.

REFERENCES

1. K. McSweeney and I. J. Jansen, "Soil Structure and Associated Rooting Behavior in Minesoils," Soil Science Society of America Journal, Vol. 48, No. 3, 1984, pp. 607-612.

2. R. I. Barnhisel and J. M. Hower, "Coal Surface Mine Reclamation in the Eastern United States: The Revegetation of Disturbed Lands to Hayland/Pasture or Cropland," Advances in Agronomy, Vol. 61, 1997, pp. 233-275.

3. A. Bradshaw, "Restoration of Mined Lands—Using Natural Processes," Ecological Engineering, Vol. 8, No. 4, 1997, pp. 255-269.

4. R. Lal, J. M. Kimble and R. F. Follett, "Land Use and Soil Carbon Pools in Terrestrial Ecosystems," In R. Lal, J. M. Kimble and R. F. Follett, Eds., Management of Carbon Sequestration in Soils, CRC Lewis Publisher, Boca Raton, 1998, pp. 1-10.

5. S. J. Indorante, I. J. Jansen and C. W. Boast, "Sorface Mining and Reclamation: Initial Changes in Soil Character," Journal of Soil and Water Conservation, Vol. 36, No. 6, 1981, pp. 347-351.

6. P. J. Thompson, I. J. Jansen and C. L. Hooks, "Penetrometer Resistance and Bulk Density as Parameters for Predicting Root System Performance in Mine Soils," Soil Science Society of America Journal, Vol. 51, No. 5, 1987, pp. 1288-1293.

7. R. I. Barnhisel, "Fertilization and Management of Reclaimed Lands," In R. L. Hosner, Ed., Reclamation of SurfaceMined Lands, CRC Press, Boca Raton, 1988, pp. 1-15.

8. M. K. Shukla, R. Lal, J. Underwood and M. Ebinger, "Physical and Hydrological Characteristics of Reclaimed Minesoils in Southeastern Ohio," Soil Science Society of America Journal, Vol. 68, No. 4, 2004, pp. 1352-1359.

9. R. E. Dunker and R. I. Barnhisel, "Cropland Reclamation," In R. I. Barnhisel, R. G. Darmody and W. L. Daniels, Eds., Reclamation of Drastically Disturbed Lands, Soil Science Society of America, Madison, 2000, pp. 323-369.

10. J. L. Powel, "Revegetation Options," In L. R. Hossner and L. R. Hossner, Eds., Reclamation of Surface Mined Lands, CRC Press, Boca Raton, 1988, pp. 49-91.

11. W. A. Dick, R. L. Blevins, W. W. Frye, S. E. Peters, D. R. Christenson, F. J. Pierce and M. L. Vitosh, "Impacts of Agricultural Management Practices on C Sequestration in Forest-Derived Soils of the Eastern Corn Belt," Soil & Tillage Research, Vol. 47, No. 3-4, 1998, pp. 235-244. doi:10.1016/S0167-1987(98)00112-3

12. G. B. Triplett, Jr. and W. A. Dick, "No-Tillage Crop Production: A Revolution in Agriculture!" Agronomy Journal, Vol. 100, No. 3, 2008, pp. S153-S165.

13. R. Lal, A. A. Mahboubi and N. R. Fausey, "Long-Term Tillage and Rotation Effects on Properties of a Central Ohio Soil," Soil Science Society of America Journal, Vol. 58, No. 2, 1994, pp. 517-522.

14. J. Six, E. T. Elliott, and K. Paustian, "Soil Macroaggregate Turnover and Microaggregate Formation: A Mechanism for C Sequestration under No-Tillage Agriculture," Soil Biology & Biochemistry, Vol. 32, No. 14, 2000, pp. 2099-2103.

15. V. A. Akala and R. Lal, "Soil Organic Carbon Pools and Sequestration Rates in Reclaimed Minesoils in Ohio," Journal of Environmental Quality, Vol. 30, No. 6, 2001, pp. 2098-2104.

16. E. S. Bendfeldt, J. A. Burger and W. L. Daniels, "Quality of Amended Mine Soils after Sixteen Years," Soil Science Society of America Journal, Vol. 65, No. 6, 2001, pp. 1736-1744.

17. D. A. N. Ussiri, R. Lal, and P. A. Jacinthe, "Post-Reclamation Land Use Effects on Properties and Carbon Sequestration in Minesoils of Southeastern Ohio," Soil Science, Vol. 171, No. 3, 2006, pp. 261-271.

18. U. Fettweis, O. Bens and R. F. Huttl, "Accumulation and Properties of Soil Organic Carbon at Reclaimed Sites in the Lusatian Lignite Mining District Afforested with Pinus Sp," Geoderma, Vol. 129, No. 1-2, 2005, pp. 81-91.

19. R. K. Shrestha and R. Lal, "Ecosystem Carbon Budgeting and Soil Carbon Sequestration in Reclaimed Mine Soil," Environment International, Vol. 32, No. 6, 2006, pp. 781- 796.

20. M. Sperow, "Carbon Seqnestration Potential in Reclaimed Mine Sites in Seven East-Central States," Journal of Environmental Quality, Vol. 35, No. 4, 2006, pp. 1428- 1438.

21. P.-A. Jacinthe and R. Lal, "Carbon Storage and Minesoil Properties in Relation to Topsoil Applications Techniques," Soil Science Society of America Journal, Vol. 71, No. 6, 2007, pp. 1788-1795.

22. D. A. N. Ussiri, R. Lal and P. A. Jacinthe, "Soil Properties and Carbon Sequestration of Afforested Pastures in Reclaimed Minesoils of Ohio," Soil Science Society of America Journal, Vol. 70, No. 5, 2006, pp. 1797-1806.

23. H. Insam and K. H. Domsch, "Relationship between Soil Organic Carbon and Microbial Biomass on Chronosequences of Reclamation Sites," Microbial Ecology, Vol. 15, No. 2, 1988, pp. 177-188.

24. R. K. Shrestha, D. A. N. Ussiri and R. Lal, "Terrestrial Carbon Sequestration Potential in Reclaimed Mine Ecosystems to Mitigate Greenhouse Effect," In R. Lal and R. F. Follet, Eds., Soil Carbon Sequestration and Greenhouse Effect, Soil Science Society of America, Madison, 2009, pp. 321-346.

25. P. D. Stahl, S. E. Williams and M. Christensen, "Efficacy of Native Vesicular Arbuscular Mycorrhizal Fungi after Severe Soil Disturbance," New Phytologist, Vol. 110, No. 3, 1988, pp. 347-354.

26. D. L. Mummey, P. D. Stahl and J. S. Buyer, "Microbial Biomarkers as an Indicator of Ecosystem Recovery Following Surface Mine Reclamation," Applied Soil Ecology, Vol. 21, No. 3, 2002, pp. 251-259.

27. D. L. Mummey, P. D. Stahl and J. S. Buyer, "Soil Microbiological Properties 20 Years after Surface Mine Reclamation: Spatial Analysis of

Reclaimed and Undisturbed Sites," Soil Biology & Biochemistry, Vol. 34, No. 11, 2002, pp. 1717-1725.

28. USDA-NRCS, "Soil Survey of Tuscarawas County, Ohio," USDA-NRCS, Washington, 1986.

29. R. B. Grossman and T. G. Reinsch, "Bulk Density and Linear Extensibility," In J. H. Dane and G. C. Topp, Eds., Methods of Soil Analysis: Part 4: Physical Methods, Soil Science Society of America Inc., Madison, 2002, pp. 201- 225.

30. R. E. Yoder, "A Direct Method of Aggregate Analysis of Soils and a Study of the Physical Nature of Erosion Losses," Journal of American Society of Agronomy, Vol. 28, No. 5, 1936, pp. 337-351.

31. J. R. Nimmo and K. S. Perkins, "Aggregate Stability and Size Distribution," In J. H. Dane and G. C. Topp, Eds., Methods of Soil Analysis. Part 4: Physical Methods, Soil Science Society of America Inc., Madison, 2002, pp. 317- 328.

32. E. D. Vance, P. C. Brookes, and D. S. Jenkinson, "An Extraction Method for Measuring Soil Micribial Biomass C," Soil Biology & Biochemistry, Vol. 19, No. 6, 1987, pp. 703-707.

33. W. R. Horwarth and E. A. Paul, "Microbial Biomass," In G. T. Weaver, J. A. Angle and P. S. Bottomley, Eds., Methods of Soil Analysis. Part 2. Microbiological and Boichemical Properties, Soil Science Society of America, Madison, 1994, pp. 753-773.

34. C. Varela, C. Vazquez, M. V. Gonzalezsangregorio, M. C. Leiros and F. Gilsotres, "Chemical and Physical Properties of Opencast Lignite Minesoils," Soil Science, Vol. 156, No. 3, 1993, pp. 193-204.

35. D. A. N. Ussiri and R. Lal, "Carbon Sequestration in Reclaimed Minesoils," Critical Reviews in Plant Sciences, Vol. 24, No. 3, 2005, pp. 151-165.

36. A. J. Franzluebbers, F. M. Hons and D. A. Zuberer, "Tillage-Induced Seasonal Changes in Soil Physical Properties Affecting Soil CO_2 Evolution under Intesive Cropping," Soil & Tillage Research, Vol. 34, No. 1, 1995, pp. 41-60.

37. B. D. Kay and A. J. VandenBygaart, "Conservation Tillage and Depth Stratification of Porosity and Soil Organic Matter," Soil & Tillage Research, Vol. 66, No. 2, 2002, pp. 107-118.

38. J. M. Tisdall and J. M. Oades, "Organic Matter and Water-Stable Aggregates in Soils," Journal of Soil Science, Vol. 33, No. 2, 1982, pp. 141-163.

39. P. Loveland and J. Webb, "Is There a Critical Level of Organic Matter in the Agricultural Soils of Temperate Regions: A Review," Soil & Tillage Research, Vol. 70, No. 1, 2003, pp. 1-18.

40. J. Balesdent, A. Mariotti and D. Boisgontier, "Effects of Tillage on Soil Organic Carbon Minrealization Estimated from C-13 Abundance in Maize Fields," Journal of Soil Science, Vol. 41, No. 4, 1990, pp. 587-596.

41. D. A. Angers, R. P. Voroney and D. Cote, "Dynamics of Soil Organic Matter and Corn Residues Affected by Tillage Practices," Soil Science Society of America Journal, Vol. 59, No. 5, 1995, pp. 1311-1315.

42. J. D. Anderson, L. J. Ingram and P. D. Stahl, "Influence of Reclamation Management Practices on Microbial Biomass Carbon and Soil Organic Carbon Accumulation in Semiarid Mined Lands of Wyoming," Applied Soil Ecology, Vol. 40, No. 2, 2008, pp. 387-397.

43. R. F. Turco, A. C. Kennedy and M. D. Jawson, "Microbial Indicators of Soil Quality," In J. W. Doran, et al., Eds., Defining Soil Quality for a Sustainable Environment, Soil Science Society of America, Madison, 1994, pp. 73-90.

44. P.-A. Jacinthe and R. Lal, "Tillage Effects on Carbon Sequestration and Microbial Biomass in Reclaimed Farmland Soils of Southwestern Indiana," Soil Science Society of America Journal, Vol. 73, No. 2, 2009, pp. 605-613. doi:10.2136/sssaj2008.0156

Chapter 7

Estimating CO_2 Sequestration by Forests in Oita Prefecture, Japan, by Combining LANDSAT ETM+ and ALOS Satellite Remote Sensing Data

Kazadi Sanga-Ngoie [1,*], Kotaro Iizuka [2] and Shoko Kobayashi [1]

[1] *Graduate School of Asia Pacific Studies, Ritsumeikan Asia Pacific University, 1-1 Jumonjibaru, Beppu, Oita 874-8577, Japan*

[2] *Graduate School of Science, Chiba University, 1-33 Yayoi-cho, Inage-ku, Chiba, Chiba 263-8522, Japan*

ABSTRACT

CO_2 sequestration of the forests in Oita Prefecture, Japan, was estimated using satellite remote sensing data. First, hybrid classification of the optical LANDSAT ETM+ data was performed using GIS to produce a detailed land cover map. CO_2 sequestration for each forest type was calculated using the sequestration rates per unit area multiplied by the forest areas obtained from the land cover map This results in 3.57 $MtCO_2$/yr for coniferous, 0.77 $MtCO_2$/yr for deciduous broadleaf, and 2.25 $MtCO_2$/yr for evergreen broadleaf, equivalent to a total of 6.60 $MtCO_2$/yr for all the forest covers in Oita. Then, two different methodologies were used to improve these estimates by considering tree ages: the Normalized Difference Vegetation Index (NDVI) and the stem volume methods. Calculation using the NDVI method shows the limitation of this method in providing detailed estimations for trees older than 15 years, because of NDVI saturation beyond this age. In the stem volume method, tree ages were deduced from stem volume values obtained by using PALSAR backscattering data. Sequestration based on tree age forest subclasses yields 2.96 $MtCO_2$/yr (coniferous) and 0.31 $MtCO_2$/yr (deciduous broadleaf forests). These results show the importance of using not only detailed forest types, but also detailed tree age information for more realistic CO_2 sequestration estimates. In so doing, overestimation of the sequestration capacity of forests could be avoided, and the information on the status and location of forest resources could be improved, thereby leading to sounder decision making in sustainable management of forest resources.

Keywords: LANDSAT; ALOS; land use/land cover maps; CO_2 sequestration; GIS

1. INTRODUCTION

Global warming is known as one of the biggest environmental issues of the 21st century. Many studies have been made and conferences held, emphasizing the importance of this issue and the associated problems. World nations are working together on policies, not only for preventing the problem, but mostly for mitigating its impacts [1–3]. Japan is one of the top nations involved in activities related to this issue, and the Kyoto Protocol (adopted at the Kyoto Conference in 1997), provided new approaches to finding solutions [4], among which are the Clean Development Mechanism (CDM), Emission Trading, Joint Implementation (JI) and Carbon Sink. Carbon sink is the focal point in the present research context, as 68.9% of land in Japan are covered by forests [5]. Japan is considering the forests to be the top candidate for its CO_2 reduction process [6]. Many prefectures in Japan have been researching carbon sequestration by forests and have come up with various policies for reducing CO_2. However, some prefectures have not yet been involved in such processes focusing on CO_2 reduction through carbon sink, notwithstanding the large amount of their forest cover. Oita Prefecture is one of them.

The Oita Prefecture government has so far focused on energy conservation as the main mechanism for effectively reducing its CO_2 emissions [7]. Oita Prefecture's global warming measure implementation started in 2005; it aimed at reducing energy consumption and wastes, and introducing more energy-efficient (cogeneration) systems. The result of this approach (Figure 1, [8]) does not seem to be as effective as claimed by the government. There is thus the need to reconsider this regulation and the tools used for its implementation toward the goals. Thus far, Oita Prefecture has not focused on the concept of carbon sinks for reducing its CO_2 emissions, and information is only marginally available. This is more likely because of the lack of sound scientific data needed for objective decision making about which process to use for effectively implementing the CO_2 reduction policy in the long term.

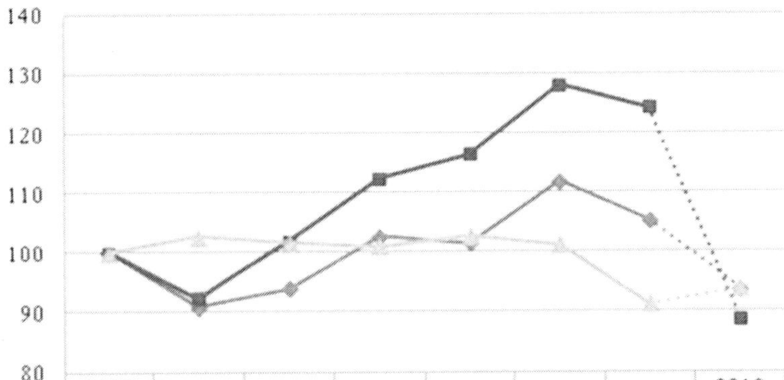

	2002 Base	2003	2004	2005	2006	2007	2008	2010 Target
Household	100	90.9	93.9	102.6	101.5	111.8	105.2	93.7
Service	100	92.3	101.8	112.3	116.4	128	124.3	88.6
Transport	100	102.7	101.6	100.9	102.6	101.2	91.3	93.9

Figure 1. Changes in CO_2 emissions from three targeted sectors in Oita Prefecture, with 2002 as the reference year (100%).

Traditional methods of estimating carbon/CO_2 sequestration through site-based methods [9,10] or using eddy covariance flux tower are known to be expensive, time consuming and limited in area coverage. Expanding the area or implementing continuous monitoring is even more costly and time consuming. Remotely sensed satellite observations have provided the scientists with an alternative method for studying the earth's biosphere (atmosphere, vegetation, etc.). As for the vegetation, it has been demonstrated that the reflected RGB (red, green, blue), and mostly the NIR (near infrared), wavelengths contain considerable information about plants biomass [11], from which precise knowledge about the vegetation can be extracted. This gives us the possibility of biomass monitoring at lower costs and with less time loss. Today, with further research advancement, remote sensing is becoming a common analysis tool, not only for producing maps needed for categorizing the land cover type of the surface and for allocating or managing the earth's resources [12], but also for analyzing the changes and their impacts for future land use/land cover (LU/LC) developments [13].

Using remote sensing data for evaluating regional carbon/CO_2 sequestration has been implemented in various ways, among which is combining the land cover information with the averaged carbon sequestration values of different land cover types [14]. The strength of using land cover information based on remotely sensed images is that it covers areas of regional or even global scales, making it possible to extend analysis over sites that are difficult to access on the ground. Moreover, it allows the implementation of continuous monitoring of the area, making it possible to analyze the temporal changes of the land cover.

Estimates of sequestration by forests in Japan are usually implemented at the national scale (whole Japan), focusing mostly on identifying forests as natural or planted forests. Hiroshima and Nakajima [15] focused on the potential carbon sinks in Japanese planted forests during the first commitment period of the Kyoto Protocol. Using subsidies and forest workers' wages as predictor variables, they estimated that the privately planted forests were sequestering 8.16–8.87 MtC/yr, depending on the different scenarios of forest management and silvicultural practices. Sasaki and Kim [16] estimated both the potential and the eligible sequestrations by the forests in Japan under the condition of the Marrakesh Accords. Using the land use and the carbon stock growth models, they estimated that the forests in Japan were likely to sequester 20.1 MtC/yr (planted forests: 15.3 MtC/yr; natural forests: 4.8 MtC/yr), while under the Marrakesh Accords scheme, this was estimated to be 10.2 MtC/yr (planted forests: 7.3 MtC/year; natural forests: 2.9 MtC/yr). In general, few studies have estimated the potential of CO_2 sequestration by the forests in Japan, and only fewer considered to perform the analysis at local or regional scales, focusing more on detailed forest types. This can be easily understood: the Kyoto Protocol deals with national level estimates, so many studies are involved in the estimates at that scale.

However, we have to note that the estimates by those authors mentioned above are based on modeling, while the information of where and how much of those resources exist is not considered. This is a serious weakness, especially when forest resources management is concerned. Moreover, we also have to consider that forests are more diverse than just being natural or planted forests. Because both tree types and tree ages make differences in CO_2 sequestration [9,17,18], and for a more realistic estimation of the sequestration, we emphasize that forests have to be analyzed on a local or regional scale in order to get more precise and detailed classes for each forest category (tree type) as well as for each tree age range. Based on this detailed information about the forests, precise estimates of the forests sink capacity can be obtained. This is also a way of providing reliable information for sound decision making for sustainable forest management and development.

The objective of this research is to estimate CO_2 sequestration by forests type in Oita Prefecture, based on precise evaluation of forests extent in Oita. This will make it possible to quantitatively estimate their potential storage capacity, and therefore maximize the usage of this environmental resource as a CO_2 sink. These findings will help the Oita Prefecture g to devise better scenarios for mitigating global warming. Local and regional scale analysis of the land cover and estimation of CO_2 sequestration will eventually contribute to the updated information of Japan's CO_2 reduction target achievement.

The study site, datasets and methodology are presented in Sections 2–4, respectively. Geographic Information Systems (GIS) and remote sensing technology are used for the analysis. The analysis results are given in Section 5, followed by discussion in section 6 and concluding remarks in Section 7.

2. STUDY SITE

The research area, Oita Prefecture, is located on the Kyushu Island in the southwestern part of Japan (Figure 2), between 130.81°E and 132.11°E, 32.72°N and 33.75°N, with a total land area of 6,339 km^2. Official figures show that 4,540 km^2 (or 72%) of Oita Prefecture's lands are covered by forests [19].

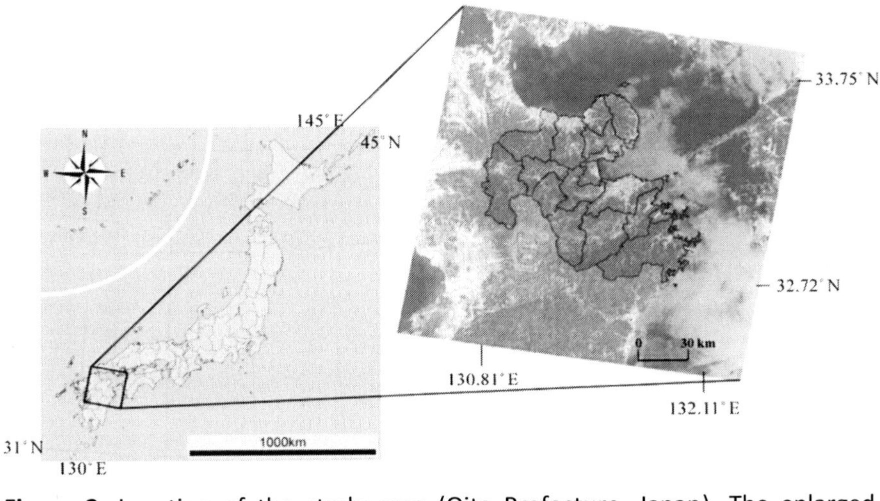

Figure 2. Location of the study area (Oita Prefecture, Japan). The enlarged image of Oita Prefecture with its administrative units (cities, towns and villages) is shown on the right.

Oita Prefecture is also known for its rich and diversified Onsen (hot springs), found especially in Beppu City. A volcanic belt (the Kirishima Volcanic Range), with several mountains ranges, culminating up to 1,758 m in altitude, runs from north to south through the prefecture, contributing to the existence of the many hot springs.

Climatic conditions divide the prefecture in two main areas: (1) the area affected by the monsoon winds coming from the coast of the Japan Sea, and bringing in large amounts of rainfall and large number of rainy days in winter, and (2) the area characterized by large amounts of rains in summer resulting from the moisture flowing in from the Pacific Ocean, and by dry and fine weathers in winter. Because of these wet climate conditions, annual rainfall in Oita Prefecture ranges from 1,500–1,600 mm over coastal areas, to 2,500–3,000 mm in the mountainous areas.

3. DATA SETS

Two optical satellite data sets from the Land Remote Sensing Satellite Enhanced Thematic Mapper, Plus (LANDSAT ETM+) with 30 m spatial resolution published by the United States Geological Survey (USGS) and two microwave satellite data sets from the Advanced Land Observing Satellite (ALOS) Phased Array L-band Synthetic Aperture Radar (PALSAR) provided by Japan Aerospace Exploration Agency (JAXA), were selected for our analytical objectives. The selection of the data sets was done by considering the following three critical issues. First, we had to choose a cloud-free image for analysis. In fact, clouds block the solar radiation from reaching the ground, rendering the covered area unavailable for the optical observing platform. Second, the Scan Line Corrector (SLC) failure that occurred after 31 May 2003 makes the LANDSAT ETM+ line of sight trace a zigzag pattern along the satellite ground track. As a result, a gap fill has to be performed in order to correct the data so as to become usable for analysis [20]. Third, we had to select an image old enough for us to create a digital LU/LC map (or digital vegetation model, here after DVM) of Oita Prefecture that could be used later on as a reference digital map for further quantitative LU/LC change analysis in this area. As a result, we selected the cloud-free LANDSAT ETM+ images captured on 25 May 2002, and 30 December 2006.

For the microwave data, we chose the level 1.5 product of ALOS PALSAR L-band data of 15 September 2009 and 14 October 2009, to cover the study area with an ascending Fine Beam Dual (FBD) Polarization, characterized by 34.3 degree of off nadir angle and 12.5 m of ground range pixel spacing, as the best fit for our analysis purposes.

The National Survey on the Natural Environment [21], published by the Ministry of Environment of Japan, provides the status of lands, surface waters and coastal areas all over the nation. The survey results are compiled and published in the form of written reports, downloadable maps, etc. The vegetation map so provided is utilized for identifying specific LU/LC sites for ground truth training data needed for land features classification and for the accuracy assessment of the classified images. Field observations were also performed for collecting additional ground truth data and for the identifying land features for the selected sites all over the prefecture.

For the correction of the remote sensing data, we utilized the digital model of terrain surface published by the Earth Remote Sensing Data Analysis Center (ERSDAC) available from the Advanced Spaceborne Thermal Emission and

Reflection Radiometer (ASTER) Global Digital Elevation Model (GDEM) with a spatial resolution of 30 m and an elevation accuracy of ±7 m. The data were downloaded from the websites and concatenated so as to cover the study area and to make them ready for further use in the analysis.

4. ANALYSIS METHOD

First, gap fill was performed before the correction of the images for the LANDSAT data using the NASA Goddard Space Flight Center frame and fill tool [22]. Then, the Integrated Radiometric Correction (IRC) method developed by Kobayashi and Sanga-Ngoie [23] was applied to the selected LANDSAT ETM+ images in order to correct the radiometric distortions caused by both the topography and the atmosphere, so as to obtain the actual (corrected) irradiance value. The IRC method is a comprehensive radiometric correction method aimed at simultaneous correction of atmospheric, solar, and topographic effects inherent to remote sensing data. The IRC makes it possible to retrieve the at-surface reflectance and radiance values as if the atmosphere were totally transparent and the underlying surface absolutely flat, using

$$L_{\mathrm{g}} = (L_{\mathrm{s}}^{*} - L_{\mathrm{h}}) \frac{A}{0.5(1 + T_{\mathrm{r}})T_{\mathrm{r}}T_{\mathrm{w}}^{2}} \tag{1}$$

where, L_g is the corrected radiance from the surface; L_s^* is the at-satellite observed spectral radiance from a sloped terrain and L_h is the upwelling atmospheric spectral radiance also known as haze; A is the topographic correction factor; T_r and T_w are atmospheric transmittance functions: Rayleigh scattering transmittance function (T_r) and water vapor transmittance function (T_w). The Dark Object Subtraction Method (DOSM) was used for the removal of haze (L_h). Haze values were computed for each band using the reflectance values over deep water areas selected as areas of known zero reflectance in the image [24]. Table 1 shows the computed haze values for each band. The decrease of haze values with increasing wavelengths is remarkable.

A hybrid classification method [25,26] is used here in the classification process for producing the digital LU/LC map of Oita. Sanga-Ngoie [27] stressed on the need to produce a detailed digital LU/LC map or DVM, especially for the vegetation categories, in order to obtain more precise estimations of CO_2 sequestration values. Here, the May data will be the main data set for the analysis while the December data and the PALSAR data will be supplemental data for additional information to the classification/reclassification process so as to produce more detailed land cover maps.

For more precise estimation of CO_2 sequestration, the PALSAR data were used for estimating stem volume, from which the information about tree age could be deduced. The PALSAR Digital Number (DN) values were converted to the backscattering intensity (σ^0) in decibel units using Equation (2),

$$\sigma^0 = 10\log_{10}DN^2 + CF \tag{2}$$

where CF is the calibration factor for the PALSAR products, for the level 1.5 product CF = −83.0 dB [28]. For a correct interpretation of the backscatter signatures, corrections for the effect of local incidence angle and normalization for the true pixel area were necessary. The corrected backscatter in gamma nought (γ^0) format can be obtained from the sigma nought (σ^0) value (Equation (3)) according to Castel et al.[29] and Ulander [30]. This approach has also been applied by other authors [31–34].

$$\gamma^0 = \sigma^0 \frac{A_{flat}}{A_{slope}} \left(\frac{\cos\theta_{ref}}{\cos\theta_{loc}}\right)^n \tag{3}$$

where θ_{loc} and θ_{ref} represent the local incidence angle and a reference angle for the normalization of the backscatter (e.g., off nadir angle), respectively. A_{slope} and A_{flat} represent the true pixel area and the local pixel area for a theoretically flat terrain, respectively. For bare surfaces, the exponent n is equal to 1. For vegetated surfaces, n expresses the variation of the scattering mechanism due to the presence of a volume on the sloped terrain. It is thus being related to the optical depth of the vegetation, and is estimated to be lower than or equal to 0.36 for the Japanese Earth Resources Satellite-1 (JERS-1) L-band SAR backscatter. We opted to use the 0.36 value, because the scattering properties of the forests should be taken into account. For future works, more precise estimates should be reconsidered, especially in the case of different polarizations.

The overall flowchart of the methodology is given in Figure 3, and consists of four main steps described hereafter. The Clark Lab's IDRISI Taiga GIS platform [24] was used for all the analysis.

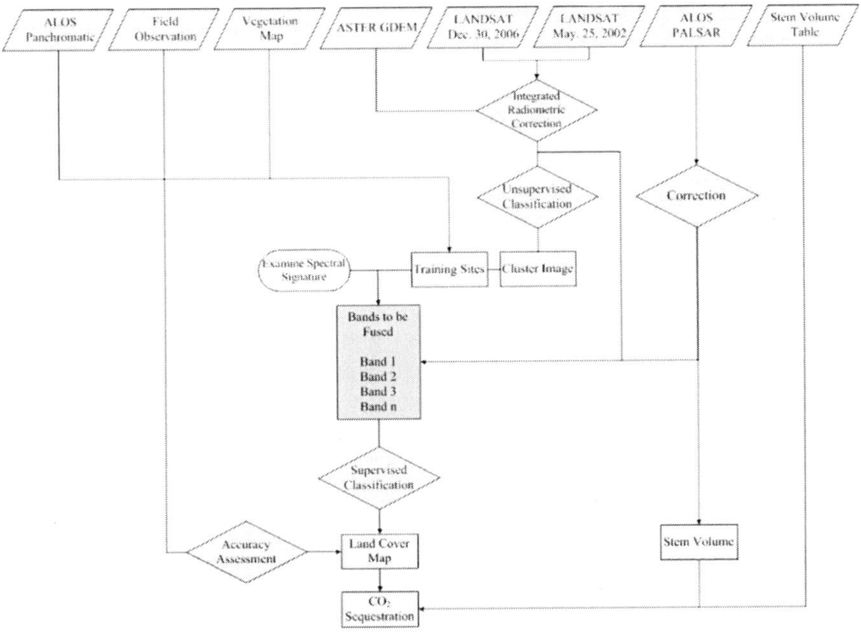

Figure 3. The analysis flowchart.

First, unsupervised classification [24,25,35] was performed to obtain a cluster image to be used for identifying the ground cover types. The IDRISI Taiga ISOCLUST module was used for the clustering process, yielding 15 distinct clusters.

Training sites were then constructed upon the cluster image and used for the supervised classification process based on the Maximum Likelihood algorithm. Data from the National Survey on the Natural Environment [21], together with our own field investigation data, were used as ground truth references for land features identification. The strength of this method is that the need of delineating unimodal spectral classes beforehand can be overcome based upon a representative subset of image data [25]. The IDRISI Taiga MAXLIKE (maximum likelihood) module and the LANDSAT ETM+ 12 bands (bands 1–5 and 7 (May and December)) along with Normalized Difference Vegetation Index (NDVI) data of May and December were used in order to perform the supervised classification. This yielded a classified image with 12 categories in total. These were reclassed later on to give our final product: a classified image with 10 LU/LC categories, designed with a special attention to the forests types found over the analysis area.

In the second step, accuracy assessment was made, based on the final image. Stratified random sampling of the image yielded 29,000 sample points used for the accuracy assessment. Using the ERRMAT module of IDRISI Taiga, the error matrix was built in reference to the land features on the vegetation map from The National Survey on the Natural Environment data [21] as ground truth.

In the third step, the information obtained from the reclassed image was used together with the CO_2 sequestration rates per unit area [9] for the estimation of the sequestration values by each forest type. For the coniferous and deciduous broadleaf forests, we performed the estimations not only by forest type but also by breaking these forest categories down into categories per tree age in the final steps of the work.

The classification of the coniferous forests per tree age was done by calculating the NDVI values using Equation (4), where IR stands for the infrared band (band 4) and R for the red band (band 3). The relationship between NDVI and tree age was deduced by Ishii [36] based on the Forest Tree Age growth curve and multitemporal NDVI data. Referring to this growth curve, we could thus classify the coniferous forests area into separate tree age categories.

$$\mathrm{NDVI} = (IR - R)/(IR + R) \tag{4}$$

Furthermore, considering the fact that NDVI saturates with age, especially for trees older than 15 years, we attempted to improve the classification of forests area (especially coniferous and deciduous broadleaf forests) into separate tree age categories, using the backscattering intensity information derived from the PALSAR data. This was implemented for comparing the results with the tree age estimation based on the NDVI method. The relationship between backscattering intensity and stem volume, or biomass, has been analyzed by various researchers [31,34,37–42]. These various studies state that the backscattering signature correlates with the forest parameters, and that SAR images can thus be used to extract information related to the forest biomass. Because we did not collect ground data of the stem volume in our study area, we opted to use the polynomial model developed by Wijaya [40] (Equation (5)) to estimate the stem volume (m^3/ha).

where, V is the stem volume (m^3/ha) and σ is the backscattering intensity. Notwithstanding the fact that this expression was developed for forests in Indonesia, we decided to apply this model because Wijaya [40] used it for very similar data to ours (PALSAR L-band FBD 34.3°HH/HV specifications). The stem volume map produced using this model was then classified into ages among the forests, referring to the relationship of the tree age and stem volume information provided by the Oita Prefectural Government. The classification was done for the coniferous and deciduous broadleaf forests, but not for the evergreen broadleaf forests for which no stem volume data were available. Sugi (cedar) tree data and Kunugi (saw tooth oak) tree data were used as representative for the coniferous forests and deciduous broadleaf forests, respectively.

Table 1. Computed haze (L_h) values for each band using the Dark Object Subtraction Method (measured in $Wm^{-2}\cdot\mu m^{-1}\cdot sr^{-1}$).

	Band 1	Band 2	Band 3	Band 4	Band 5	Band 7
May	40	11.8	5	1.3	0.13	0.05
December	26	11.8	5	1.3	0.13	0.04

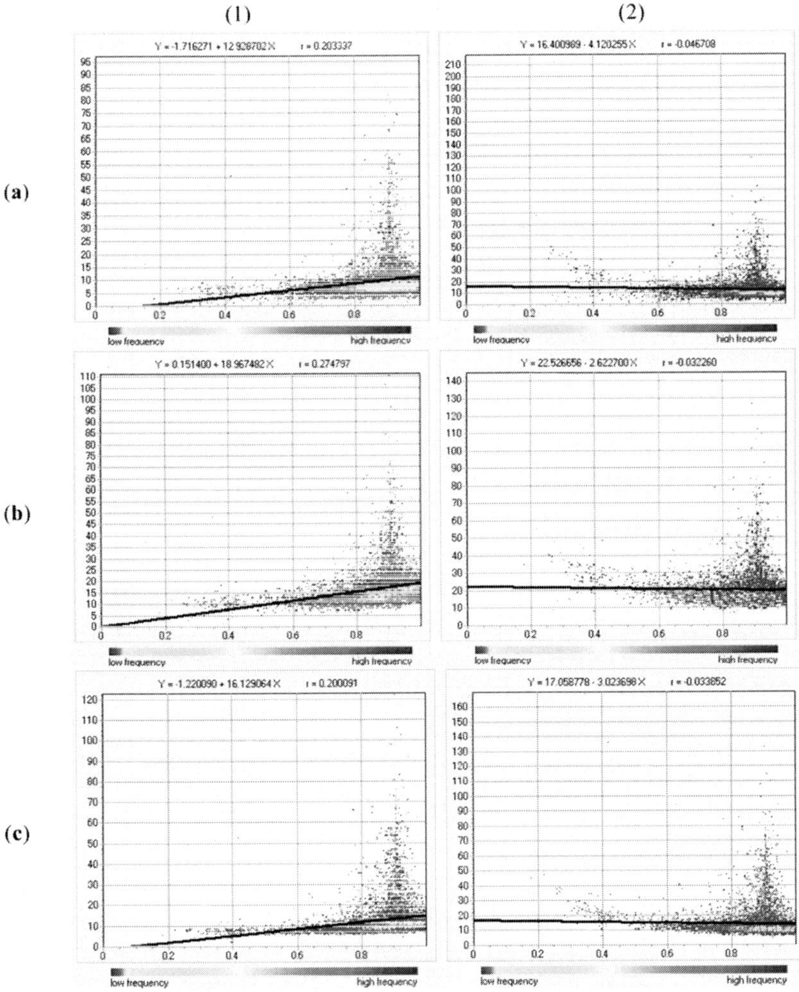

Figure 4. Regression analysis between cos i on x-axis, and the radiance values on y-axis, for (**a**) band 1, (**b**) band 2 and (**c**) band 3, for the noncorrected (1) and for the corrected radiance data (2), respectively.

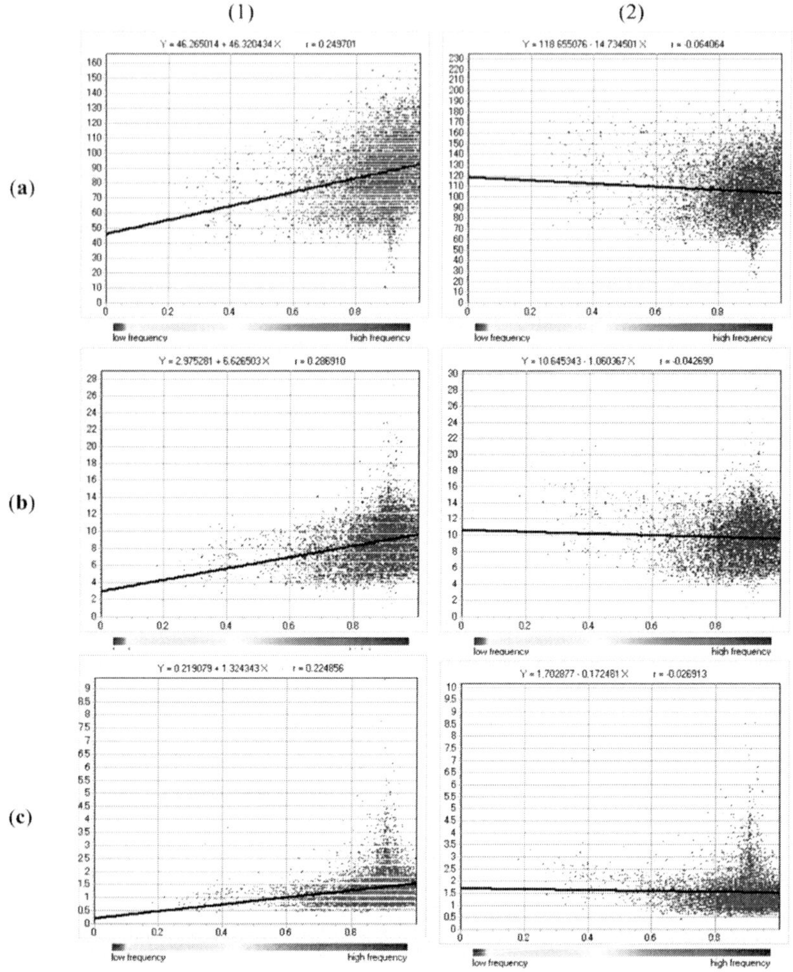

Figure 5. Same as in Figure 4, but for (**a**) band 4, (**b**) band 5 and (**c**) band 7, for the noncorrected (1) and for the corrected radiance data (2), respectively.

5. RESULTS

5.1. Radiometric Correction

Quantitative evaluation of the performance of the correction is done by assessing the absence of correlation between cos i (i: solar incidence angle), which is a measure of the solar illumination on the sloped area, and the corrected radiance data. The regression analysis between cos i and noncorrected data (Figures 4 and 5(1)) shows that these parameters are highly correlated: higher radiances for higher cos i values, because of the various inherent errors, including the atmospheric and topographic effects. However, lower correlation, or mostly no correlation, between those two parameters is obtained for radiometrically corrected radiances (Figures 4 and 5 (2)). Figure 6 shows an example of the difference of images before and after correction, respectively, for the band 4 image of May data. The zoomed-in area is a mountainous location, showing that the radiance values are corrected in these mountainous areas, while almost no changes are detected over the flat residential areas on the top left side of the inset.

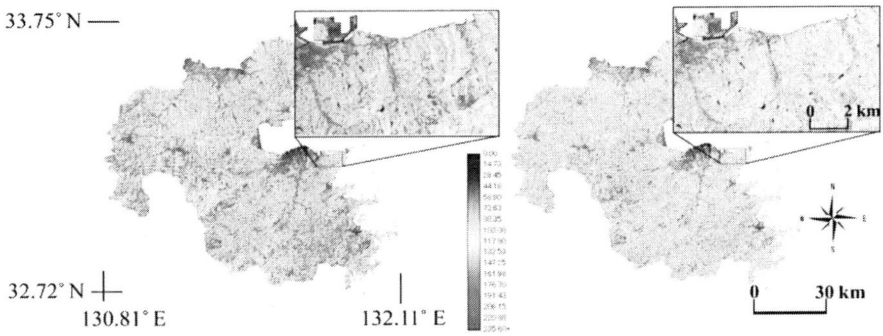

Figure 6. Original near-infrared band (band 4) image (**left**) and IRC corrected image (**right**) over the study area.

The use of radiometrically corrected data ensures that more precise land use/land cover classification of optical satellites data will be performed, with minimum perturbations related to atmospheric effects (haze, transmittance, etc.) or topographic effects (hillshading).

5.2. Oita LU/LC Digital Map

The LANDSAT ETM+ image captured on the date of 25 May 2002 was classified referring to the Natural Survey on the Natural Environment vegetation map [21] and our own field investigation records as ground truth, into the following 12 land cover types (Table 2(a)): cedar–cypress (Cryptomeria japonica–Chamaecyparis obtusa), saw tooth oak (Quercus acutissima), live oak (Castanopsis Cyclobalanopsis), silver grass, grasslands, bare lands, water bodies, crop lands, paddy fields, residential areas, deciduous broadleaf forests and evergreen broadleaf forests. These detailed classes were then reclassed into the following final 10 LU/LC categories (Table 2(b)): coniferous forests, silver grass, grass lands, bare lands, water bodies, crop lands, paddy fields, residential areas, deciduous broadleaf forests and evergreen broadleaf forests (Figure 7).

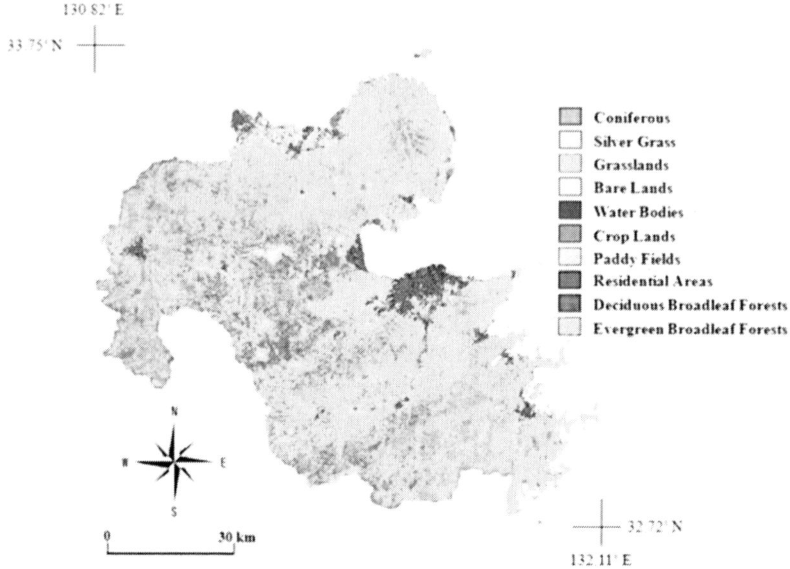

Figure 7. Classified LU/LC types in Oita Prefecture obtained by using the maximum likelihood algorithm.

Table 2. Land cover classes: (**a**) Original outputs; (**b**) The final product.

(a)	
Class Num.	**Class Type**
Class 1	Cedar-Cypress (*Sugi-Hinoki*)
Class 2	Saw Tooth Oak (*Kunugi*)
Class 3	Live Oak (*Shii-Kashi*)
Class 4	Silver Grass (*Susuki*)
Class 5	Grass Lands
Class 6	Bare Lands
Class 7	Water Bodies
Class 8	Crop Lands
Class 9	Paddy Fields
Class 10	Residential Areas
Class 11	Deciduous BL Forests
Class 12	Evergreen BL Forests
Class Num.	Class Type

(b)	
Class Num.	**Class Type**
Class 1	Coniferous Forests
Class 2	Silver Grass (*Susuki*)
Class 3	Grasslands
Class 4	Bare Lands
Class 5	Water Bodies
Class 6	Crop Lands
Class 7	Paddy Fields
Class 8	Residential Areas
Class 9	Deciduous BL Forests
Class 10	Evergreen BL Forests
Class Num.	Class Type

Table 3. The error matrix for the classified LU/LC map. Area for ground truth categories, together with the Commission Errors (ErrCom), are given in the columns, while the areas for the classified categories and the Omission Errors (ErrOm) are given in the rows.

Classified \ Reference	Coniferous	Silver Grass	Grasslands	Bare lands	Water bodies	Crop lands	Paddy fields	Residential	Deciduous-Bl	Evergreen-Bl	Total	ErrCom	Users A
Coniferous	**9,305**	100	52	16	61	125	158	39	1,086	1,543	12,485	0.2547	74.53%
Silver Grass	6	**365**	48	7	2	7	3	1	17	0	456	0.1996	80.04%
Grasslands	69	192	**211**	17	8	86	71	26	65	18	763	0.7235	27.65%
Bare lands	5	10	14	**41**	12	5	7	59	3	2	158	0.7405	25.95%
Water bodies	5	9	2	12	**279**	0	8	6	1	14	336	0.1696	83.04%
Crop lands	179	29	67	52	28	**235**	237	197	42	71	1,137	0.7933	20.67%
Paddy fields	309	128	129	109	223	216	**1,339**	349	120	105	3,027	0.5576	44.24%
Residential	5	15	40	55	70	34	104	**651**	7	9	990	0.3424	65.76%
Deciduous-Bl	1,190	363	129	18	4	85	73	12	**1,788**	500	4,162	0.5704	42.96%
Evergreen-Bl	1,734	77	89	25	16	257	108	44	1,123	**1,729**	5,202	0.6676	33.24%
Total	12,807	1,288	781	352	703	1,050	2,108	1,384	4,252	3,991	**28,716**		
ErrOm	0.2734	0.7166	0.7298	0.8835	0.6031	0.7762	0.3648	0.5296	0.5795	0.5668		0.4448	
Producers A	72.66%	28.34%	27.02%	11.65%	39.69%	22.38%	63.52%	47.04%	42.05%	43.32%			

Overall Accuracy
15,943/28,716 = 55.52%

Table 4. Fuzzy Error Matrix [43,44] following the same methodology as the traditional error Non-diagonal cells in the matrix contains accepted and poor columns. The numbers in the accepted estimating fuzzy accuracy, the numbers in the poor columns are considered as an error. The fuzzy percentage of sites where the acceptable reference labels matched the classified label.

	Reference																				Users Accuracy	
	Coniferous		Silver Grass		Grasslands		Bare lands		Water bodies		Croplands		Paddy fields		Residential		DBL		EBL		Total Fuzzy	Percent Fuzzy
	Accept	Poor	Accept	Poor	Accept	Poor	Accept	Poor	Accept	Poor	Accept	Poor	Accept	Poor	Accept	Poor	Accept	Poor	Accept	Poor		
Coniferous	**9,305**	0	0	100	0	52	0	16	0	61	0	125	0	158	0	39	**642**	444	**911**	632	10,858/12,485	86.97%
Silver Grass	0	6	**365**	0	**27**	21	0	7	0	2	**4**	3	0	3	0	1	0	17	0	0	396/456	86.78%
Grasslands	0	69	**108**	84	**211**	0	0	17	0	8	**47**	39	**39**	32	0	26	0	65	0	18	405/763	53.1%
Bare lands	0	5	0	10	0	14	**41**	0	0	12	0	5	**4**	3	**25**	34	0	3	0	2	70/158	44.07%
Water Bodies	0	5	0	9	0	2	0	12	**279**	0	0	0	**3**	5	0	6	0	1	0	14	282/336	84.07%
Croplands	0	179	**22**	7	**43**	24	0	52	0	28	**235**	0	**199**	38	**182**	15	0	42	0	71	681/1,137	59.92%
Paddy fileds	0	309	**99**	29	**98**	31	**79**	30	**140**	83	**185**	31	**1,339**	0	**290**	59	0	120	0	105	2,229/3,027	73.62%
Residential	0	5	0	15	0	40	**23**	32	0	70	**22**	12	**66**	38	**651**	0	0	7	0	9	762/990	76.97%
DBL	**934**	256	0	363	0	129	0	18	0	4	0	85	0	73	0	12	**1,788**	0	**443**	57	3,165/4,162	76.04%
EBL	**1,194**	540	0	77	0	89	0	25	0	16	0	257	0	108	0	44	**999**	124	**1,729**	0	3,922/5,202	75.39%
Producers	11,433/12,807		594/1,288		379/781		143/352		419/703		493/1,050		1,650/2,108		1,148/1,384		3,429/4,252		3,083/3,991		Fuzzy Overall Accuracy	
Accuracy	89.27%		46.12%		48.53%		40.63%		59.60%		46.95%		78.27%		82.95%		80.64%		77.25%		22,771/28,716 = 79.30%	

Table 3 shows the results of the error matrix of the accuracy assessment obtained by using the ERRMAT module of the IDRISI Taiga GIS software, while Table 4 shows the fuzzy error matrix modified from the conventional error matrix based on the concept of fuzzy accuracy assessment developed by Green and Congalton [43,44]. The error matrix so obtained tabulates the errors of omission and commission, as well as the marginal and total errors. The overall accuracy of the produced LU/LC map is 55.5% and the fuzzy overall accuracy is 79.3%.

5.3. CO_2 Sequestration by Forest Types

The three dominant forest types found in Oita Prefecture are: coniferous, deciduous broadleafs and evergreen broadleaf forests. Estimations of CO_2 sequestrations per forest type are obtained by multiplying the observational CO_2 sequestration rates [9,27,45] by the total area of each forest type (Table 5). The following results are obtained for the total CO_2 sequestrations per forest type: 3.57 $MtCO_2$/year for the coniferous forests, 0.78 $MtCO_2$/year for the deciduous broadleaf forests, and 2.25 $MtCO_2$/year for the evergreen broadleaf forests. This sums up into a total CO_2 sequestration of 6.60 $MtCO_2$/year for all the forested areas in Oita Prefecture (Table 5). These findings emphasize the fact that more precise sequestration estimates can be obtained by using detailed LU/LC maps (with detailed forests categories and precise areas) than by just dividing the forest types into natural and planted forests only as recommended by the IPCC [46].

Table 5. CO_2 sequestration by forest types in Oita Prefecture using the IPCC Method.

Forest Type	CO_2 Sequestration (tCO_2/ha/yr) [9,47]	Area (ha)	Total CO_2 Sequestration (tCO_2/yr)	Total C Sequestration (tC/yr)
Coniferous	13.5 ± 4.2	264,152	3,566,052	972,560
Deciduous-BL	8.7 ± 3.0	89,208	776,109	211,666
Evergreen-BL	18.1 ± 3.0	124,494	2,253,341	614,548
Total		477,854	6,595,502	1,798,774

Carbon sequestration equivalents are simply obtained by multiplying the CO_2 sequestrations by 12/44, where 12 and 44 are the molecular masses of the carbon and the CO_2, respectively. Our aim is to evaluate atmospheric CO_2 sequestration by forests. We also present in the tables the carbon sequestration equivalents, for informational purposes.

5.4. CO_2 Sequestration by Tree Ages (NDVI Method)

Further estimation of the CO_2 sequestration was considered for the coniferous forests. These forests were classified into four different categories (Figure 8), based on the NDVI values [36], each category indicating the range of the tree age for the coniferous forests: 0–5, 5–10, 10–15 and over 15 years old categories. Based on this detailed classification, a total CO_2 sequestration of 2.35 $MtCO_2$/yr was obtained for coniferous forests in Oita (Table 6).

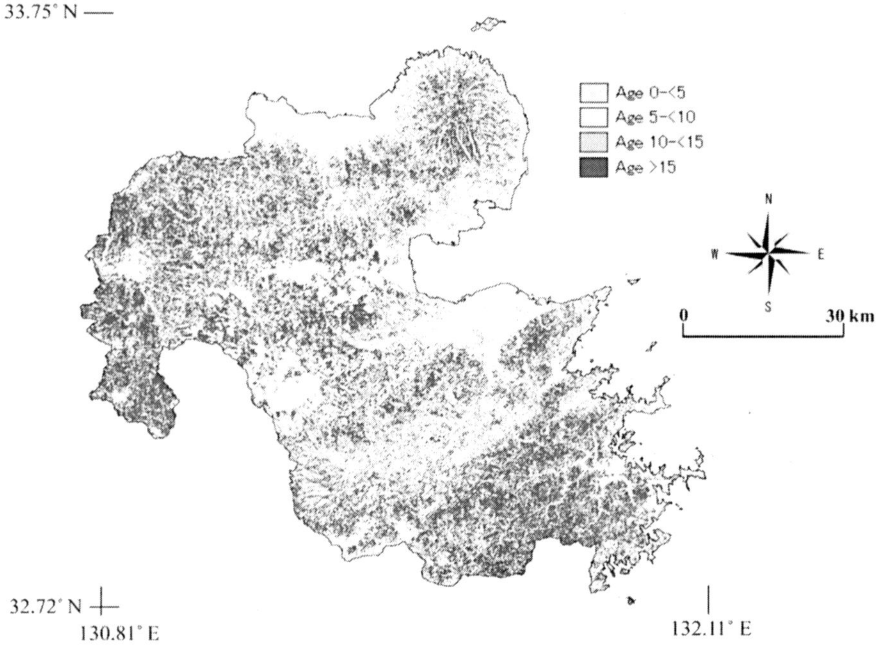

Figure 8. Area of coniferous forests (ceda–cypress) divided per tree age using the NDVI values.

Table 6. CO_2 sequestration by coniferous forests (cedar–cypress) per tree age using the NDVI Method.

Coniferous Age	CO_2 Sequestration (tCO_2/ha/yr) [47]	Area (ha)	Total CO_2 Sequestration (tCO_2/yr)	Total C Sequestration (tC/yr)
0–5	0.6	15	9	2
5–10	4.8	1,316	6,317	1,723
10–15	13.5	11,772	158,922	43,342
15–50	8.7	251,048	2,184,118	595,669
Total		264,151	2,349,366	640,736

The averaged sequestration value used for planted forests following the traditional (IPCC) methodology [46] is 13.5 tCO_2/ha/yr. We have to note that in Table 6 (sequestration value per unit area per tree age), a similar value of 13.5 tCO_2/ha/yr is shown to be valid only for trees 10–15 years old. This age range is also the period of time when the coniferous trees seem to have the highest growth rate. There is therefore a strong possibility of overestimating the sequestration results if the estimates are made using these averaged values (which turn out to be the values for the tree age category with the highest sequestration rate) for all the tree age categories. In fact, the comparison of the total sequestration values obtained for coniferous forests by using the averaged value (3.57 $MtCO_2$/yr) with our results in Table 6 (2.35 $MtCO_2$/yr) shows a 34.2% overestimation in CO_2 sequestration for the former methodology.

5.5. CO_2 Sequestration by Tree Age using PALSAR Data (Stem Volume Method)

Using Equation (5) and the backscattering intensity derived from the PALSAR data, we calculated stem volumes for forests in Oita, as shown in the stem volume map (Figure 9). The stem volume values for various tree ages as provided by the Oita Prefectural Government are shown by the graphs in Figure 10. Then, the so-obtained stem volume map (Figure 9) is reclassified into range of ages, by considering the different growth rates among the various regions of the study area, based on Figure 10.

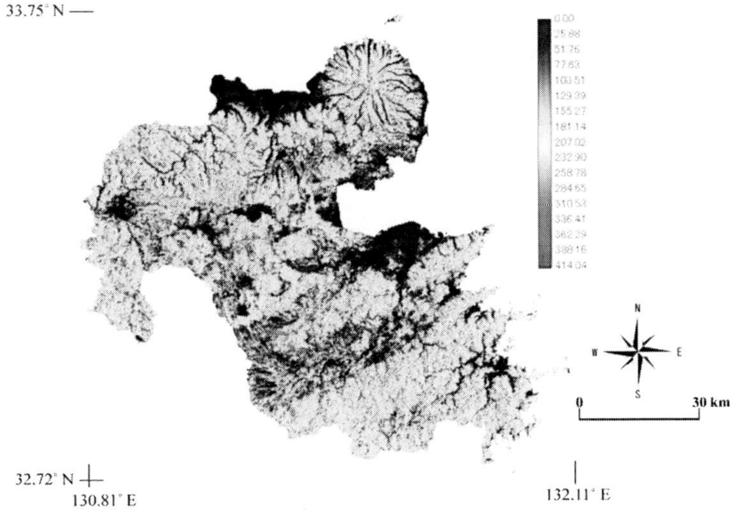

Figure 9. Stem volume map of the study area (m^3/ha).

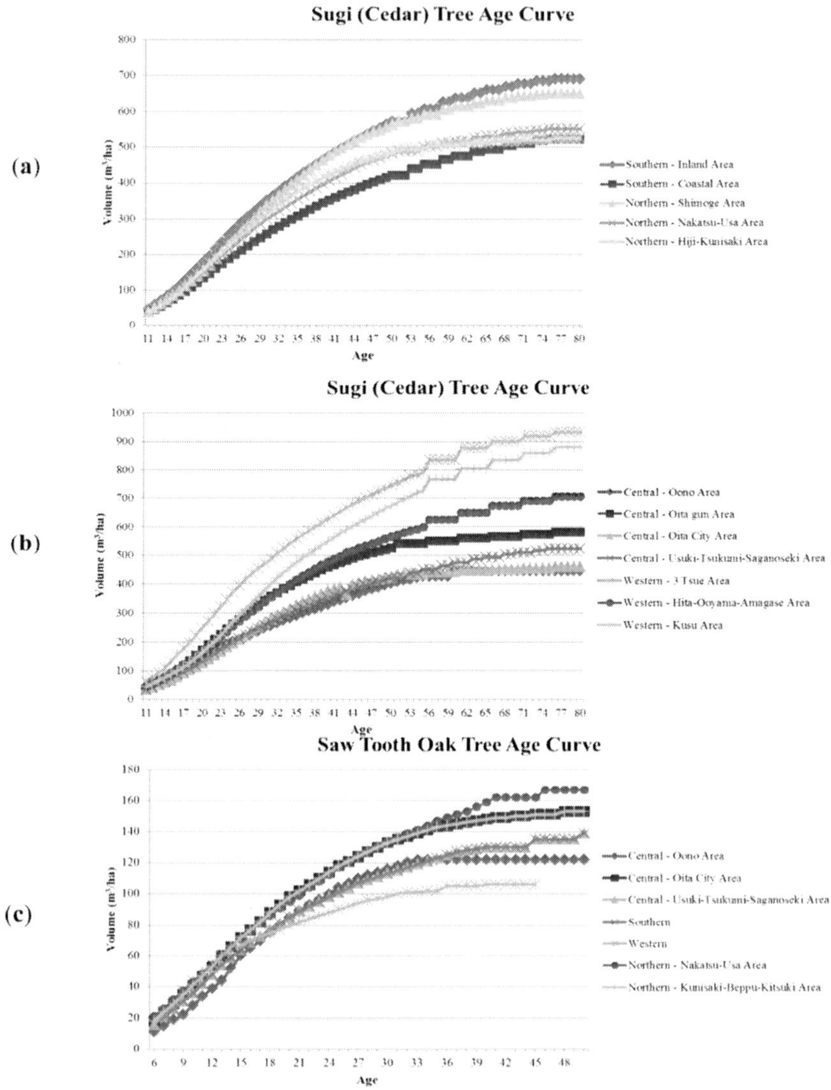

Figure 10. Stem volume as a function of tree ages for (**a**,**b**) the Sugi (cedar) and (**c**) the saw tooth oak, in Oita Prefecture. The difference of stem volume growth among the different regions in Oita is remarkable.

Coniferous forests data were analyzed and the results compared with those obtained by using the NDVI Method. The same was done for the deciduous broadleaf forests. Here, obtained forest ages are more detailed or more specific than those used for the NDVI method, because of the difference between the saturation points for the NDVI and the stem volume methods. In fact, the NDVI values saturate after 0.7, showing similar values for all the ages above, therefore the difficulty in classifying trees of more than 15 years of age. For the stem volume method, the same can be said except that here, the saturation point is higher than for the NDVI, as is widely discussed in the literature [40,47].

Table 7 shows the new total CO_2 sequestration estimates for coniferous and broadleaf forests per age category, respectively. In general, the new total CO_2 sequestration for coniferous (2.96 $MtCO_2$/yr) is found to be larger than that obtained by using the NDVI method (2.35 $MtCO_2$/yr) (Table 6). As for the deciduous broadleaf forests, a total sequestration of 0.38 $MtCO_2$/ha/year is obtained (Table 7) in comparison with the 0.78 $MtCO_2$/ha/yr (Table 5) by using the traditional averaged method. These differences come from the fact that the sequestration values, which were averaged for all the trees of the same category (traditional averaged method), or for all the trees of more than 15 years of age due to NDVI saturation (NDVI method), are now broken down into detailed age categories up to the year 50 (stem volume method).

Table 7. CO_2 sequestration of coniferous and deciduous broadleaf forests per tree age using the PALSAR data and the stem volume method.

Coniferous Age	CO_2 Sequestration (tCO_2/ha/yr) [47]	Area (ha)	Total CO_2 Sequestration (tCO_2/yr)	Total C Sequestration (tC/yr)
0–5 5–10	2.9	3,184	7,323	1,997
10–15	13.5	8,941	112,657	30,725
15–20	14.8	90,441	1,293,306	352,720
20–25	11.3	147,084	1,500,257	409,161
25–30	10.2	5,358	46,615	12,713
30–50	6.4	662	3,972	1,083
Total	11.59	255,640	2,964,130	808,399
Deciduous-Bl Age				
0–5	5.5	310	1,705	465
5–10	9.4	614	5,772	1,574
10–15	8.2	1,205	9,881	2,695
15–20	8.2	1,557	12,767	3,482
20–25	5.9	2,168	12,791	3,488
25–30	5.8	2,709	15,712	4,285
30–50	3.2	78,044	249,741	68,111
Total	3.56	86,607	308,369	84,100

In fact, by using the detailed sequestration rates per tree ages (stem volume method), sequestration overestimates that implicitly occur with the averaged method [46] for coniferous trees of less than 10 years of age and for trees of more than 20 years of age, as well as the underestimation for 15–20 years old trees, can altogether be avoided. The same can also be said, mutatis mutandis, for deciduous broadleaf forests for which the 5–10 years category is shown to have the highest sequestration rate (9.4 tCO_2/ha/yr) according to the stem volume method (Table 7), as compared to the averaged values for all ages (8.7 tCO_2/ha/yr) in Table 5. In general, the highest sequestration rates are shown for 10–25-year-old coniferous forests, and for 5–20-year-old deciduous broadleaf forests.

6. DISCUSSION

6.1. Land Cover Map (DVM)

Focus has been put on producing a land cover map in which land cover types are defined not just simply as forested and nonforested areas [46], but more in detailed forest subclasses and other nonforested areas. Sanga-Ngoie [27] had stressed the need to produce a detailed digital land cover map, especially for the vegetation categories, in order to allow for more precise estimations of CO_2 sequestration values.

Various combinations of approaches and data sets (noncorrected vs. corrected, single vs. multitemporal) were examined in order to produce the best results for a highly accurate land cover map (Figure 7). Difficulty did arise when using single temporal datasets for producing land cover maps, especially when one has to classify the vegetation types in detail. For example, the spectral signatures (mean radiance values per training site for each band) for forested land cover types given in Figure 11 show that it is very hard to differentiate the evergreen broadleaf forests and deciduous broadleaf forests categories in the May data, because of similar signatures for most of the bands in this period of the year. However, it is shown that clear discrimination among all the different forest types can be obtained from the December data (bands W1–W5 and W7). In general, although the radiance values in December (winter season) are lower than those in May (spring season) for all the forests types, and coniferous clearly differentiated from other broadleaf forests in both seasons, it can be said that, when seasonal changes in the forests canopy are considered: (1) clear differentiation among the deciduous and evergreen subgroups of broadleaf forests can be obtained with larger radiance values for deciduous broadleaf forests than for evergreen broadleaf forests, (2) discrimination within each broadleaf forest subgroup can be obtained, with radiance values (in $Wm^{-2}\cdot\mu m^{-1}\cdot sr^{-1}$) for saw tooth oak and live oak forests larger than those for other deciduous and evergreen broadleaf forests, respectively. Similar improvements are also established in terms of accuracy assessment, showing by far larger accuracy values (72.3%, 42.1% and 43.3%) in Table 3 for multitemporal corrected images than the 40%, 14% and 22% obtained for coniferous, deciduous and evergreen broadleaf forests, respectively, based on the analysis of single and noncorrected bands. These findings firmly establish the importance of considering seasonal differences (multitemporal data sets) in order to improve the land cover map results, because they make it possible to distinguish various land cover categories normally difficult to classify from one single temporal data sets.

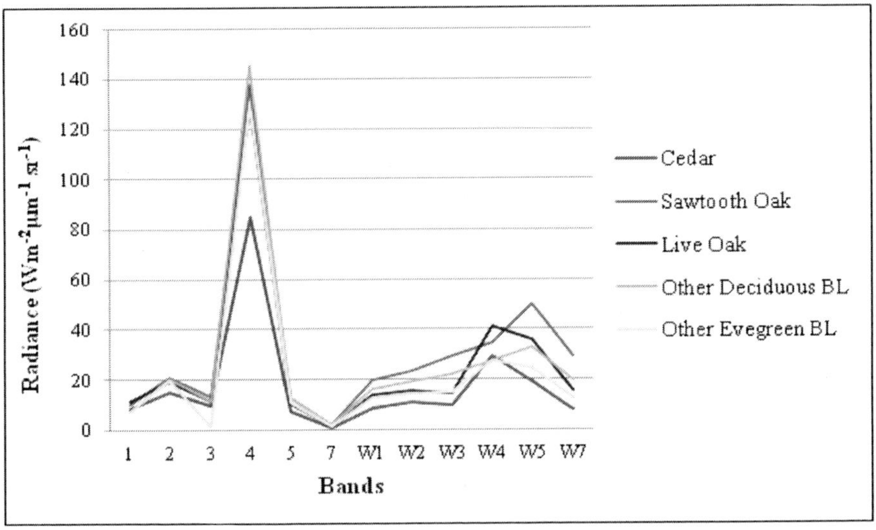

Figure 11. Signatures comparison between the mean radiance values (in $Wm^{-2} \cdot \mu m^{-1} \cdot sr^{-1}$) per band for the pixels within the training sites for each forested land cover type. May data mean values for bands 1–5 and 7 are marked by 1–5 and 7, while December data mean values for bands 1–5 and 7 are marked by W1–W5 and W7.

6.2. Accuracy Assessment

It is a common practice in the remote sensing society to use a conventional error matrix for checking the reliability of the classified land cover map. Detailed information from within the error matrix is often overlooked because people mostly focus only on the overall accuracy. Congalton and Green [48] recommended to consider the error matrix as a jumping-off point for identifying not simply the error itself, but mostly the sources of confusion that might come from the differences between the map created from remotely sensed data and the reference data, or from erroneous human interpretation, for example when defining these features from aerial photos. The confusion might thus lead to errors in the process of producing reference maps, which would in their turn produce errors when used later on as a ground truth references. Accuracy assessment should thus be a process for identifying errors and understanding the problem of the misclassification so as to prevent them. We stress that more attention has to be paid to this issue instead of concentrating on the overall accuracy alone.

In this study, fuzzy accuracy assessment [43,44,48] was applied along with the conventional accuracy assessment process in order to compare how the accuracy varies when we fuzz the boundaries of the classes. Vegetation Condition Index (VCI) [49] was used to compute the degree of acceptance for each class. Considering the complexity of the real environment, we suggest that fuzzy accuracy assessment should also be implemented in the same way as the conventional accuracy assessment for any sound classification scheme.

6.3. NDVI, PALSAR data and CO_2 Sequestration Estimates

The overall sequestration results have been summarized in Table 8 so as to enhance our findings in this paper. As can be seen in this table, if the conventional IPCC method is used alone, coniferous and deciduous broadleaf forest sequestrations are set to be 13.5 tCO_2/ha/yr and 8.7 tCO_2/ha/yr, respectively. The NDVI methodology made it possible to break the Coniferous sequestration down to each age category, showing very clearly the range of overestimation implied when the IPCC method (with no special attention to tree age) is used, especially for trees younger than, or older than, 10–15 years of age. However, because of the saturation effect for the NDVI values, sequestration rates could not be discriminated for trees older than 15 years with the NDVI method.

Table 8. Summary of CO_2 sequestration values and total sequestrations for the different methods of estimation.

Estimation Method	Coniferous (tCO_2/ha/year)		Deciduous-BL (tCO_2/ha/year)		Evergreen-BL (tCO_2/ha/year)	Total CO_2 Sequestration (tCO_2/year)	Total C Sequestration (tC/year)
Conventional	13.5		8.7		18.1	6,595,502	1,798,774
NDVI	Age 0–5	0.6	8.7		18.1	5,378,816	1,466,950
	5–10	4.8					
	10–15	13.5					
	15–50	8.7					
Stem Volume	Age 0–5	2.9	Age 0–5	5.5	18.1	5,525,840	1,507,047
	5–10		5–10	9.4			
	10–15	13.5	10–15	8.2			
	15–20	14.8	15–20	8.2			
	20–25	11.3	20–25	5.9			
	25–30	10.2	25–30	5.8			
	30–50	6.4	30–50	3.2			

The use of PALSAR data made this discrimination possible, yielding sequestration rates per tree age, of up to trees of 50 years of age, not only for the coniferous, but also for the Deciduous broadleaf forests. It is thus clearly established that sequestration rates of 13.5 tCO_2/ha/yr (cfr IPCC Method) are valid only for 10–15-year-old coniferous forests, while deciduous broadleaf forests reach, or go beyond, the 8.7 tCO_2/ha/yr threshold only during the 5–10-year-old range. Any use of these IPCC rates for trees of different ages from the classes mentioned here above will lead to sequestration overestimation. The same is also true for the results obtained using the NDVI method for the coniferous, except for the underestimation that occurs for the trees between the 15 and 20-year range.

It is very important to note that, the averaged sequestration value for the coniferous forest type matches the sequestration value per tree ages obtained here for the 10–15-year-old coniferous trees. This is also the category of the most fast-growing period for coniferous trees. The same is also true for the deciduous broadleaf forests for the 5–10-year-old trees. This strongly implies that, if estimation is made based on the averaged values alone, it will round up the values for other trees in different growth stages, giving way to an overall overestimation of the carbon sink capacity, both for natural or planted forests [46]. Our results emphasize the fact that sequestration differs among tree ages: this has to be considered, otherwise erroneous values of forests' sequestration capacity will be obtained, and a number of negative consequences in the further use of these values can be foreseen.

This demonstrates and emphasizes the importance of using PALSAR data (stem volume method), based on the estimation of the stem volume computed from the PALSAR backscattering intensity data by applying the model developed by Wijaya [40] for Indonesian forests. Nowadays, many works attempt to analyze the correlation between backscatter with different polarizations and the stem volume or biomass for forests with the same tree types [41,42]. They do not look through the backscattering signatures among different vegetation types for different polarizations. It is known that the backscattering signature gets affected by the soil moisture or the topography and it can also be argued on how well Wijaya's model applies to the study area in Oita because of different climate conditions and vegetation. The impacts, or the eventual limitations, on the applicability of Wijaya's expression to forests in Oita will be one of our most urgent tasks to deal with in the future.

6.4. CO_2 Sequestration Estimates by Oita Government

Recently, the Oita Prefectural Government published its sequestration estimates by forests of Oita [8]. They divided Oita forests into natural and planted forests, and calculated the sequestration values as prescribed in the IPCC Guideline [46], using the following equation:

$$\text{Annual Carbon Sequestration (per ha)} = \text{Annual Growth (m}^3\text{/ha)} \times \text{Magnification Factor} \times \text{Bulk Density} \times \text{Carbon Content} \quad (6)$$

where the magnification factor is a factor for expanding the stem volume into an estimate of the whole volume of the tree, including the volume of all leaves, branches and roots together. A magnification factor of 1.7 and 1.8 is proposed for coniferous trees and broadleaf trees, respectively. Bulk Density represents the weight per cubic meter of volume. We have to note that, although a common value is used everywhere in Japan, we presume that the values of these two factors should vary among trees according to, not only their types, but also their ages and the regions of growth. The carbon content is usually set to be 0.5 [46], meaning that the composition of wood contains roughly 50% of carbon. CO_2 sequestration is simply obtained by multiplying the C sequestrations by 44/12, where 12 and 44 are the molecular masses of the carbon and the CO_2, respectively. Finally, in order to obtain the total carbon sequestration by the forests, Equation (6) will have to be multiplied by the forested total area (in ha).

Published Oita Prefecture's CO_2 sequestration results [8] using the IPCC Method [46] are still not very conclusive and need some more work and systematic streamlining. Strong discrepancies are found among the few results published so far, more likely due to both the difference in forest classification and in the sequestration assumptions used for the calculations. We have to stress here that these prefectural values are obtained without consideration to the differences in sequestration rates by tree type or tree age, they are also relatively low compared to the values in the literature [9,27,45], and to our present results (Tables 6–8) in which we emphasized the diversity in sequestrations among tree type and age.

In summary, it is shown that Oita Prefectural Government is underestimating its forests sequestration capacity. We would like to recommend the prefecture to consider the methodology developed here in the case that the carbon sink concept is to be considered as part of the prefectural CO_2 countermeasure plan; because the government's present estimates could lead to incorrect decision making which ought to be avoided.

6.5. Decision Making in Forests Management

Our results have shed new light on the role and the potentials of forests in Oita Prefecture concerning CO_2 sequestration. If well implemented, the present methodology is more likely to help the prefecture in devising better scenarios for forest resource management, and for global warming mitigation, as well. Through this methodology, resources could be localized and their areal covers quantitatively estimated using the satellite optical remote sensing data, while the status of the resources could be analyzed using the SAR data.

By comparing with the observed sequestration rates per tree age, it was shown that the highest sequestration rates are obtained for 10–25-year-old trees for coniferous, and for 5–20-year-old for deciduous broadleaf forests. Tree harvesting time and afforestation plans can thus be pinpointed for specific forested areas, taking into consideration the important information about sequestration rates per tree type and per tree age. Planted trees will thus need to be checked if they can yield good timber products at around 20–25 years or 15–20 years of age for coniferous or for deciduous broadleaf forests, respectively. Bibilis et al.[50] emphasize the importance of having more precise information about timber products, because the percentage of acceptance to grade requirements is known to increase with tree age. Altogether, these findings can be a strong information source in support of sound decision making and actions not only for the Oita Prefectural Government, but also for other prefectures in Japan, or all over the world.

7. CONCLUDING REMARKS

This research has employed multitemporal LANDSAT ETM+ and ALOS PALSAR data together with field data collected in Oita Prefecture, Japan, to estimate the CO_2 sequestration potential of the forests in Oita. We have shown that sequestrations obtained by using averaged values for natural or planted forests without forest-type and tree-age discrimination can lead to large overestimations of the sink capacity of forests. It has been made clear here that, by combining the optical and SAR remote sensing data, it became possible to obtain more realistic CO_2 sequestration amounts by the forests, based on detailed discrimination of forest type and tree age. Moreover, the detailed information about the spatial location and the status of the forest resources is more likely to provide the decision makers with a sound basis for more effective policies and actions aiming at the sustainable use and conservation of the environmental resources.

In this study, the retrieval of tree ages using the stem volume information extracted from SAR data was implemented only for coniferous and deciduous broadleaf forests, taking cedar (Sugi) and saw tooth oak (Kunugi) trees as representatives of those forest types, respectively. With this method, we have obtained more realistic sequestration amounts than with the other methods. This methodology can thus help to investigate more in-depth the CO_2 sequestration values for trees of different ages and types, especially for the coniferous or other evergreen or deciduous broadleaf forests, which are found over the study area. We foresee that this would lead to even more realistic estimation of the CO_2 sequestration of the real environment.

The correct positioning of PALSAR images is another general and important issue, especially when these images are to be fused with optical data. This problem can lead to errors, especially when constructing the training sites. Because of the positioning problem, the training sites coordinates in the optical and the SAR data will not match, corrupting thus the spectral signatures information for each land class. Orthorectification is usually used to deal with this issue, but it was not implemented in this research. The correction of the PALSAR image using the methods by Castel et al.[29] does not account for layover and shadowing effects. This leads to some basic problems with the SAR data, especially in Japan, a country with many mountains and slopes. Small [51] has developed a new method for improving the normalization of the SAR image, in which the layover and shadowing effects are considered. We intend to include this method in our future work in order to overcome these issues and improve the results obtained in the present work. In so doing, we will have room for more options for the satellite data to be acquired, such as using images for ascending or descending modes, or for any other modes available.

ACKNOWLEDGMENTS

We are very grateful to the students and the faculty members of the Laboratory of Environmental Geoscience, Graduate School of Asia Pacific Studies, Ritsumeikan Asia Pacific University, and the members of the ENVOL Program, for the informative remarks and suggestions during the elaboration of this work. We also thank the Department of Agriculture Affairs, Forest Management Division in the Oita Prefectural Government, for providing the stem volume data.

REFERENCES

1. Jain, A.K.; Hayhoe, K.A.A. Global Air Pollution Problems. In Handbook of Atmospheric Science: Principles and Applications; Hewitt, C.N., Jackson, A.V., Eds.; Blackwell: Padstow, UK, 2003; pp. 368–370.
2. Chiras, D.D. Environmental Science, 7th ed.; Jones and Bartlett: Massachusetts, UK, 2006; p. 458.
3. IPCC. Climate Change 2007: Mitigation of Climate Change; Cabridge University Press: Cambridge, UK, 2007.
4. UNFCCC. Kyoto Protocol to the United Nations Framework Convention on Climate Change, Available online: http://unfccc.int/resource/docs/convkp/kpeng.pdf (accessed on 28 June 2011).

5. OECD. Environmental Data Compendium Part 1. The State of the Environment, Available online: http://www.oecd.org/document/40/0,3746,en_2649_34283_39011377_1_1_1_1,00.html (accessed on 10 April 2011).

6. Ministry of Economy Trade and Industry. Kyoto Protocol Target Achievement Plan, Available online: http://www.kantei.go.jp/jp/singi/ondanka/kakugi/050428keikaku.pdf (accessed on 12 July 2011).

7. Oita Prefectural Government. Oita Prefecture Global Warming Measures 2006, Available online: http://www.pref.oita.jp/uploaded/life/111018_123085_misc.pdf (accessed on 28 June 2011).

8. Oita Prefectural Government. Oita Prefecture Global Warming Measures 2010, Available online: http://www.pref.oita.jp/soshiki/13020/ondan-keikaku.html (accessed on 28 June 2011).

9. Tadaki, Y.; Hachiya, K. Forest Ecosystems and Their Productivity; Ringyo Kagakugijutsu Shinkosho: Toyko, Japan, 1968.

10. Bird, M. Monitoring the Carbon Cycle in Tropical Forests. Proceedings of the ENVOL 2010 International Symposium, Beppu, Japan, 22 January 2011; pp. 39–91.

11. Tucker, C.J. Red and photographic infrared linear combinations for monitoring vegetation. Remote Sens. Environ **1979**, 8, 127–150.

12. Lo, C.P.; Choi, J. A hybrid approach to urban land use/cover mapping using Landsat 7 enhanced thematic mapper plus (ETM+) images. Int. J. Remote Sens **2004**, 25, 2687–2700.

13. Chen, X.W. Using remote sensing and GIS to analyse land cover change and its impacts on regional sustainable development. Int. J. Remote Sens **2002**, 23, 107–124.

14. Janssens, I.A.; Freibauer, A.; Ciais, P.; Smith, P.; Nabuurs, G.J.; Folberth, G.; Schlamadinger, B.; Hutjes, R.W.A.; Ceulemans, R.; Schulze, E.D.; et al. Europe's terrestrial biosphere absorbs 7 to 12% of european anthropogenic CO_2 emissions. Science **2003**, 300, 1538–1542.

15. Takuya, H.; Tohru, N. Estimation of sequestered carbon in Article-3.4 private planted forests in the first commitment period in Japan. J. For. Res **2006**, 11, 427–437.

16. Sasaki, N.; Kim, S. Biomass carbon sinks in Japanese forests: 1966–2012. Forestry **2009**, 82, 113–123.

17. Ministry of Agriculture, Forestry and Fisheries. Function of the Sequestration of Forests/Trees, Available online: http://www.rinya.maff.go.jp/kinki/hyogo/mori-grow/mori-CO2.html (accessed on 10 April 2011).

18. Turner, D.P.; Guzy, M.; Lefsky, M.A.; Ritts, W.D.; Tuyl, S.V.; Law, B.E. Monitoring forest carbon sequestration with remote sensing and carbon cycle modeling. Environ. Manage **2004**, 33, 457–466.

19. Oita Prefectural Government. Oita Prefecture Forestry Figures, 2003. Available online: http://www.pref.oita.jp/site/toukei/h1506.html (accessed on 10 April 2011).

20. USGS. SLC-Off Products: Background. Available online: http://landsat.usgs.gov/products_slcoffbackground.php (accessed on 10 April 2011).

21. Ministry of Environment, Japan. The National Survey on the Natural Environment (6th (1999–2004), 7th (2005–Present)), Available online: http://www.vegetation.jp/ (accessed on 10 April 2011).

22. Irish, R.; Scaramuzza, P. SLC-Off Gap Filled Product Generation, Available online: http://igett.delmar.edu/Resources/Remote%20Sensing%20Technology%20Training/SLC-off%20GapFilled_Jun09.ppt (accessed on 10 June 2010).

23. Kobayashi, S.; Sanga-Ngoie, K. The integrated radiometric correction of optical remote sensing imageries. Int. J. Remote Sens **2008**, 29, 5957–5985.

24. Eastman, J.R. Idrisi Taiga Guide to GIS and Image Processing (Manual Version 16.02); Clark Labs: Worcester, MA, USA, 2009.

25. Richards, J.A.; Jia, X. Remote Sensing Digital Image Analysis: An Introduction, 3rd ed.; Springer: New York, NY, USA, 1999.

26. Sanga-Ngoie, K.; Kobayashi, S. A Perspective on the Integration of the Land-Use/Land-Cover Classification Results Using GIS and LANDSAT Data. Proceedings of 2003 Annual Conference of the Mie University GIS Society: GIS and Remote Sensing-Powerful Tools for Eco-Climatic Analysis, 23 April 2003; Mie, Japan; 3, pp. 69–74.

27. Sanga-Ngoie, K. Quantitative Estimation of CO_2 Sequestration by Forest Systems in Mie Using GIS and Remote Sensing. Proceedings of 2003 Annual Conference of the Mie University GIS Society: GIS and Remote Sensing-Powerful tools for Eco-Climatic Analysis, Mie, Japan, 23 April 2003; 3, pp. 75–78.

28. JAXA. PALSAR Calibration Factor Updated, Available online: https://auig.eoc.jaxa.jp/auigs/en/doc/an/20090109en_3.html (accessed on 8 November 2012).

29. Castel, T.; Beaudoin, A.; Stach, N.; Stussi, N.; Le Toan, T.; Durand, P. Sensitivity of space-borne SAR data to forest parameters over sloping terrain. Theory and experiment. Int. J. Remote Sens **2001**, 22, 2351–2376.

30. Ulander, L.M.H. Radiometric slope correction of synthetic-aperture radar images. IEEE Trans. Geosci. Remote Sens **1996**, 24, 1115–1122.

31. Chen, E.; Li, Z.; Ling, F.; Lu, Y.; He, Q.; Fan, F. Forest Volume Density Estimation Capability of ALOS PALSAR Data Over Hilly Region. Proceedings of 4th International Workshop on Science and Applications of SAR Polarimetry and Polarimetric Interferometry—PolInSAR 2009, Frascati, Italy, 26–30 January 2009.

32. Santoro, M.; Fransson, J.E.S.; Eriksson, L.E.B.; Magnusson, M.; Ulander, L.M.H.; Olsson, H. Signatures of ALOS PALSAR L-band backscatter in Swedish forest. IEEE Trans. Geosci. Remote Sens **2009**, 47, 4001–4019.

33. Armston, J.; Carreiras, J.; Lucas, R.; Shimada, M. ALOS PALSAR Backscatter Mosaics for Queensland, Australia. Proceedings of the 15th Australasian Remote Sensing & Photogrammetry Conference, Alice Springs, Australia, 13–17 September 2010.

34. Santoro, M.; Beer, C.; Cartus, O.; Schmullius, C.; Shvidenko, A.; McCallum, I.; Wegmüller, U.; Wiesmann, A. Retrieval of growing stock volume in boreal forest using hyper-temporal series of envisat asar scansar backscatter measurements. Remote Sens. Environ **2011**, 115, 490–507.

35. Kato, M. Forestry Remote Sensing-Basics to Application; Japan Forestry Investigation Committee: Tokyo, Japan, 2004.

36. Ishii, T. Estimation of Forest Leaf Area Index Using Remote Sensing Data and Its Applications to Predicting Forest Carbon Sequestration and Water Budget. Ph.D. Thesis, Waseda University, Tokyo, Japan, 2007..

37. Kuplich, T.M.; Salvatori, V.; Curran, P.J. Jers-1/SAR backscatter and its relationship with biomass of regenerating forests. Int. J. Remote Sens **2000**, 21, 2513–2518.

38. Champion, I.; Dubois-Fernandez, P.; Guyon, D.; Cottrel, M. Radar image texture as a function of forest stand age. Int. J. Remote Sens **2008**, 29, 1795–1800. [Google Scholar]

39. Karjalainen, M.; Pyysalo, U.; Karila, K.; Hyyppa, J. Forest Biomass Estimation Using ALOS PALSAR Images in Challenging Natural Forest Area in Finland. Proceedings of the ALOS PI 2008 Symposium, Rhodes, Greece, 3–7 November 2008.

40. Wijaya, A. Evaluation of ALOS-PALSAR mosaic data for estimating stem volume and biomass: A case study from tropical rainforest of central Indonesia. Jurnal Geografi **2009**, 2, 14–21.

41. Kobayashi, S.; Widyorini, R.; Kawai, S.; Omura, Y.; Sanga-Ngoie, K.; Supriadi, B. Backscattering characteristics of L-band polarimetric and optical satellite imagery over planted acacia forests in Sumatra, Indonesia. J. Appl. Remote Sens **2012**, 6, 063519–063525.

42. Kobayashi, S.; Omura, Y.; Sanga-Ngoie, K.; Widyorini, R.; Kawai, S.; Supriadi, B.; Yamaguchi, Y. Characteristics of decomposition powers of L-band multi-polarimetric SAR in assessing tree growth of industrial plantation forests in the tropics. Remote Sens **2012**, 4, 3058–3077.

43. Green, K.; Congalton, R. An Error Matrix Approach to Fuzzy Accuracy Assessment: The NIMA Geocover Project; CRC Press: Boca Raton, FL, USA, 2004.

44. Congalton, R.G.; Green, K. Assessing the Accuracy of Remotely Sensed Data; Principles and Practices, 2nd ed.; CRC Press: New York, NY, USA, 2009.

45. Shida, N. A GIS Analysis of Forests Growth and CO_2 Sequestration by Mie Prefecture forests. Bachelor Thesis, Mie University, Tsu, Japan, 2007..

46. IPCC. Revised 1996 IPCC Guidelines for National Greenhouse Gas Inventories; Meteorological Office: Bracknell, UK, 1996.

47. Luckman, A.; Baker, J.; Honzák, M.; Lucas, R. Tropical forest biomass density estimation using jers-1 SAR: Seasonal variation, confidence limits, and application to image mosaics. Remote Sens. Environ **1998**, 63, 126–139.

48. Congalton, R.G.; Green, K. A practical look at the sources of confusion in error matrix generation. Photogramm. Eng. Remote Sensing **1993**, 59, 641–644.

49. Kogan, F.N. A typical pattern of vegetation conditions in Southern Africa during El Niño years detected from AVHRR data using three-channel numerical index. Int. J. Remote Sens **1998**, 19, 3688–3694.

50. Bibilis, E.J.; Brinker, R.; Carino, H.F.; Mckee, C.W. Effect of stand age on flexural properties and grade compliance of lumber from loblolly pine plantation timber. Forest Prod. J **1993**, 43, 23–28.

51. Small, D. Flattening gamma: Radiometric terrain correction for SAR imagery. IEEE Trans. Geosci. Remote Sens **2011**, 49, 3081–3093.

Chapter 8

Estimation of CO_2 Sequestration by the Forests in Japan by Discriminating Precise Tree Age Category using Remote Sensing Techniques

Kotaro Iizuka [1,*] and Ryutaro Tateishi [2]

[1] *Research Institute for Sustainable Humanosphere, Kyoto University, Gokasho, Uji, Kyoto 611-0011, Japan*

[2] *Center for Environmental Remote Sensing (CEReS), Chiba University, 1-33 Yayoi-cho, Inage-ku, Chiba 263-8522, Japan*

ABSTRACT

This study estimates CO_2 sequestration by forests in Japan using Land Remote Sensing Satellite (Landsat) Operational Land Imager (OLI) and the Advanced Land Observing Satellite (ALOS) Phased Array L-band Synthetic Aperture Radar (PALSAR) remote sensing data for the in-depth retrieval of forest growth stages (tree age). Landsat imagery was used to develop a detailed forest cover map, while the PALSAR data were used to estimate the volume information. The volume was converted to tree age information for each of the three forest types in Japan. An estimation of CO_2 sequestration values for each forest type and for each tree age from the forest inventory data was made. The forest cover map results in four classes, and the overall accuracy yields approximately 74%. For the volume estimation, Root Mean Square Error (RMSE) was computed with the ground reference information resulting in 105.58 m^3/ha. The final result showed that total CO_2 sequestration in Japan based on tree age forest subclasses yields 85.0 Mt·CO_2 (coniferous), 4.76 Mt·CO_2 (evergreen broadleaf) and 21.61 Mt·CO_2 (deciduous broadleaf), which in total is 111.27 Mt·CO_2. Using remote sensing techniques to quantitatively estimate CO_2 sequestration in Japanese forests has been shown both to have advantages and to offer further possibilities.

Keywords: forest; sequestration; CO_2; PALSAR; Landsat; tree age; Japan

1. INTRODUCTION

Forest resources are essential to life on Earth because forests play a large role in providing organic matter through photosynthesis, which is important for countless species of plants and animals, and in processing atmospheric gasses such as CO_2, which is of crucial environmental importance and among the most important issues faced in the 21st century: Global Warming/Climate Change. Research is performed and conferences are held to investigate this issue and its associated problems, and many nations are working together on policies to prevent or mitigate the associated problems and impacts [1,2]. Japan has been at the forefront of nations involved in the many activities related to this issue. The Kyoto Protocol and the Reducing emissions from deforestation and forest degradation mechanisms (REDD, REDD+) provided a new approach to finding solutions, giving the role of forests more importance, especially when focusing on how forests function as a carbon sink. Forest carbon sinks are highly important in the current research context, as 67% of the land in Japan is covered by forests [3]. Japan considers forests to be the top candidate for its CO_2 reduction process [4]. Prefectural Governments and private companies in Japan highly value forest carbon sequestration research; however, some Prefectures have not yet been involved in such processes even when they have a great amount of forest cover [5], which may be because they lack the equipment or effective methodology to quantitatively estimate carbon sequestration.

The estimation implemented in Japan mostly focuses on identifying forests as either natural or planted. Hiroshima and Nakajima [6] focused on the potential carbon sinks in Japanese plantation forests during the first commitment period of the Kyoto Protocol. They estimated that planted forests were expected to sequester 8.16–8.87 Mt·C/yr, depending on various scenarios of forest management based on silvicultural practices, and employed subsidies and forest workers' wages as predictor variables. Sasaki and Kim [7] estimated potential CO_2 sequestration by forests in Japan and the eligible sequestration under the condition of the Marrakesh Accords. Using the land use model and the carbon stock growth model, it has been estimated that the forests in Japan are likely to sequester 20.1 Mt·C/yr (planted forests: 15.3 Mt·C/yr, natural: 4.8 Mt·C/yr), and under the conditions of the Marrakesh Accords, it is estimated at 10.2 Mt·C/yr (planted forests: 7.3 Mt·C/yr, natural: 2.9 Mt·C/yr). Recently, the National Institute for Environmental Studies (NIES) Center for Global

Environmental Research (CGER) calculated a result for the IPCC reports on the issue of forest carbon sinks, estimating the CO_2 sequestration by the forests of Japan [8]. Using the forest registration data from each prefectural government, and differentiating the forests by tree species, they have calculated the total CO_2 amount sequestered both aboveground and in litter and soil. Their results showed an estimate of 77.67 Mt·CO_2 for the year 2012.

Utilizing satellite remote sensing data for quantitative C/CO_2 estimations is challenging but well-known method for improving broad scale estimations compared to conventional ground-based observations [9]. The techniques can be separated into two major methodologies. The first is the indirect measure of C/CO_2, which calculates several parameters related to the function of the forest system, such as Leaf Area Index (LAI), Photosynthetically Active Radiation (PAR) and others, and then models multiple parameters to estimate either Gross Primary Product (GPP) or Net Primary Product (NPP) [10,11]. The second method integrates detailed land cover information with ground observations of forest inventories [5]. Alternatively, both methods may be integrated together [12]. All these methods have advantages and disadvantages. For example, one popular product by the former method is the Moderate Resolution Imaging Spectroradiometer (MODIS) NPP [13], which can be collected globally with a fine temporal resolution. Depending on the need, such products can be unsatisfying because the spatial resolution of 1 km by 1 km is too coarse; when we want to focus more on a regional-scale analysis, a spatial resolution of 1 km cannot delineate the actual local environment accurately. The latter method is usually implemented in a regional scale analysis using fine resolution imagery, such as Landsat, to discriminate details in land cover/land use types to estimate accurate values for C/CO_2 content. This can also be conducted by indirectly estimating modeling parameters, but it is often performed with the associated forest inventory data collected from other ground observation sources and then integrated with the land cover information for the estimation. In comparison with MODIS, Landsat provides a much finer spatial resolution (30 m), making it possible to allocate the heterogeneity of the natural environment; however, there are again issues that must always be faced, such as the fundamental problem of utilizing optical imagery due to cloud cover. Fine resolution imagery has lower temporal resolution, and it is sometimes difficult to collect cloud free imagery through only a few observations. Sanga-Ngoie et al. [5] has utilized synthetic aperture radar (SAR) data to extract the forest parameters related to CO_2 sequestration. This method has shown the usefulness of implementing SAR to observe forest extent, which could be broadly usable in conditions where there is frequent cloud cover over the area of interest. This method made it clear that not only land cover type discrimination, but also tree age, or the growth stages of the forests, plays an important role. Taking these variables into account provides more accurate estimations for the overall result, because each forest growth stage differs in its sequestration rate [5,14]. Utilizing both optical and microwave satellite imagery makes this possible.

Most forest carbon sequestration capacity estimates within Japan are performed at a national scale, and detailed analyses at a regional scale are still rare; therefore, such estimates are either rough estimations on a national scale or detailed analysis on a local scale. Unfortunately, local-scale studies cannot be expanded to a national scale; they are only for smaller regions and do not apply to the country as a whole. The studies mentioned usually use a methodology involving a modeling base, site observation and statistics. However, detecting and locating forest resources does not take place. Locating precise forest resources of both forest types and their growth stages requires not only consideration of realistic estimation of sequestration [5,10,14,15,16] but also an understanding of the current status of the forest resources, which may be the most important information to gather for long-term sustainable plans to manage forests in a strategic way. Consequently, the objective of this research is to reassess the estimation of CO_2 sequestration (i.e., NPP) by forest cover in Japan based on a precise evaluation of forest extent by both forest type and age, together with the use of remote sensing data, to quantitatively estimate the CO_2 sequestration potential.

2. STUDY AREA

Our area of study focuses on mainland Japan (except Okinawa and small islands) (Figure 1). Japan is made up of over 7000 contiguous islands from the northeast to the southwest, with a total length of approximately 3000 km. Total land area is approximately 377,900 km^2. Over 60% of the land is recognized as mountainous (over 2000 m in altitude) and 12% as hilly. The total forested area makes up approximately 67% [3], the majority of which is located around the aforementioned mountain or hill areas. Climatic conditions in Japan vary widely due to its geographical location; its lands stretch from a southern latitude of 30°N (Okinawa omitted) to a northern latitude of 45°N. This range, together with the ocean currents, gives each region a different climate. Because of the variety in topography and climatic conditions, Japan has a great diversity of vegetation both horizontally and vertically. For example, there are over 6000 species of Pteridophytes, each of which has evolved and adapted to its specific location [17]. The dominant coniferous tree types in Japan are the Japanese cedar (Cryptomeria japonica) and cypress (Chamaecyparis obtusa) trees, which occur widely from the north to the south. As a whole, coniferous and broadleaf forests are distributed in approximately a 50/50 ratio, while 80% of coniferous forests tend to be planted forests that were part of a large plantation plan that lasted from the 1950s to the 1970s. On the other hand, the broadleaf forests are primarily natural (over 90%), and few deciduous or evergreen broadleaf trees have been planted. Broadleaf hardwoods are dominant in the central region, while mixed deciduous evergreen forests or evergreen broadleaf forests can be observed in the temperate south zones.

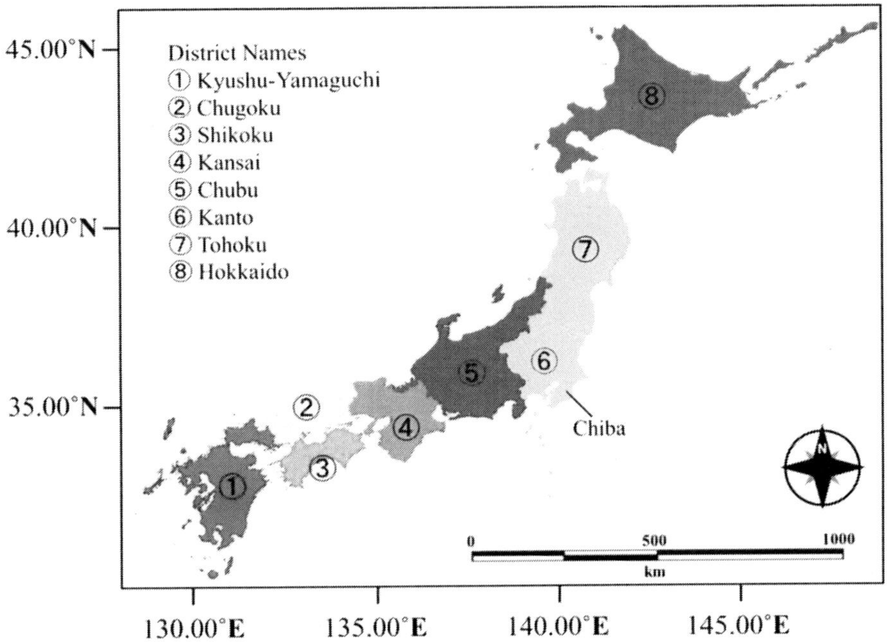

Figure 1. Study Area, Japan and its eight major districts.

Remote sensing data for Chiba Prefecture was used for a detailed analysis of statistical data, so a brief explanation is given. Chiba is located on the east coast of Japan along the Pacific Ocean, just east of the Tokyo metropolitan area, where the peninsula juts out. Climatic conditions include a warm oceanic climate, which is a condition with high humidity and high precipitation in summer, and low humidity and low precipitation in the winter. However, comparing the southern and northeastern regions of Chiba, the climate is warm throughout the year in contrast to the inland area, which shows more diversity such as larger temperature drops in the winter [18]. Annual precipitation is highest in the southern area, with more than 2000 mm, and is next highest in the northern area, which has approximately 1400 mm to 1600 mm. The difference in precipitation clearly shows the distribution of the forests, which are more dense in the south and sparse in the north [18]. A characteristic of Chiba is that it is one of the flattest regions in Japan; the average land altitude is 46 m above sea level, and there are no mountains higher than 500 m in altitude.

3. METHODS AND DATA

The overall flowchart of the methodology is shown in Figure 2. A three-step process is used: (1). Estimating the volume of the forests in Japan using SAR and converting these data to tree age information data; (2). Developing a detailed land cover (forest cover) map to allocate the precise forest types across the land area of Japan; and (3). Utilizing the obtained data from the preceding steps together with the CO_2 sequestration rate per unit area to estimate the sequestration values by each forest type and age.

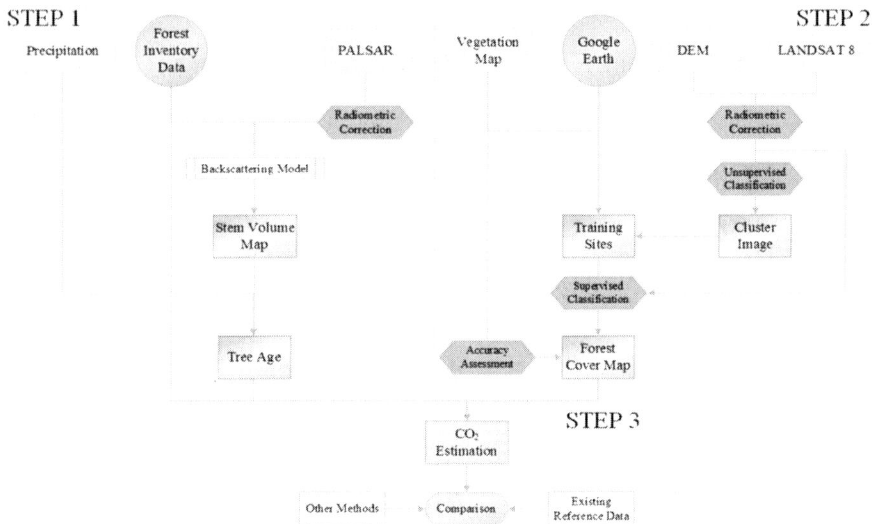

Figure 2. Overall flowchart of the methodology.

3.1. L-Band PALSAR for Volume Mapping

In the first step of the study, a simple relationship analysis was performed between the biophysical parameters collected from field observations by the Chiba Prefectural Government and the L-band backscattering intensity data collected by the Advanced Land Observing Satellite (ALOS) Phased Array L-band Synthetic Aperture Radar (PALSAR) provided by the Japan Aerospace Exploration Agency (JAXA). This step involved extracting the biophysical parameters (volumes) of the forested areas and mapping them for the whole of Japan. The use of SAR for detecting forest parameters works because the radar waves are longer than the optical waves, so the radar beam penetrates through the top layer of canopy and scatters from the stems and branches of the trees when using L-band radar [19,20,21,22,23]. A number of studies have been conducted to develop models related to those studies' specific regions of interest. Those studies confirm that the backscattering signature correlates with forest parameters and that SAR images can be used to extract information related to forest biomass. While most of the papers present and acknowledge the possibility of extracting forest parameters using the L-band SAR information, some studies show difficulties in data interpretation [24], mostly due to distortions caused by local topography, which is also one of the challenges for this work.

To meet this topography challenge for the analysis, Chiba Prefecture was selected due to its relatively flat terrain, which limits this critical and fundamental topography problem when utilizing SAR imagery [25].

The 25 m PALSAR Global Mosaic (PGM) image used in this study was converted from the provided Digital Number (DN) values to the backscattering intensity information, also known as backscattering coefficients, using Equation (1). The PALSAR specification is in ascending Fine Beam Dual (FB) mode characterized by a 34.3 degree of off-nadir angle, calculated as:

$$\sigma^0 = 10 \times \log_{10}(\mathrm{DN}^2) + \mathrm{CF} \tag{1}$$

where σ^0 is the backscattering intensity represented in decibel units (dB) and CF is the calibration factor for the data obtained, depending on the observation period and polarization [26]. The image is already processed for ortho-rectification and slope correction [27], so further analysis uses these data directly without any additional correction.

A statistical analysis for formulating relationships among the field observation data of the forest stand characteristics and the remotely sensed microwave satellite data was carried out. The process analyzes the relationships between SAR backscattering intensity (σ^0) and the stand characteristics (volume) for the 25 m pixel spacing PALSAR image. The PALSAR image used here is the mean value of the years 2009 and 2010, to minimize local variations from such things as precipitation. Using the vegetation continuous field data [14], the sample plots were selected from an area of less than 70% forest cover to avoid uncertainties associated with scattering, as Iizuka and Tateishi [28] explain that attenuation of the radar beam is caused by high dense forested areas, resulting in lower backscattering values compared to the peak areas (70%–75% forest cover), and causing difficulties when interpreting the actual forest trend.

Using the result, the model function was applied to the 50 m PALSAR mosaic product (which can be downloaded free from the JAXA web page [29]) to obtain volume information for the forests across Japan. The 50 m PALSAR data were selected because of the spatial coverage and availability (it is freely available and covers the whole of Japan). The equation obtained from the relationship analysis was assumed to be acceptable to apply to the downscaled data, because several papers confirm an increase in accuracy and higher correlations when the pixel spacing is reduced [30,31].

DN values of the data from 2007 and 2009 were averaged to avoid small changes from backscattering caused by local environmental factors at that particular observation time (e.g., precipitation) and then converted to dB units. The data were then slope corrected, because they had been previously ortho-rectified but not slope corrected, resulting in enhancements of the backscatter from the rugged hills facing towards the sensor direction [28].

A correction method proposed by Zhou et al. [32] was modified and applied to the 50 m PALSAR mosaic product. The correction of the SAR imagery is performed by applying a slope correction factor to the original SAR data. The equation for the correction is as follows:

$$\sigma_c^0 = SCF^q \times \sigma^0 \tag{2}$$

where σ^0_c is the corrected sigma naught backscattering value, σ^0 is the original sigma naught backscattering data, q is the diffuse scattering, which is polarization dependent, and SCF is the slope correction factor for the correction of the topography, expressed as:

$$SCF = \frac{\cos i}{\sqrt{1 + \left(\frac{\tan \alpha \times \sin \alpha \times \sin 2\beta}{2}\right)^2}} \tag{3}$$

where cos i is the cosine of the incidence angle, α is the slope and β is the aspect. The HV polarization image of PALSAR was corrected using this method. The value for the diffuse scattering depends on the environment and was unknown in this case due to the large study area with differing forest structures, so here it has been assumed that $q = 1$. The volume map for the whole of Japan was obtained by applying the volume model equation to the corrected 50 m PALSAR mosaic product.

3.2. Land Cover Mapping of Japan

We selected and analyzed 35 scenes from the Land Remote Sensing Satellite (Landsat) Operational Land Imager (OLI) data published by the United States Geological Survey (USGS) to develop a land cover map of Japan that takes forest types into account. Each scene was processed with an Integrated Radiometric Correction [33] method to reduce effects from the atmosphere and topography, resulting in fewer errors throughout the classification process. Landsat OLI 5 bands (bands 2–6) along with the Normalized Difference

Vegetation Index (NDVI), Normalized Difference Water Index (NDWI) [34], and Green Red Vegetation Index (GRVI) [35] were used to perform the supervised classification using a maximum likelihood algorithm. This yielded a final land cover product with 12 classes. However, our interests here are in the forest classes only, so the number of classes was reorganized into three forest type classes (coniferous, deciduous broadleaf, and evergreen broadleaf) plus one additional class (others), resulting in a total of four classes. These were utilized for further procedures along with the accuracy assessment, which is fundamental for evaluating the resulting categorical map [36]. The date of the acquired scenes included two from March, four from April, four from May, six from June, five from August, five from September and nine from October. Selection of the images was made by considering a general and critical issue for such optical images: that they were cloud-free. Not all scenes were completely cloud free. In such cases, selections were made from the clearest images that could be found. The specifications of the chosen data are listed in Table 1.

Table 1. Specification of the Landsat Imagery used.

	Landsat OLI Data Sets				Landsat OLI Data Sets		
Region	Path	Row	Date	Region	Path	Row	Date
Kyushu Yamaguchi	112	36	2 May 2014	Tohoku	106	33	8 May 2014
	112	37	13 April 2013		107	31	29 April 2014
	112	38	13 April 2013		107	32	19 October 2013
	113	37	23 April 2014		107	33	17 September 2013
	113	38	29 October 2013		108	31	20 April 2014
Chugoku Shikoku Kansai	110	36	17 March 2014		108	32	4 June 2013
	110	37	17 March 2014		108	33	4 June 2014
	111	35	24 May 2013		108	34	4 June 2014
	111	36	24 May 2013	Hokkaido	105	30	2 June 2014
	111	37	11 May 2014		106	29	25 June 2014
	112	35	19 August 2013		106	30	25 June 2014
Chubu	108	35	7 August 2013		106	31	28 October 2013
	108	36	7 August 2013		107	29	19 October 2013
	109	35	14 August 2013		107	30	19 October 2013
	109	36	14 August 2013		108	28	10 October 2013
	109	37	17 October 2013		108	29	10 October 2013
	110	35	17 May 2013		108	30	10 October 2013
Kanto	107	34	17 September 2013				
	107	35	17 September 2013				
	107	36	17 September 2013				
	109	34	14 August 2013				

3.3. Estimating CO_2 Sequestration from the Forests of Japan

Estimation of the sequestration from Japanese forests was obtained by utilizing the two data sets described above (i.e., the volume map and land cover map), along with an additional data set of forest inventory data provided by two Prefectural Governments and the precipitation data published by the Ministry of Land, Infrastructure, Transport and Tourism (MLIT) of Japan. CO_2 sequestration per unit area data published by the Tochigi Prefectural Government was used to estimate the sequestration values not only for each forest type but also for each tree age. This was achieved by transforming the volume map into ages among the different forest types, referring to the volume-tree age relationship data provided by the Oita Prefectural Government. The precipitation data were also used to help discriminate between growth rates for each region, which is an important step, because even with the same stem volume, the forests' per unit areas could have different tree ages due to local site productivity [37].

The stem volume-age relationship data provided by the Oita Prefectural Government is based on observations made on ground-level plots at multiple sites within Oita Prefecture. The curves were averaged for each growth region and computed as high, moderate and low growth rate areas. Each location within the study area that lies on a matching growth-rate region was determined to have a particular growth curve for that forest; the tree age conversion is based on that curve. To differentiate the growth rate of the study area, a relationship analysis was made using the precipitation data along with the total volume of growth made between the two periods of time. The sugi (cedar) and kunugi (sawtooth oak) data were used as representative of the coniferous and deciduous broadleaf forests, respectively. Unfortunately, there is a lack of data for the evergreen broadleaf forests, so the growth rate is considered only for the first two classes. Here, the precipitation data were used to determine whether precipitation would be one of the limiting factors for forest growth.

The sequestration values were modified from the reference data published by the Tochigi Prefectural Government by taking a mean value of each tree type at each tree age range for deciduous and evergreen broadleaf forests, using the tree data for sawtooth oak, Japanese beech, Japanese zelkova and hornbeam for deciduous broadleaf forests, and using evergreen oak and chinquapin oak for the evergreen broadleaf forests. For the coniferous forests, a weighted mean value of the sugi (cedar) and hinoki (cypress) trees was taken as 7:3 because the distribution of those trees approximates that ratio for the whole of Japan. Together with the estimation, an in-depth analysis of the uncertainties of the estimated results was assessed by considering possible errors that could generate inaccuracies in the results.

4. RESULTS

4.1. Stem Volume Modeling and Mapping

Figure 3 shows the trends of the stem volume vs. the backscattering intensity (dB) for both HH and HV polarizations. An increasing trend can be observed with the increasing parameter and increasing backscattering intensity, which is consistent with many of the other related studies [19,20,21,22,23,30,31]. According to these results, a second polynomial function from the HV polarization was selected and applied to obtain the stem volume information for the whole of Japan, because it has higher correlation compared to HH (HH: $R^2 = 0.2123$, HV: $R^2 = 0.2198$).

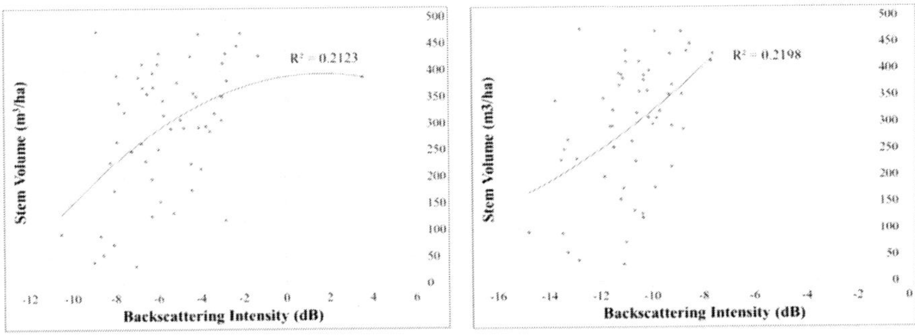

Figure 3. Relationship between Stem Volume (m^3/ha) and backscattering for HH (**left**) and HV (**right**) polarization at mean 25 m pixel spacing data from 2009 and 2010 (HH: $R^2 = 0.2123$, HV: $R^2 = 0.2198$).

The equation of the function is described as:

$$V = 1.4192\sigma^{0^2}_{HV} + 60.472\sigma^0_{HV} + 773.8 \quad (4)$$

where V is the stem volume and σ^0_{HV} is the backscattering intensity from the HV polarization. Validation of the backscattering model was assessed by using the remaining observation plot data that were not used in the modeling process. Using the plot data as ground truth information, the root mean square error (RMSE) between the estimated results and the reference data was calculated. The RMSE result is 105.58 m^3/ha, and $R^2 = 0.164$ as shown in Figure 4. This model will be applied to obtain the stem volume information for the forested areas within Japan.

Figure 5 shows the interpretation of the original and the corrected imagery of the PALSAR data. In addition to the interpretation, a quantitative evaluation of the performance of the correction was conductedby assessing the non-correlation between cos i and the corrected data. Comparing the correlation coefficients, the value shows $r = 0.4$ and $r = 0.06$ for non-corrected and corrected images, respectively. From these results, it is presumed that the correction of the PALSAR was performed successfully.

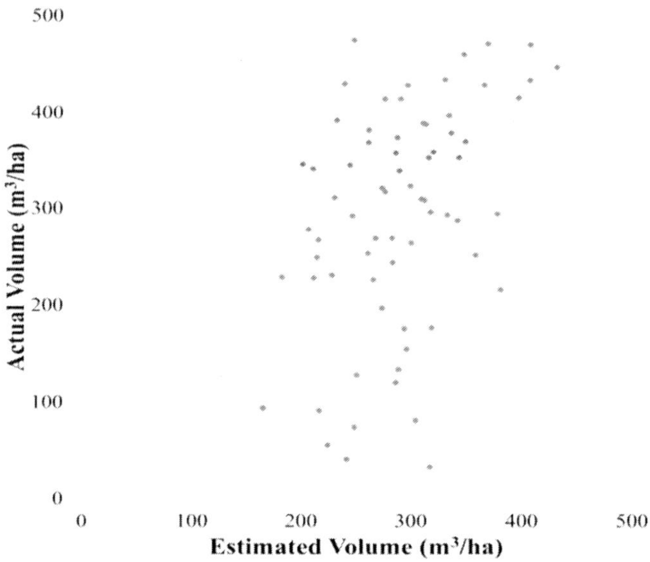

Figure 4. Validation of the backscattering model. The X-axis is the estimated volume (m³/ha) using the model, and the Y-axis is the reference data from the observation plot. R^2 = 0.164 Root Mean Square Error (RMSE) = 105.58 m³/ha.

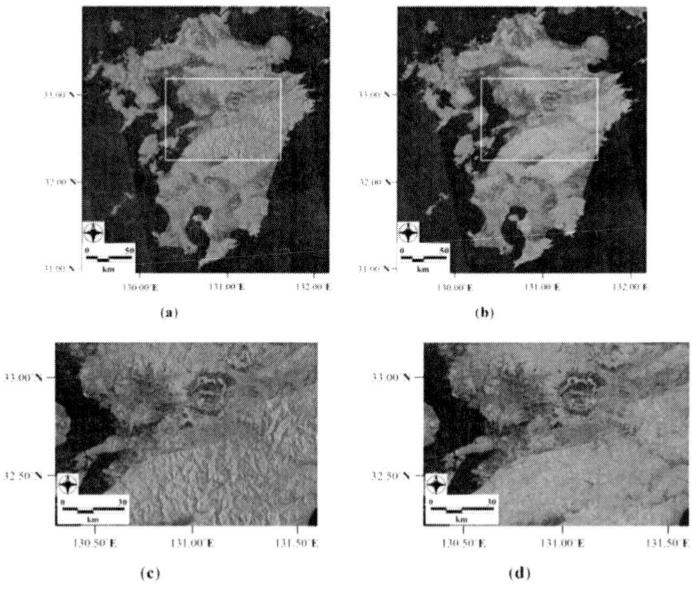

Figure 5. Visual interpretation of the Phased Array L-band Synthetic Aperture Radar (PALSAR) 50 m at the Kyushu Island region, for (**a**) the original image and (**b**) the slope-corrected image of the HV polarization data. The (**c**,**d**) images show the zoomed areas from the (a,b) images, respectively.

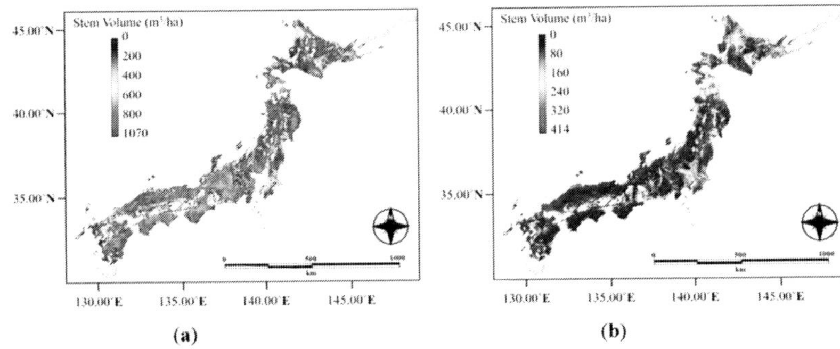

Figure 6. Stem volume map of Japan for the (**a**) coniferous forests and (**b**) deciduous and evergreen broadleaf forests. Forest cover data were overlaid with the stem volume map to extract only the forest types needed for each stem volume map.

The formula (Equation (4)) obtained through the model was applied to the corrected PALSAR imagery to obtain stem volume information for the entire forest area of Japan. By doing this, the spatial distribution of the estimated stand volume for Japanese forests can be shown. However, this model is specific to the coniferous stands in Japan, and the possibility that backscattering characteristics of different forest types would give different trends due to the structure of the stand and leaf size should be taken into account. Instead, for the broadleaf forests, a polynomial model developed by Wijaya [38] from the evergreen trees of Indonesia was applied to fit the situation (Equation (5)). Utilizing the two different models specific to those forest types, the stem volume distribution across Japan was obtained, as seen in Figure 6a,b. Using the model developed from the results of this study, the maximum stand volume of the coniferous forests reaches up to 1069 m^3/ha; for the broadleaf forests, it reaches up to 414 m^3/ha. Although the coniferous stand volume gives a high maximum value, most of the distributed areas lie near the 400–500 m^3/ha range and rarely exceeds 700 m^3/ha. The original stem volume information was distributed over all land cover types, so the exact location of each forest was extracted using the land cover map. The white areas in Figure 6 indicate non-coniferous (a) and non-broadleaf forests (b) areas.

$$V = 256.85 - (65.458\sigma^0) - (6.8{\sigma^0}^2) \quad (5)$$

4.2. Land Cover Map and ItsAccuracy

The Landsat OLI images captured on multiple dates from years 2013 to 2014 were classified into the following 12 land cover types: water bodies, urban, paddy fields, crop lands, orchards, grasslands, silver grass, bare lands, coniferous forests, evergreen broadleaf forests, deciduous broadleaf forests and snow/clouds. These detailed classes were then reclassified into the three general forest classes and one other-than-forest class, resulting in a total of four land cover categories. Only the forest categories are of interest; the accuracy assessment was also performed with this map. The final product of the land cover map (before reclassification) of Japan is shown in Figure 7.

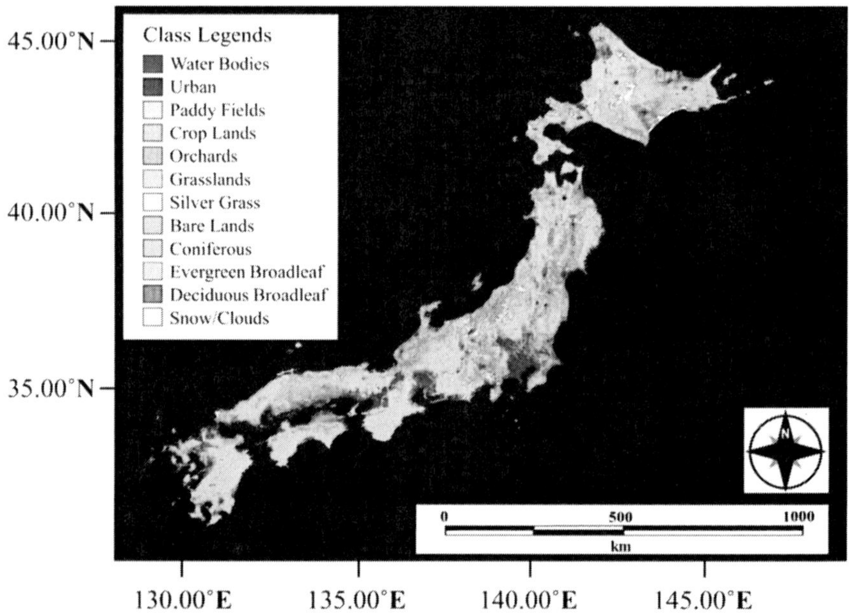

Figure 7. Land Cover map of Japan developed using 35 Landsat Operational Land Imager (OLI) scenes yielding 12 classes in total.

The ground truth points obtained from Google Earth together with the reference vegetation map [39] were used in the resulting land cover map to perform an accuracy assessment by outputting an error matrix (Table 2). The final number of ground truth points yields 3550 points. However, the number of points was normalized for each class to 1000 points, to make calculations and interpretation easier in the accuracy assessment. The overall accuracy of the land cover (forest cover) map result was 73.93%, and the overall Kappa Index of Agreement (KIA) result was 0.65. KIA is a statistical measure adapted for accuracy assessment in the remote sensing fields by Congalton and Mead [40]. KIA values range from 0 to 1, with zero representing completely random chance and 1 meaning a perfectly true agreement. In this study, KIA = 0.65, which could be said to show that the map is 65% likely to be the result of true agreement.

Table 2. Error Matrix showing the reference data (ground truth) versus the image classification. Overall accuracy shows approximately 74% and Kappa Index of Agreement (KIA) as 0.65.

	KIA = 0.65	Reference					
		Coniferous	Evergreen Broadleaf	Deciduous Broadleaf	Others	Total	ErrorC
Classified	Coniferous	**852**	253	149	24	1278	0.3334
	Evergreen Broadleaf	41	**533**	113	30	717	0.2566
	Deciduous Broadleaf	59	200	**680**	54	993	0.3152
	Others	48	14	58	**892**	1012	0.1186
	Total	1000	1000	1000	1000	4000	
	ErrorO	0.1480	0.4670	0.3200	0.1080		0.2607

4.3. CO_2 Sequestration

4.3.1. CO_2 Sequestration by Forest Types and Ages

The stem volume-age relationship data were modified from the original Oita data [5] to compute high, moderate and low growth curves for coniferous and deciduous broadleaf forests (Figure 8), which were used in turn to convert the so-obtained stem volume map (Figure 6) into tree age classes. Figure 9 shows the analysis between annual precipitation and tree growth for coniferous and deciduous broadleaf forests.

The results show a clear trend in the relationship between precipitation and tree growth. For the sugi trees (coniferous), increasing precipitation results in larger tree growth. On the other hand, for the sawtooth oak (deciduous broadleaf), the opposite trend can be observed, where tree growth increases with decreasing precipitation. The negative trends in the deciduous broadleaf forests might be caused by the precipitation data used. The precipitation data are an annual value, which means that it was not taken into account during the time when deciduous trees are likely to be photosynthesizing. If the precipitation falls mostly in the fall or winter, the trees may not be able to take advantage of it and thus would have lower than expected growth rates.

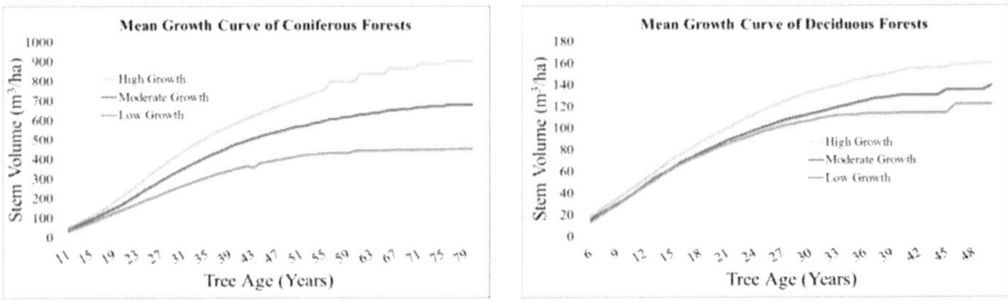

(a) (b)

Figure 8. Mean growth curve of (**a**) coniferous and (**b**) deciduous broadleaf forests in each growth rate region, computed from the original Oita data.

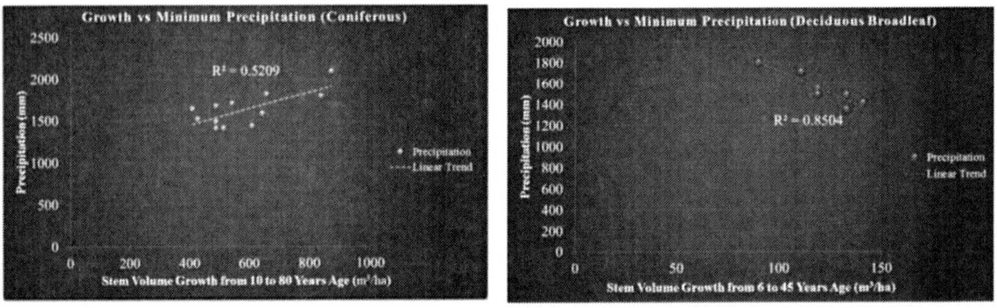

(a) (b)

Figure 9. Relationship analysis between annual precipitation and tree growth for (**a**) sugi (cedar) trees and (**b**) kunugi (sawtooth oak) trees.

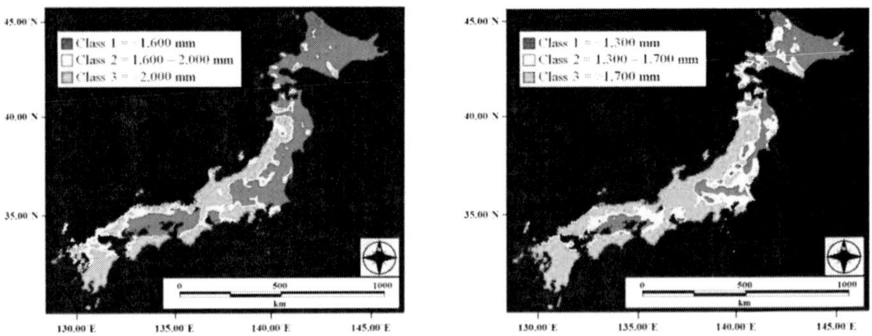

(a) (b)

Figure 10. Precipitation data classified using the threshold values for evaluating the growth rate of (**a**) coniferous forests and (**b**) deciduous broadleaf forests.

The distribution of the precipitation classes and the thresholds is seen in Figure 10. For the coniferous forests, Class 1 is the area with a low growth rate, Class 2 is moderate, and Class 3 is a high growth rate. For deciduous broadleaf forests, Class 1 is high growth rate, Class 2 moderate, and Class 3 is low growth rate. Finally, Figure 11 shows a graph of the distribution of tree age overlaid with the forest cover data to extract the exact forest types. For the coniferous and deciduous broadleaf forests, the growth rates are considered for the age conversion, however for the evergreen broadleaf it was not, due to the lack of reference data, so that has been converted directly from the stem volume map using the stem volume-age relationship information provided by Sasaki et al. [41].

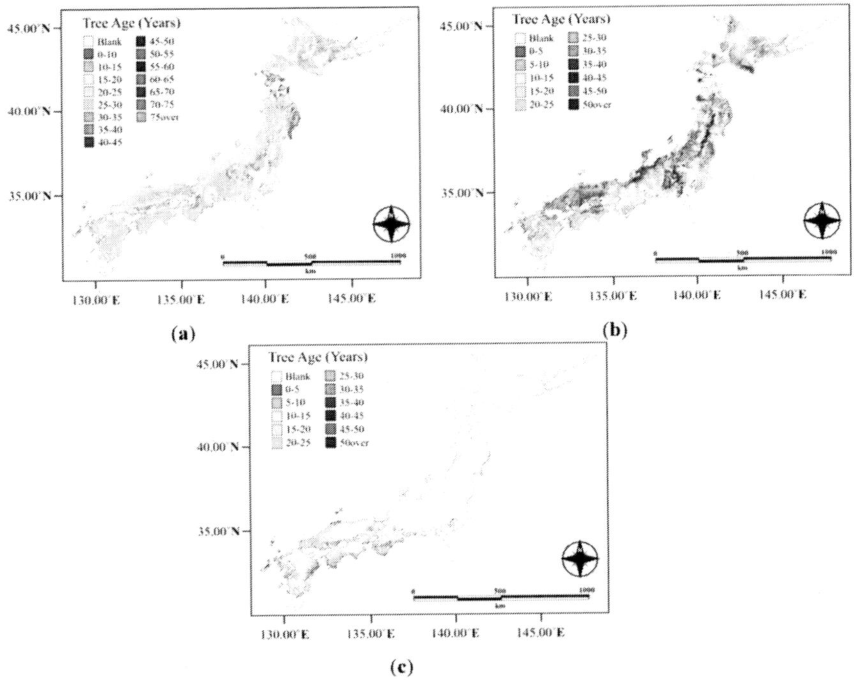

Figure 11. Tree age map transformed from the stem volume map for the (**a**) coniferous, (**b**) deciduous broadleaf and (**c**) evergreen broadleaf forests.

The statistical distribution of forest tree ages can be seen in Figure 12. Coniferous forests can be widely distributed among the different tree ages. The majority lie in the 25–30 year age range for Site 1 (high growth rate regions), 30–35 years range for Site 2 (moderate growth rate regions) and 35–40 years range for Site 3 (low growth rate regions). We can also confirm some sizable areas over 75 years of age, which can be understood to be old mature forested areas left naturally and still growing without any management, while the areas with lower tree ages are places where plantations or forest management is in operation. However, future work is needed to validate the accuracy of this concept. Compared to coniferous forests, the majority of deciduous and evergreen broadleaf forests can be observed to have tree ages of over 50 years. These forests include few to no areas that are actively managed; they are just left to grow naturally, resulting in a majority of the areas being in highly matured age classes.

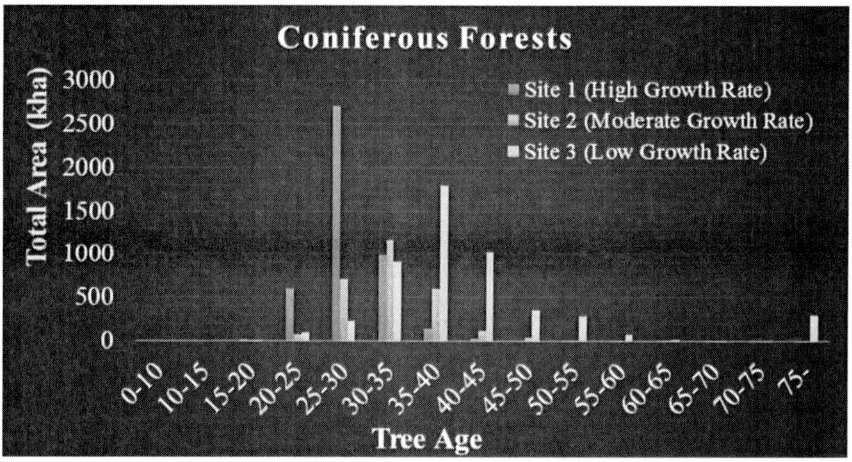

(a)

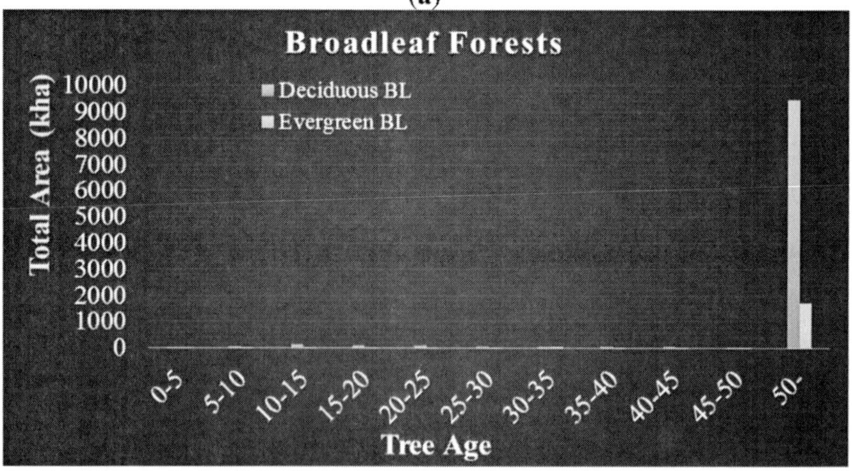

(b)

Figure 12. Statistical distribution of forest area by tree age for (**a**) coniferous and (**b**) deciduous/evergreen broadleaf forests in Japan.

4.3.2. CO_2 Sequestration by the Forests of Japan

Estimations of CO_2 sequestration by forest type were obtained by multiplying the observed CO_2 sequestration rates from the Prefectural data by the total area of each forest type's tree age category. Table 3, Table 4 and Table 5 show the total CO_2 sequestration estimates for coniferous, deciduous broadleaf, and evergreen broadleaf forest age categories, respectively.

Figure 13 shows graphs of the sequestration values among each tree age for (a) coniferous and (b) broadleaf forests. Together with the calculated sequestration value, a mean sequestration value of each forest type from the reference by Tadaki and Hachiya [15] has been plotted. The coniferous forests show a peak value at approximately 15–20 years of age and then gradually decrease to a value similar to that of the 0–10 year range when they reach 75 years. Each site dependency for the growth rate clearly shows a difference in the total sequestration amount, where compared with trees in low growth rate areas, high growth rate area trees have approximately a 10 $t{\cdot}CO_2/ha/yr$ difference at the peak value. For the broadleaf forests in this study, the results do not show a large difference between evergreen and deciduous broadleaf forests. The peak value can be observed at approximately 15–20 years of age, which gradually decreases, similar to the coniferous forests; however, the total sequestration value is much smaller than for coniferous forests. It is obvious that Japan has operated a large-scale plantation operation using sugi and hinoki trees because of their fast growth characteristics, which these values reflect. By using the detailed sequestration values for tree ages, sequestration overestimates that implicitly occur with the estimation method using averaged values [15] for coniferous trees of less than 10 years of age and for trees of more than 20 years of age, as well as the underestimation for the peak tree ages (15–20 years), can be avoided altogether. Similarly, for deciduous and evergreen broadleaf forests, compared to the averaged values for all ages, the values for the tree age categories show lower results.

Finally, Figure 14 shows the overall output of the final work, visually illustrating CO_2 sequestration values for forests in Japan.

4.4. CO_2 Sequestration Estimates and Errors

4.4.1. CO_2 Estimates and Potential Errors

The total CO_2 from the forests of Japan has been calculated by taking into consideration both forest type and tree age. It should be understood that determining the accuracy of this result is quite difficult, because there is no known "true value" for this information. It is also necessary to note that there are always sources of error when attempting to calculate such estimations. Possible errors that could generate inaccuracies in the results have been identified and listed here:

Table 3. CO_2 sequestration of coniferous forests by tree age using the PALSAR data and the stem volume method.

Forest Type	Age (years)	t·CO_2/ha/yr			Area (ha)			Total Mt·CO_2		
		Site 1	Site 2	Site 3	Site 1	Site 2	Site 3	Site 1	Site 2	Site 3
Coniferous	0–5 5–10	3.38	2.56	1.70	0	0	0	0	0	0
	10–15	13.23	9.57	6.20	0	0	0	0	0	0
	15–20	16.11	11.45	6.76	22.826	6.575	7.332	0.37	0.08	0.05
	20–25	12.21	8.49	5.83	592.573	69.564	97.357	7.24	0.59	0.57
	25–30	10.05	7.83	5.34	2.700.597	705.043	230.650	27.15	5.52	1.23
	30–35	8.16	6.74	5.21	989.309	1.168.617	904.797	8.07	7.87	4.72
	35–40	6.42	5.87	4.89	137.687	600.268	1.791.110	0.88	3.52	8.76
	40–45	5.61	5.07	4.41	31.225	122.804	1.023.606	0.18	0.62	4.51
	45–50	5.41	4.63	3.78	7.835	38.352	349.890	0.04	0.18	1.32
	50–55	4.84	4.20	3.30	2.179	5.265	287.697	0.01	0.02	0.95
	55–60	4.08	3.61	2.91	684	7.685	78.352	0.00	0.03	0.23
	60–65	3.35	2.92	2.29	183	877.37	15.174	0.00	0.00	0.03
	65–70	2.47	1.78	1.40	1.57	1.986	9.501	0.00	0.00	0.01
	70–75	2.17	1.41	1.03	0.79	190.96	9.136	0.00	0.00	0.01
	over 75	1.18	0.42	0.42	30.57	1.098	296.715	0.00	0.00	0.13
Total					4.485.131	2.728.325	5.101.317	43.95	18.44	22.52
Overall						12.314.773			85.0	

Table 4. CO_2 sequestration of deciduous broadleaf forests by tree age using the PALSAR data and the stem volume method.

Forest Type	Age (years)	t·CO_2/ha/yr	Area (ha)	Total Mt·CO_2
Deciduous Broadleaf	0–5	2.91	38.756	0.11
	5–10	6.33	58.850	0.37
	10–15	6.50	151.418	0.98
	15–20	6.89	91.892	0.63
	20–25	5.96	90.147	0.54
	25–30	5.20	78.751	0.41
	30–35	5.09	62.852	0.32
	35–40	3.79	48.437	0.18
	40–45	3.52	57.704	0.20
	45–50	3.13	36.219	0.11
	over 50	1.87	9.475.569	17.74
Total			10.190.332	21.61

Table 5. CO_2 sequestration of evergreen broadleaf forests by tree age using the PALSAR data and the stem volume method.

Forest Type	Age (years)	$t \cdot CO_2/ha/yr$	Area (ha)	Total $Mt \cdot CO_2$
Evergreen Broadleaf	0–5	2.46	35,540	0.09
	5–10	6.43	33,957	0.22
	10–15	6.57	15,752	0.10
	15–20	6.93	17,987	0.12
	20–25	6.27	20,771	0.13
	25–30	5.32	23,800	0.13
	30–35	5.38	68,927	0.37
	35–40	3.93	31,164	0.12
	40–45	3.74	36,252	0.14
	45–50	3.23	41,367	0.13
	over 50	1.87	1,716,706	3.21
Total			2,042,228	4.76

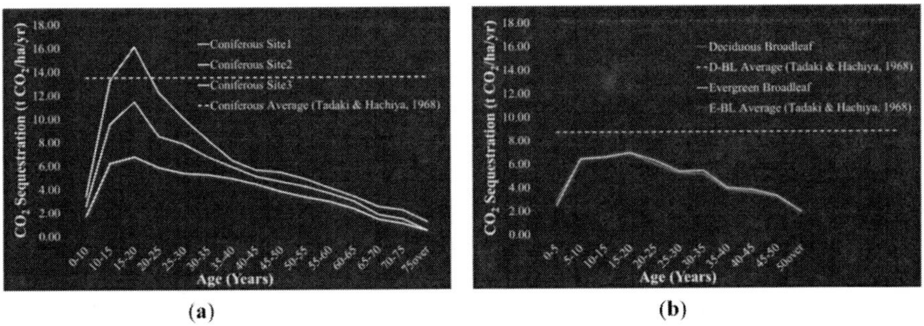

Figure 13. Trend of CO_2 values by tree age classes for (**a**) coniferous and (**b**) deciduous/evergreen broadleaf forests.

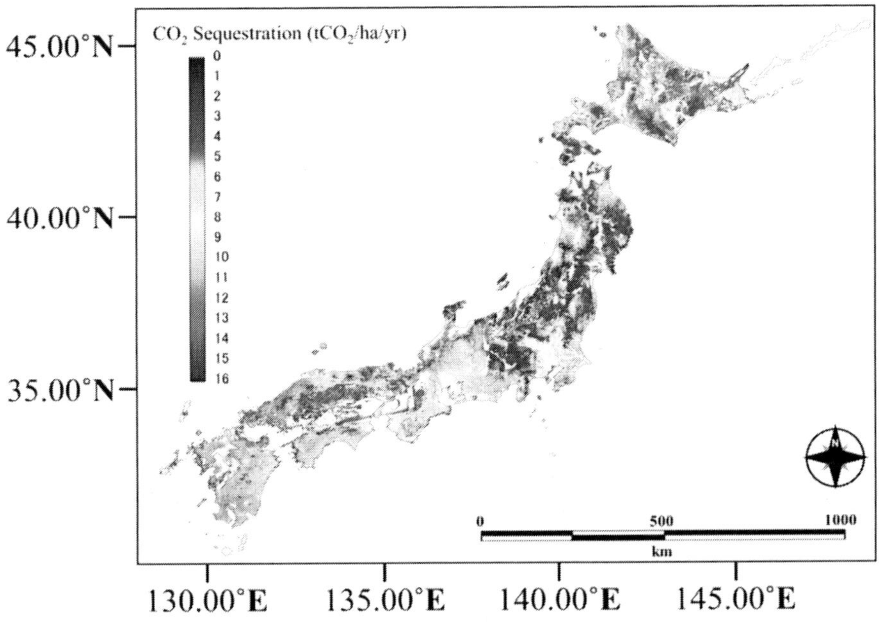

Figure 14. CO_2 sequestration map of Japan.

- Accuracy of the land cover map (forest cover map)
- Dependency on the backscattering model for stem volume estimation
- Stem volume-age relationship curve
- The sequestration value itself

Number 1 refers simply to errors resulting from the misclassification of each forest type. As acknowledged in the work, the total area size of a forest type is an important factor in calculating total CO_2 sequestration, because the sequestration value per unit area is multiplied by the total area of each forest type. Figure 15 illustrates the relationship between accuracy and errors for each forest type in reference to the error matrix (Table 2).

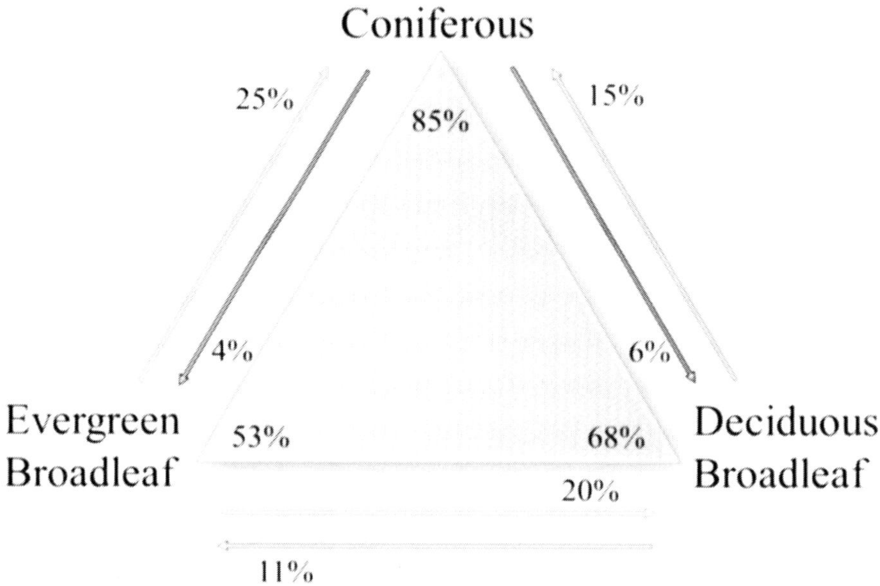

Figure 15. Triangular relationship of the forest classes and their accuracy and errors.

The figure shows the relationship between the classes used in the classification scheme. It indicates the accuracy (percentage value inside the triangle) and errors of omission (outside the triangle). The arrows indicate the misclassification of the classes that might have been classified into one class (beginning of the arrow) but may have instead been classified into a different class (end of the arrow). For example, coniferous forests may have had an 85% accuracy rate in classification, while 4% were misclassified into evergreen broadleaf, and 6% were misclassified into deciduous broadleaf forests. The same can be said for the other two classes. Thus, an 85% accuracy rate for coniferous may be true, but there are 4% and 6% chances of errors due to misclassification into evergreen and deciduous broadleaf forests, respectively, in calculating the total CO_2 amount. The same can be shown for the other classes, mutatis mutandis. The evergreen broadleaf class shows 20% and 25% chances of error from deciduous broadleaf and coniferous forests, respectively, while deciduous broadleaf shows 11% and 15% chances of error from evergreen broadleaf and coniferous forests, respectively.

For number 2, errors in the backscattering model can be explained by the RMSE calculated from the estimated stem volume versus the ground truth volume information (Figure 4). As has been explained in an earlier section, the RMSE of the backscattering model resulted in 105.58 m^3/ha. This can lead to errors in transforming the stem volume into the tree age information, and thus become an error in the estimation of CO_2 sequestration. For example, in the coniferous trees in high growth rate areas, a 100 m^3/ha difference can miss the true tree age by ±6 years. Larger differences accrue to lower growth rate areas, for example at Site 3 (low growth rate region) 100 m^3/ha can miss the tree age by ±13 years by looking at the stem volume-age relationship data from the Oita Prefectural Government. A 6-year difference might change the estimation values; for example, the forested coniferous areas at peak ages could be classified to earlier or older age ranges and vice versa: the older age classes could change to the peak tree age classes. Therefore, this error source must be acknowledged as one possible error in the estimation.

For number 3, the tree age transformation used the reference data provided by the Oita Prefectural Government and from Sasaki et al. [41]. According to this growth curve, the stem volume information was reclassified for the whole of Japan. The issue here is whether this growth curve also has implications for other regions in Japan. This study used the reference information as a representative growth curve for each tree type, and using the precipitation data, it has been assumed that the other regions of Japan would similarly lie along one of those growth curves. To make sure, additional references were collected to compare the curve information to other regions of Japan [42,43]. According to the tree growth curve information from the Kagoshima region (Kyushu district) and Aomori region (Tohoku district), there are differences in the growth curves depending on the different site productivity areas; however, comparing the two figures, a large difference between growth curves even in opposite environmental regions (the warmer Kagoshima region, and the cooler Aomori region) cannot be confirmed [42,43]. It is known that there is a most suitable growth curve for each Prefecture in Japan, but this study considered that by making an attempt to normalize the differences of the growth curves by using the precipitation data and setting the threshold. Thus it can be said that the growth curve differences among each region are averaged out and reduced to the least possible error in this aspect. Still, this is one point that could be considered as a potential error in the estimation.

Finally, for number 4, the sequestration information was modified from the Tochigi Prefectural Government by taking the mean or weighted mean of each tree type and using that value as representative of each forest type tree age class. The Prefectural data for the CO_2 sequestration is estimated using the guidelines commonly used for estimations in the forests of Japan, based on the following equation:

$$\text{Annual } CO_2 \text{ Sequestration (per ha)} = \text{Annual Growth } (m^3/ha) \times \text{Biomass Expansion Factor} \times \text{Bulk Density} \times \text{Carbon Content} \times CO_2 \text{ Equivalent Factor} \tag{6}$$

Annual growth is measured by stem volume growth. The biomass expansion factor is used for expanding the stem volume into an estimate for the whole volume of the tree, including the volume of leaves, branches and roots together. This value usually differs among tree types and species. Bulk Density represents the weight per cubic meter of volume. Carbon content commonly uses the value of 0.5, which means that the wood is composed of approximately 50% carbon. Finally, to convert the carbon value to a CO_2 sequestration value, the CO_2 Equivalent Factor of 44/12 is multiplied, where 12 and 44 are the molecular masses of carbon and CO_2, respectively.

An error that can occur from this is the sequestration value used in multiplying the total forest area, which would be different if it were actually broken down in further detail for each specific tree type and age. Some trees capture more or less CO_2 compared to other tree types within the same forest type, but those values were averaged for this study. Another possible error source is the minimum and maximum sequestration value for each tree age class. Because it is difficult to discriminate between forests using a one-year age distribution, the tree ages were reclassified into 5-year interval ranges. This study used the mean value within that range, and there are minimum and maximum values for those categories, and those can cause errors in the sequestration estimation as well. Table 6 shows the minimum and maximum sequestration values for each forest type at each tree age class range. As shown, there are some classes that have high sequestration values at the peak age of coniferous trees (10–15 years) at site 1, which is both a little higher than the average value from the evergreen broadleaf forests of Tadaki and Hachiya [15] and also similar to the deciduous broadleaf forests. Looking at all the ages, the biggest differences between maximum and minimum can be observed in the peak aged forest classes. This point is one of the more probable errors.

4.4.2. Actual CO_2 Estimates Considering the Errors

The CO_2 sequestration amount by the forests in Japan was calculated after each error was taken into consideration as discussed in Section 4.4.1. Table 7 shows the summary of the final output of the total CO_2 amounts when these errors are considered.

Table 6. CO_2 sequestration minimum and maximum values for each forest type and tree age class.

	Coniferous ($t \cdot CO_2$/ha/yr)					
	Site 1		Site 2		Site 3	
Age (years)	Min	Max	Min	Max	Min	Max
0–5 5–10	3.26	3.49	1.95	2.75	1.23	1.84
10–15	11.75	18.43	8.70	11.78	5.38	8.64
15–20	14.23	18.08	9.90	13.15	5.99	8.88
20–25	11.13	13.23	7.70	9.49	5.29	7.31
25–30	9.55	10.93	7.50	8.20	4.68	5.98
30–35	7.99	8.71	6.39	6.98	5.03	5.95
35–40	6.30	6.71	5.63	6.61	4.55	5.27
40–45	5.19	5.91	4.61	5.40	3.85	4.78
45–50	4.37	5.89	3.79	5.01	3.46	4.51
50–55	3.44	5.75	3.23	4.94	3.02	3.78
55–60	2.79	5.82	2.72	4.55	2.58	3.78
60–65	2.08	5.28	2.01	3.65	1.71	2.93
65–70	1.32	4.35	1.04	2.77	0.97	1.98
70–75	1.32	3.89	0.97	2.39	0.97	1.22
75 over	0.59	2.06	0.22	0.79	0.24	0.53

	Deciduous Broadleaf ($t \cdot CO_2$/ha/yr)		Evergreen Broadleaf ($t \cdot CO_2$/ha/yr)	
Age (years)	Min	Max	Min	Max
0–5	1.72	4.16	1.56	3.44
5–10	3.80	9.92	3.43	10.29
10–15	5.06	8.44	5.14	8.57
15–20	5.17	8.33	5.14	8.57
20–25	5.08	7.20	5.32	7.60
25–30	3.97	5.56	4.18	5.70
30–35	3.60	5.45	3.80	5.70
35–40	2.12	4.81	2.28	4.94
40–45	1.86	4.34	1.90	4.56
45–50	1.86	3.70	1.90	3.80
50 over	0.60	3.41	1.50	3.61

In Table 7, the summary of the total CO_2 calculated from this study is listed when each of the error cases is considered. Four different cases of potential errors were examined: land cover misclassification, reliability of the backscattering model, dependency of the tree age curve, and the sequestration value. The tree age curve is listed as N/A because it is simply the errors caused by the growth trend of the trees, and giving an estimation by using other Prefectural Governments' tree age curve information will give results similar to those obtained using the Oita data. However, for the errors from land cover mapping misclassification, backscattering model and sequestration values, and after considering the backscattering model and sequestration value errors together, the estimation was calculated.

In the table, Avg_{val} Min_{val} and Max_{val} represent the average, minimum and maximum sequestration value used for the estimation, respectively. Min_{tot} and Max_{tot} represent the minimum total and maximum total CO_2 sequestration by each category, respectively. Thus, for backscattering model error cases in the coniferous category, the total CO_2 amount was estimated using the average sequestration value (the value listed in Table 3). Values in brackets written after N/A are the total sequestration outputs listed in the tables of deciduous and evergreen broadleaf forests (Table 4 and Table 5).

Errors of misclassification were considered by using the percentage of the total area of each distributed forest class modified based on the actual misclassification percentage shown in Figure 15. For example, 4% and 6% of coniferous classes were misclassified to evergreen broadleaf and deciduous broadleaf forests, respectively. Thus, 4% and 6% of the total area from those forest classes, respectively, were subtracted and added to the coniferous class (this is determined as the maximum status). At that point, the deciduous and evergreen broadleaf status was determined as the minimum status. On the other hand, 25% and 15% were misclassified to the coniferous class when they were likely to be evergreen and deciduous broadleaf forests, respectively. Similarly, those percentages of the total area were subtracted from the coniferous forests and added to the evergreen and deciduous broadleaf forest classes (this is determined as the minimum status). Again, at this point, the maximum deciduous and evergreen broadleaf statuses were determined. The dependency of the classification results to errors in estimates of sequestration were assumed in this method. As a result, the minimum total sequestration was 75.80 $Mt{\cdot}CO_2$, while the maximum was 127.11 $Mt{\cdot}CO_2$. Here, the sequestration value by tree age class for each forest type does not change, and only the total area of the forests fluctuates. We can see how the total size of the forested area is one important factor that contributes to total sequestration.

Table 7. Summary of total CO_2 sequestration estimated after considering each possible error source.

Error Types / Forest Types	Coniferous		Deciduous Broadleaf		Evergreen Broadleaf		Total CO_2 Sequestration	
	Unit: $Mt{\cdot}CO_2$							
Land Cover Mapping	Min_{tot}: 50.94	Max_{tot}: 89.67	Min_{tot}: 20.29	Max_{tot}: 25.50	Min_{tot}: 4.57	Max_{tot}: 11.94	Min_{tot}: 75.80	Max_{tot}: 127.11
Backscattering Model	(Avg_{val}) Min_{tot}: 65.46	(Avg_{val}) Max_{tot}: 99.81	N/A (21.61)		N/A (4.76)		Min_{tot}: 91.83	Max_{tot}: 126.18
Tree Age Curve	N/A		N/A		N/A		N/A	
Sequestration Value	Min_{tot}: 78.80	Max_{tot}: 90.97	Min_{tot}: 7.38	Max_{tot}: 31.94	Min_{tot}: 6.55	Max_{tot}: 14.78	Min_{tot}: 92.73	Max_{tot}: 137.69
Backscattering Model + Sequestration Value	(Min_{val}) Min_{tot}: 64.65 (Max_{val}) Min_{tot}: 75.03	(Min_{val}) Max_{tot}: 91.45 (Max_{val}) Max_{tot}: 110.64	N/A (21.61)		N/A (4.76)		(Min_{val}) Min_{tot}: 91.02 (Max_{val}) Min_{tot}: 101.4	(Min_{val}) Max_{tot}: 117.82 (Max_{val}) Max_{tot}: 137.01

Errors of the backscattering model were considered from the developed RMSE model, which resulted in 105.58 m^3/ha. According to the stem volume-tree age curve information, the difference of stem volume results in 6-year, 7-year and 10-year differences for the coniferous trees on site 1, site 2 and site 3, respectively. Using this information, the estimation of the CO_2 in coniferous forests was re-assessed by shifting the age range by the above tree age year difference. The minimum and maximum then became 65.46 and 99.81 Mt·CO_2, respectively. The total CO_2 was then calculated using these values along with the total values of other forest types (deciduous broadleaf (DBL) and evergreen broadleaf (EBL)). The estimation was made only for the coniferous forests in this case because the actual error values of the model developed by Wijaya [38] were not found.

For the sequestration value errors listed in Table 6, the minimum and maximum values were applied, multiplying by the total forested area of each forest type. Overall total CO_2 resulted in 92.73 Mt·CO_2 and 137.69 Mt·CO_2 for minimum and maximum totals, respectively. This case is easier to interpret compared to the other errors because all the forest classes can be estimated clearly.

Finally both the errors from the backscattering model and the sequestration value were added, and then the estimation was re-assessed. This was conducted only for the coniferous class, and resulted in 64.65 Mt·CO_2 and 91.45 Mt·CO_2 for minimum and maximum amounts, respectively, when the minimum sequestration value (Table 6) was used, and 75.03 Mt·CO_2 and 110.64 Mt·CO_2 for minimum and maximum amounts, respectively, when the maximum sequestration value (Table 6) was used.

5. DISCUSSION

5.1. Stem Volume Empirical Modeling

In Section 4.1, an empirical backscattering model was developed to estimate stem volume information across the forests of Japan. The challenge is that backscattering can have multiple causes within a forested area, and can be affected by species, structures, soil moisture [44], density of forests [18], etc. A simple empirical modeling method was used by taking a correlation analysis between the backscattering intensity and the ground reference data. We have avoided most of the possible distracting factors affecting the difference in the backscattering to make a more accurate trend between the parameters and SAR image, which allowed a better interpretation when compared to many studies performed in similarly rugged areas [24]. We could discuss about the limitation of L-band SAR for the estimation of biomass, due to its sensitivity which gives saturation at a certain biomass level [45], however the most important variable in our study is the topographic effect. Although various radiometric corrections have been developed to normalize the SAR imagery [19,26,27,46] the radiometric accuracies of those correction methods can always be challenged. However, without their application, the topographic effects caused by rugged terrain will remain a crucial issue, especially for the study area here in Japan, and can result in higher errors compared to not applying them. There might be solutions to improve the interpretation of the modeling, but in this first study on national-scale estimation using remote sensing, we believe that the result can serve as a standard. Note that during the relationship analysis, speckle filtering was not applied to the PGM data. This decision was made because the data were downscaled to 25 m resolution from the original, which averaged out the area, reducing the noise. Further filtering might lose the actual information from the surface objects. However, in the future, a different filtering method might be applied to evaluate whether the difference affects the results.

5.2. Land Cover Mapping of Japan

A land cover map was produced that not only defined the areas as simple forests and non-forested areas but also provided more detail (i.e., subclasses) such as forest types (coniferous, deciduous and evergreen broadleaf), so the information could be used in many different ways. The need for a detailed digital land cover (or digital vegetation model) map, especially for the vegetation categories, is important for obtaining more precise estimations of CO_2 sequestration values. For this study, only single-scene data sets were used to develop the land cover map for each area. Sanga-Ngoie et al. [5] states the usefulness of applying temporal data sets to improve land cover accuracy due to the clear difference of radiance values that change during different seasons of the year. In the case of forest classes, similar spectral signatures would be observed throughout a season after blooming finishes, but clear and distinguishable differences can be observed for different forest types, such as trees whose leaves fall in the winter. Such methods were not applied in this case because few cloud-free products were captured, which made it difficult to find a set of scenes with two different seasons for each location. Operation of Landsat 8 is still in its third year, so its scene database is still relatively small compared to Landsat 5 or 7, so such methods are currently limited to very small areas or must wait for more data to be collected over time.

However, because of the lack of temporal data scenes, the classification process was performed not by increasing the number of images but by collecting as many detailed training sites as possible. This method led to the ability to discriminate between forest classes in more detail and with more precision, because constructing more detailed and precise land cover classes resulted in greater chances of unclear pixels being classified into the forest class to which they should correctly belong. For example, if the training sites of the forest classes are specified and classified only as "coniferous, deciduous broadleaf and evergreen broadleaf", deciduous and evergreen broadleaf forest classification drop inaccuracy by 33% and 36%, respectively, compared to when more detailed training classes of those forest types are constructed. Therefore, this methodology is effective when there are few scenes available for classifying a region. Constructing detailed training sites can improve the categorical map. In the future, we recommend reprocessing the classification of areas that had lower accuracy results or regions that were covered by snow/cloud to improve and correct the land cover types used here.

5.3. Stem Volume Method for CO_2 Sequestration Estimates

5.3.1. CO_2 Sequestration Values and Trends

The conventional method [15], when used alone for estimating CO_2 sequestration, results in large errors in the total amount compared to methods that give special attention to tree age. The stem volume method for discriminating tree ages within each forest type clearly shows the importance of this variable for avoiding over- or underestimations, depending on the growth stage of the forests examined. The use of L-band SAR data rather than optical data [5] made this discrimination possible, yielding sequestration rates by tree age for trees up to 75 years of age for coniferous and 50 years of age for deciduous and evergreen broadleaf forests. It is thus clearly established that sequestration rates of 13.5 t·CO_2/ha/yr [15] are only valid for 10–15-year-old coniferous forests in high growth rate regions. In contrast, deciduous broadleaf forests reach just below the 8.7-t·CO2/ha/yr threshold only in the 15–20-year-old range. For evergreen broadleaf forests, this is a large issue when the conventional method is applied, because it shows approximately 3 times greater value than observed in the results of this study, meaning that there would be a large overestimation of the forest carbon sink capacity if such a value is applied. High sequestration values from evergreen broadleaf forests are difficult to confirm for many reasons; however, the amount of sequestration by evergreen trees in Japan is acknowledged to depend on total solar radiation and air temperature. Gross Primary Product (GPP) results from the total radiation, while the Net Primary Product (NPP) is set by the amount of respiration subtracted from the GPP [47]. In this case, higher air temperature leads to higher respiration and decreases the NPP. On the other hand, when both total radiation and air temperature are low, respiration rates decrease but NPP becomes the same or even greater compared to higher radiation periods [37]. Depending on the season and location of the observation, the average value of evergreen broadleaf forests will differ, and it might well be possible that the large value obtained by Tadaki and Hachiya [15] resulted from such causes.

It is very important to note that the averaged sequestration value for the coniferous forest type matches the sequestration value by tree ages obtained here for the 10–15-year-old coniferous trees at Site 1. This is also the category of the second fastest-growing period for coniferous trees at that site. The same is also true for deciduous broadleaf forests for 15–20-year-old trees; it is just a little lower than the average value. This strongly implies that an estimation based on the averaged values alone will round up the values for trees in different growth stages, causing an overall overestimation of the carbon sink capacity [5], both for natural and planted forests. The results emphasize the fact that sequestration differs among tree ages. This reality should be taken into consideration in all cases, otherwise, erroneous values of the forests sequestration capacity will be obtained, generating results that possibly have errors from using less accurate values.

The uncertainty assessment gave a clear view of how and why the estimations can fluctuate, where the estimated total CO_2 lies between the minimum 75.80 Mt·CO_2 and maximum 137.69 Mt·CO_2. The minimum estimate results from land cover error and the maximum from sequestration value error. After examining the results presented in this paper, it can be assumed that these two parameters are the most important factors when estimating sequestration amounts. Focusing on the results where 111.27 Mt·CO_2 was shown, the fluctuation of the estimated amount is a difference of approximately 23 to 32%. Certainly, if the errors for the DBL and EBL classes for the backscattering model errors could be obtained, some change might be observed in the results; however, it may be determined that even if this occurred, it would not result in a very large difference from the current numbers. Through the uncertainty assessment, a better understanding of the fluctuation of the CO_2 trend became clear, and thus this result leads to a strong confidence in the practical usage of this methodology.

5.3.2. Validity of the Method in Comparison

For further knowledge, other estimation methods were investigated and compared to the results of this study. Four main methods were examined and compared; one is from this study, the second is from the conventional method by Tadaki and Hachiya [15], the third is the work conducted by Sasaki and Kim [7], and the last is from the National Institute for Environmental Studies (NIES) [8] of the IPCC report for the forest carbon sink. The total CO_2 estimation is summarized in Table 8 [7,8,15] for all four methods. For the conventional method, the sequestration value was used along with the total forest areas obtained from our work.

Table 8. Summary of the total CO_2 sequestration estimated by various methodologies in the different studies.

Methods	Total CO_2 Sequestration
Conventional (Tadaki and Hachiya [15])	308.51 $Mt{\cdot}CO_2$
Sasaki and Kim [7]	73.7 $Mt{\cdot}CO_2$
NIES [8]	77.67 $Mt{\cdot}CO_2$
Stem Volume Method (our work)	111.27 $Mt{\cdot}CO_2$

The stem volume method used in this study, shows a result of 111.27 $Mt{\cdot}CO_2$ by considering both forest type and tree age information (coniferous: 85.0 $Mt{\cdot}CO_2$; deciduous broadleaf: 21.61 $Mt{\cdot}CO_2$; evergreen broadleaf: 4.76 $Mt{\cdot}CO_2$). With the conventional method [15] using the averaged value compared to the results in this study, a large overestimation of the sequestration can be observed, because the sequestration value dumps everything into one category, omitting tree age information and ignoring lower sequestration stages, leading to high overestimation of the result. Sasaki and Kim [7] and NIES [8] show similar estimation values, however for the former study, the result is obtained by using the land use model and carbon stock growth model and estimates the sequestration value classifying the forests as planted and natural forests. Planted forests showed an estimation of 56.1 $Mt{\cdot}CO_2$, while natural forests were estimated at 17.6 $Mt{\cdot}CO_2$, and then summed for a total result of 73.7 $Mt{\cdot}CO_2$. The most interesting result is from the NIES, whose approach used forest registration data from each prefectural government and differentiated the forests by tree type, collecting annual growth information for each, and then performing an estimation by multiplying the annual sequestration value by the total forested area of each tree type, which is similar to this work. They have not only made calculations for aboveground sequestration but also for the litter and soil. However, the majority of the sequestration results from the aboveground biomass (aboveground: 76.53 $Mt{\cdot}CO_2$; litter: 0.41 $Mt{\cdot}CO_2$; soil: 2.58 $Mt{\cdot}CO_2$; dead tree: −1.85 $Mt{\cdot}CO_2$). If we assume that the methods used by the NIES [8] study are likely to be the most accurate estimation of the natural environment, the remote sensing method applied in this study shows only a 43% overestimation in comparison. This difference can be explained from the errors discussed in the previous section. If methods are developed to reduce those potential errors, our results may become somewhat similar to the results by the NIES. However, even though the NIES results raise some questions, such as the uncertainty of the total forest area, which is important as mentioned, total CO_2 is calculated by multiplying the sequestration value by the total forested areas. The data from prefectural government is sometimes unclear, especially in terms of the area that the forest covers, as researchers have noted (the uncertainty rate of NIES estimates shows 5% for forest areas without discriminating by forest type [8]), so there is a possibility that the NIES is underestimating forest capacity. What is significant about this work compared to the NIES is that this method can be performed throughout various seasons in the future in less time and with lower labor costs through the use of remote sensing data. The work performed by NIES is possible in countries where a large database of forest information is compiled for use in applications, but in countries where this is not the case, it will be very difficult to implement such a method. On the other hand, the method developed in this study can be applied in any country as long as the required data sets are available. Moreover, the uncertainty of the total forest area compared to NIES can be overcome using our methodology, because the correct amounts and locations of forests are easier to discern. Overall, this study explains why the result is much lower (better) than the conventional method, which seems to result in high overestimation, while when compared to NIES results, the estimation arrived at by this study does not show serious underestimation. The true result likely falls between the two as discussed in this paper. The results achieved using the stem volume method in this study show both an improvement and the potential for obtaining even better results than other methods by using remote sensing method. As a note, the term sequestration used in this study can be described as NPP, while the NIES result can be described as Net Ecosystem Product (NEP) because it also considers decomposers. In the future, improvements in the methodology will likely take this into consideration and obtain NEP values of the forests so we can estimate a more precise value not only for use in forest management and policy but also for use in modeling the biogeochemical cycle of terrestrial ecosystems.

6. CONCLUDING REMARKS

Rapid change in climatic conditions in recent years has raised many pressing issues and problems that need to be solved. These are of global interest among researchers, companies, politicians and others. Therefore, in response, various measures are being taken, such as the Kyoto Protocol, REDD+ and others, to mitigate the impacts of climate change. Although such policies and actions are being implemented around the world, the quantitative estimation of how well they have contributed to the natural environment, such as in terms of carbon sequestration, is still under discussion. Meanwhile, this work has shown a path forward in quantitative estimates on a national scale for CO_2 sequestration by utilizing fine resolution satellite imagery through a local/regional scale analysis. Focusing on forest type and tree age information using the LANDSAT and PALSAR data, respectively, yields more accurate data. This methodology gives better interpretations and more precise estimations of CO_2 sequestration, as it is clear that without both tree type and tree age discrimination, overestimations may well result.

The CO_2 amount and other important information revealed in this study has provided important data. Do old mature trees sequestrate as much as younger trees? The answer is no when we see the trend of the sequestration as a function of tree age. Using this information, we can discuss policies such as the REDD+ procedure. Is planting trees for 100 years' time [48] for a credit really an effective and correct way to manage forest resources? Rather, it could be suggested that planting faster growing trees in a period of 20 to 30 years and then replacing them continuously might be more sustainable and effective, even in economic terms. The findings in this paper can thus suggest these types of ideas to policy makers for more sustainable development/conservation.

The methodology is not only of interest for forestry, stakeholders, and policy makers but also for researchers who address modeling of the earth's ecosystems. The knowledge of CO_2 trends in the terrestrial biosphere developed from this work contributes to a better understanding of the biogeochemical cycle and thus sheds more light on the influence of the terrestrial ecosystem and its interaction with the atmosphere. This could lead to a step forward in understanding more about climatic systems. Utilizing fine resolution satellite imagery on a more local/regional scale can be expected to improve modeling accuracy and help better understand the trends of the earth's systems.

The in-depth quantification of forest sequestration can play an important role in many fields, helping them to improve how they set strategic actions into operation. Promoting such methods will certainly support further development and discussion to make a difference in solving or mitigating the issues faced in a more effective way.

ACKNOWLEDGMENTS

The authors would like to give their heartfelt appreciation to Kazadi Sanga-Ngoie for cooperating to strengthen the methodology of this study. The authors would also like to thank the Oita Prefectural Government, Forest Management Division Department of Agriculture Affairs along with the Chiba Prefectural Government for providing the forest inventory data.

AUTHOR CONTRIBUTIONS

K.I. and R.T. conceived this research. K.I. collected the data, performed and analyzed the experiments and drafted the manuscript. R.T. supervised throughout the research and edited the manuscript.

REFERENCES

1. Hewitt, C.N.; Jackson, A. Handbook of Atmospheric Science: Principles and Applications; Blackwell: Padstow, UK, 2003.
2. Chiras, D.D. Environmental Science, 7th ed.; Jones and Bartlett: Massachusetts, UK, 2006.
3. Ministry of Agriculture, Forestry and Fisheries (MAFF) of Japan (2010). Annual Report on Forest and Forestry in Japan (Summary). Available online: http://www.rinya.maff.go.jp/j/kikaku/hakusyo /22hakusho/pdf/22_e.pdf (accessed on 6 August 2015).
4. Ministry of Economy, Trade and Industry (METI) of Japan (2005). Kyoto Protocol Target Achievement Plan. Available online: http://www.meti.go.jp/policy/energy_environment/global_warming/study_policy.html (accessed on 6 August 2015). (In Japanese).
5. Sanga-Ngoie, K.; Iizuka, K.; Kobayashi, S. Estimating CO_2 sequestration by forests in Oita Prefecture, Japan, by combining Landsat ETM+ and ALOS satellite remote sensing data. Remote Sens. **2012**, 4, 3544–3570.
6. Hiroshima, T.; Nakajima, T. Estimation of sequestered carbon in article-3.4 private planted forests in the first commitment period in Japan. J. For. Res. **2006**, 11, 427–437. [Google Scholar] [CrossRef]
7. Sasaki, N.; Kim, S. Biomass carbon sinks in Japanese forests: 1966–2012. Forestry **2009**, 82, 113–123.
8. National Institute for Environmental Studies Japan (NIES), Center for Global Environmental Research (CGER). National Greenhouse Gas Inventory Report of Japan. Available online: http://www.gio.nies.go.jp/aboutghg/nir/2014/NIR-JPN-2014-v3.0.pdf (accessed on 6 August 2015).
9. Gibbs, J.K.; Brown, S.; Niles, J.O.; Foley, J.A. Monitoring and estimating tropical forest carbon stocks: Making REDD a reality. Environ. Res. Lett. **2007**, 2, 1–13.
10. Goetz, S.J.; Prince, S.D.; Goward, S.N.; Thawley, M.M.; Small, J. Satellite remote sensing of primary production: An improved production efficiency modeling approach. Ecol. Model. **1999**, 122, 239–255.
11. Pachavo, G.; Murwira, A. Remote sensing net primary productivity (NPP) estimation with the aid of GIS modelled shortwave radiation (SWR) in a southern African Savanna. Int. J. Appl. Earth Obs. Geoinf. **2014**, 30, 217–226.

12. Zheng, G.; Chen, J.J.; Tian, Q.J.; Ju, W.M.; Xia, X.Q. Combining remote sensing imagery and forest age inventory for biomass mapping. J. Environ. Manag. **2007**, 85, 616–623.

13. Turner, D.P.; Ritts, W.D.; Cohen, W.B.; Gower, S.T.; Running, S.W.; Zhao, M.; Costa, M.H.; Kirschbaum, A.A.; Ham, J.M.; Saleska, S.R.; et al. Evaluation of MODIS NPP and GPP products across multiple biomes. Remote Sens. Environ. **2006**, 102, 282–292.

14. Turner, D.P.; Guzy, M.; Lefsky, M.A.; Ritts, W.D.; Tuyl, S.V.; Law, B.E. Monitoring forest carbon sequestration with remote sensing and carbon cycle modeling. Environ. Manage. **2004**, 33, 457–466.

15. Tadaki, Y.; Hachiya, K. Forest Ecosystems and Their Productivity; Ringyo Kagakugijutsu Shinkosho: Tokyo, Japan, 1968. (In Japanese)

16. Ministry of Agriculture, Forestry and Fisheries (MAFF) of Japan (2010). Function of the Sequestration of Forests/Trees. Available online: http://www.rinya.maff.go.jp/kinki/hyogo/mori-grow/mori-co2.html (accessed on 6 August 2015). (In Japanese).

17. Ministry of Environment (MOE) of Japan. Vegetation Distribution of Japan. Available online: http://www.vegetation.biodic.go.jp /zu/index.html (accessed on 9 September 2015). (In Japanese).

18. Iizuka, K.; Tateishi, R. Simple relationship analysis between L-band backscattering intensity and the stand characteristics of sugi (Cryptomeria japonica) and hinoki (Chamaecyparis obtusa) trees. Adv. Remote Sens. **2014**, 3, 219–234.

19. Castel, T.; Beaudoin, A.; Stach, N.; Stussi, N.; Le Toan, T.; Durand, P. Sensitivity of space-borne SAR data to forest parameters over sloping terrain. Theory and experiment. Int. J. Remote Sens. **2001**, 22, 2351–2376.

20. Dobson, M.C.; Ulaby, F.T.; Le Toan, T.; Beaudoin, A.; Kasischke, E.S.; Christensen, N. Dependence of radar backscatter on coniferous forest biomass. IEEE Trans. Geosci. Remote Sens. **1992**, 30, 412–415.

21. Harrell, P.; Kasischke, E.S.; Bourgeau-Chavez, L.L.; Haney, E.; Christensen, N.L. Evaluation of approaches to estimating aboveground biomass in southern pine forests using SIR-C data. Remote Sens. Environ. **1997**, 59, 223–233.

22. Saatchi, S.S.; Moghaddam, M. Estimation of crown and stem water content and biomass of boreal forest using polarimetric SAR imagery. IEEE Trans. Geosci. Remote Sens. **2000**, 38, 697–709.

23. Sandberg, G.; Ulander, L.M.H.; Fransson, J.E.S.; Holmgren, J.; Le Toan, T. L- and P-Band backscatter intensity for biomass retrieval in hemiboreal forest. Remote Sens. Environ. **2011**, 115, 2874–2886.

24. Chen, E.; Li, Z.Y.; Ling, F.L.; Lu, Y.; He, Q.S.; Fan, F.Y. Forest volume density estimation capability of ALOS PALSAR data over hilly region. In Proceedings of 4th International Workshop on Science and Applications of SAR Polarimetry and Polarimetric Interferometry (PolInSAR), Frascati, Italy, 26–30 January 2009.

25. Small, D. Flattening gamma: Radiometric terrain correction for SAR imagery. IEEE Trans. Geosci. Remote Sens. **2011**, 49, 3081–3093.

26. Japan Aerospace Exploration Agency (JAXA). PALSAR Calibration Factor Updated. Available online: http://www.eorc.jaxa.jp/en/about /distribution/info/alos/20090109en_3.html (accessed on 6 August 2015).

27. Shimada, M. Ortho-Rectification and slope correction of SAR data using DEM and its accuracy evaluation. IEEE J. Sel. Topics Appl. Earth Obs. **2010**, 3, 657–671. [Google Scholar] [CrossRef]

28. Iizuka, K.; Tateishi, R. Analysis of backscattering characteristics of L-band SAR over the mountainous region of Chiba Japan, using 50 m PALSAR mosaic product. In Proceedings of International Symposium on Remote Sensing, Chiba, Japan, 22–26 April 2013.

29. Japan Aerospace and Exploration Agency (JAXA). PALSAR 50 m Orthorectified Mosaic Product. Available online: http://www.eorc.jaxa.jp/ALOS/en/kc_mosaic/kc_mosaic.htm (accessed on 6 August 2015).

30. Cartus, O.; Santoro, M.; Kellndorfer, J. Mapping forest aboveground biomass in the northeastern United States with ALOS PALSAR dual-polarization L-band. Remote Sens. Environ. **2012**, 124, 466–478.

31. Santoro, M.; Beer, C.; Cartus, O.; Schmullius, C.; Shvidenko, A.; McCallum, I.; Wegmüller, U.; Wiesmann, A. Retrieval of growing stock volume in boreal forest using hyper-temporal series of Envisat ASAR ScanSAR backscatter measurements. Remote Sens. Environ. **2011**, 115, 490–507.

32. Zhou, Z.-S.; Lehmann, E.; Wu, X.; Caccetta, P.; McNeill, S.; Mitchell, A.; Milne, A.; Tapley, I.; Lowell, K. Terrain slope correction and precise registration of SAR data for forest mapping and monitoring. In Proceedings of International Symposium for Remote Sensing of the Environment, Sydney, NSW, Australia, 10–15 April 2011.

33. Kobayashi, S.; Sanga-Ngoie, K. The integrated radiometric correction of optical remote sensing imageries. Int. J. Remote Sens. **2008**, 29, 5957–5985.

34. Li, W.; Du, Z.; Ling, F.; Zhou, D.; Wang, H.; Gui, Y.; Sun, B.; Zhang, X. A comparison of land surface water mapping using the Normalized Difference Water Index from TM, ETM+ and ALI. Remote Sens. **2013**, 5, 5530–5549.

35. Motohka, T.; Nasahara, K.N.; Oguma, H.; Tsuchida, S. Applicability of green-red vegetation index for remote sensing of vegetation phenology. Remote Sens. **2010**, 2, 2369–2387. [Google Scholar] [CrossRef]

36. Congalton, R.G.; Green, K. Assessing The Accuracy of Remotely Sensed Data: Principles and Practices, 2nd ed.; CRC/Taylor & Francis: New York, NY, USA, 2009.

37. Perry, D.A.; Oren, R.; Hart, S.C. Forest Ecosystems, 2nd ed.; The Johns Hopkins University Press: Baltimore, MD, USA, 2008.

38. Wijaya, A. Evaluation of ALOS-PALSAR mosaic data for estimating stem volume and biomass: A case study from tropical rainforest of central Indonesia. J. Geogr. **2009**, 2, 14–21.

39. Ministry of Environment (MOE) of Japan. National Survey on the Natural Environment. Available online: http://www.biodic.go.jp/english/J-IBIS.html (accessed on 6 August 2015).

40. Congalton, R.G.; Mead, R.A. A quantitative method to test for consistency and correctness in photointerpretation. Photogramm. Eng. Remote Sens. **1983**, 49, 69–74.

41. Sasaki, H.; Hayashi, H.; Ise, S. Considerations on the selective logging updating of evergreen broadleaf forests. Tokushima Prefect. For. Res. Cent. Rep. **1989**, 27, 22–29. (In Japanese).

42. Forestry Agency. Aomori Region Sugi Tree Rinbun-Shukaku-Hyou (1962). Available online: http://www.ffpri.affrc.go.jp /labs/shukakushiken/02gyoken/02gyoken_28.pdf (accessed on 6 August 2015). (In Japanese).

43. Forestry Agency. Kagoshima Region Sugi Tree Rinbun-Shukaku-Hyou (1965). Available online: http://www.ffpri.affrc.go.jp/labs /shukakushiken/02gyoken/02gyoken_33.pdf (accessed on 6 August 2015). (In Japanese).

44. Ulaby, F.T.; Sarabandi, K.; McDonald, K.; Whitt, M.; Dobson, C. Michigan microwave canopy scattering model. Int. J. Remote Sens. **1990**, 11, 1223–1253.

45. Baghdadi, N.; Le Maire, G.; Bailly, J.S.; Osé, K.; Nouvellon, Y.; Zribi, M.; Lemos, C.; Hakamada, R. Evaluation of ALOS/PALSAR L-band data for the estimation of Eucalyptus plantations aboveground biomass in Brazil. IEEE J. Sel. Top. Appl. Earth Obs. Remote Sens. **2015**, 8, 3802–3811.

46. Ulander, L.M.H. Radiometric slope correction of Synthetic-Aperture Radar images. IEEE Trans. Geosci. Remote Sens. **1996**, 34, 1115–1122.

47. Matsumoto, M. Carbon stock and carbon sequestration by the forests of Japan. Shinrin Kagaku. **2001**, 33, 30–36. (In Japanese).

48. Forestry and Forest Products Research Institute. Guidelines for Implementing REDD-plus ver. 1.1. Available online: http://www.ffpri.affrc.go.jp/reddrdc/en/reference/guideline/REDD+Guidelines_E_ver.1.1_20140327.pdf. (accessed on 6 August 2015).

Chapter 9

Modeling The Impacts of Climate Variability and Hurricane on Carbon Sequestration in a Coastal Forested Wetland in South Carolina

Zhaohua Dai[1], Carl C. Trettin[2], Changsheng Li[1], Ge Sun[3], Devendra M. Amatya[2], Harbin Li[2]

[1]*Earth Science Research Center, EOS, University of New Hampshire, Durham, USA; zdai@fs.fed.us*

[2]*Center for Forested Wetland Research, USDA Forest Service, Cordesville, USA*

[3]*Eastern Forest Environmental Threat Assessment Center, Southern Research Station, USDA Forest Service, Raleigh, USA*

ABSTRACT

The impacts of hurricane disturbance and climate variability on carbon dynamics in a coastal forested wetland in South Carolina of USA were simulated using the Forest-DNDC model with a spatially explicit approach. The model was validated using the measured biomass before and after Hurricane Hugo and the biomass inventories in 2006 and 2007, showed that the ForestDNDC model was applicable for estimating carbon dynamics with hurricane disturbance. The simulated results indicated that Hurricane Hugo in 1989 substantially influenced carbon storage immediately after the disturbance event. The simulated net ecosystem exchange (NEE) for the 58-year period (1950-2007) indicated that the hurricane reduced CO_2 sequestration due primarily to the increased decomposition of a large amount of litter and woody debris, including fallen trees (over 80% of pre-hurricane trees), debris and branches, and dead roots. The inter-annual fluctuation of soil CO_2 flux showed that the climate variability interfered substantially soil carbon dynamics in the forest. The results showed that there were substantial spatial and temporal differences in CO_2 flux (3.2 - 4.8 $Mg \cdot C \cdot ha^{-1}$) and wood biomass due to the differences in physical and biogeochemical characteristics in the forest.

Keywords: Greenhouse Gas; Forest-DNDC; Biogeochemical Model; Biomass; Climate Change

1. INTRODUCTION

The mean global surface air temperature has increased by 0.8˚C in last hundred years [1,2], and is increasing at a rate of 0.2˚C per decade in recent years [2]. The increase in surface air temperature may be mainly caused by increasing greenhouse gases (GHG) emissions with respect to human activities, such as land use change and fossil fuel consumptions. Forest ecosystems, including wetland ecosystems, are believed to be large carbon sinks [3-6]. Future climate change is likely to influence forest carbon sink strength [4]. Changes in temperature and/or precipitation can influence the forest hydrology regulating C balance in forest ecosystems, especially in forested wetland ecosystems where the hydrology is one of the most important elements impacting on C accumulation and consumption [3,7-9]. Understanding the interactions of wetland hydrology and carbon balances is highly needed to assess the role of forested wetlands in regulating global CO_2 that directly contribute to global warming.

Forested wetland ecosystems on Atlantic Coastal Plain in South Carolina, especially those forests in the firstorder headwater watersheds, are characteristic of the hydrology highly regulated by precipitation and evapotranspiration. However, observed climate warming and precipitation variability in the subtropical area have showed their potential in altering watershed hydrology [10,11]. For example, climate change can decrease water table level of the first-order forested wetland watersheds in southeast Atlantic coastal area due to an increase in evapotranspiration caused by temperature rise [12]. Water table decrease can lead to changes in vegetation species which have been observed in South Carolina [13], and the decrease can influence soil C storage in forested wetland ecosystems [3,7]. Pietsch et al. [7] found that water table decrease did not influence C sequestration to timber production in their forested wetlands. However, they found the forest soil C loss to water table decrease. This result indicates that global warming can influence soil C storage in the first-order forested watersheds dominated by a shallow water table, regulated by precipitation and evapotranspiration, on South Carolina coastal plain. Understanding the impact of climate variability on C dynamics in forested wetland ecosystems is needed to estimate the role of the ecosystems in mitigating global warming.

Except for anthropogenic factors affecting forest C balance and storage, many natural disasters, such as strong storms, flooding and wild fire, can substantially disrupt forest C dynamics. Although, it is not clear if global warming has increased the frequency and/or intensity of hurricanes, tropical storms and/or tropical cyclones [14], many scientists argued that global warming could affect those storms, including increasing their frequency and/or rising the energy [15-18] due to sea surface temperature increased by global warming. Those disastrous storms (categorized as 3, 4 and 5) can destroy forest. For example, Hurricane Hugo severely destroyed the forests on the Francis Marion National Forest in South Carolina in 1989 [19], the woody biomass reduced by over 80% [19]. Hurricane Katrina in 2005 destroyed forests along the Gulf coast [20-23], led to wood biomass loss 50 - 370 $m^3 \cdot ha^{-1}$ in swamp forests along the Gulf coast in Louisiana and Mississippi as reported by Middleton [23]. Accordingly, those disastrous storms can substantially influence C storage in forests in a short time period, such as within several years. Generally, only the direct impact of the disastrous storms has been considered as assess their damage. However, to understand a longer term impact of the storms may be valuable for estimating C balance in the forested ecosystems.

Computer models (e.g., MAESTRO [24], Forest-DNDC [25], BIOME-BGC [26], CABALA [27]) are very useful tools to assess C dynamics in forest ecosystems. However, some models, such as CABALA and Forest-DNDC, can perform better than some models [5,28]. ForestDNDC [25,29] is a process-based biogeochemical model used to simulate forest growth, C and N dynamics in forest ecosystems, including trace gas emissions (e.g., CO_2, N_2O, NO, NH_3, and CH_4), based on the balance of water, energy and nutrition in forest ecosystems [5,25, 29]. It integrates photosynthesis, decomposition, nitrification-denitrification, carbon storage and consumption, and hydro-thermal balance in forest ecosystems. The model has been tested to simulate C and N dynamics in forest ecosystems across different climatic regions, including boreal, temperate, subtropical and tropical [29- 34]. Recently this model has been modified such that it becomes a spatially explicit modeling tool for estimating C and N dynamics in forest ecosystems; the new version can utilize spatiotemporally physical and biogeochemical characteristics of catchments or regions [35], including soils, vegetation, hydro-geology and climate. The modified model is a flexible tool to assess carbon dynamics on watersheds with complex physical and biogeochemical characteristics. The size and shape of simulating units are flexible, ranging from a single plant-soil profile to watersheds [35]. The detailed modeling algorithm and parameterization can be found in some publications [5,25, 29-34,36-38].

This study used the Forest-DNDC with spatially explicit modeling framework [35] to assess 58-year (1950- 2007) C dynamics in a forested wetland landscape on the Santee Experimental Forest on coastal South Carolina plain by using onsite climatic and hydrologic observations, which was used to estimate the impacts of climatic variability and hurricane disturbance on the C sequestration in the forest. The spatially explicit physically-based hydrologic model MIKE SHE [39] was linked to the Forest-DNDC such that the differences in the spatiotemporally hydrologic and biogeochemical features in the catchment can be considered. The 60-year onsite climate observation showed a substantial climate fluctuation with an increase in air temperature at an average rate of 0.19˚C per decade and without substantially changing precipitation from 1950-2007 in this forest [11]; and a natural disaster, Hurricane Hugo categorized as IV storm, severely destroyed the forest of the watershed in 1989 [19,40]. These conditions were good for testing whether or not the model can capture the impacts of climate variability and hurricanes on C sequestration to this forest. The simulated results were compared to observations, which can be used to project the impact of future climate change on C storage in this type of forested watersheds.

2. MATERIALS AND METHODS

2.1. Study Watershed

Watershed WS80, located at 33.15˚N, 79.8˚W on the Santee Experimental Forest (SEF), 55 km northwest of Charleston, South Carolina, was used for this study (Figure 1). WS80 is first-order forested watershed with a mosaic landscape consisting of uplands and wetlands. Historically this forest was a part of agricultural wetlands (rice paddies) along Cooper River [41]. However, it was converted to a forested wetland in 19th century. The earlier trees were clearly cut in earlier 20th century [42]. The land was acquired by U.S. government in 1930's and used as forests. In 1938, SEF was established for the studies on forest and water. Thereby, it has gauging records since 1967.

WS80 serves as the control watershed for a paired watershed system within the second-order watershed (WS79; Figure 1) draining into Huger Creek, a tributary of East Branch of the Cooper River. The climate of this area is characteristic of the subtropical region of the Atlantic Coast with short, warm and humid winters and long and hot summers; the annual average temperature is 18.5˚C, and the mean annual precipitation (1946-2008) is 1370 mm [11]. The topography is planar, and the slope is less than 4%. The elevation is between 4 - 10 m above mean sea level. The watershed has a shallow water table, and over 23% of the watershed is classified as wetlands [43].

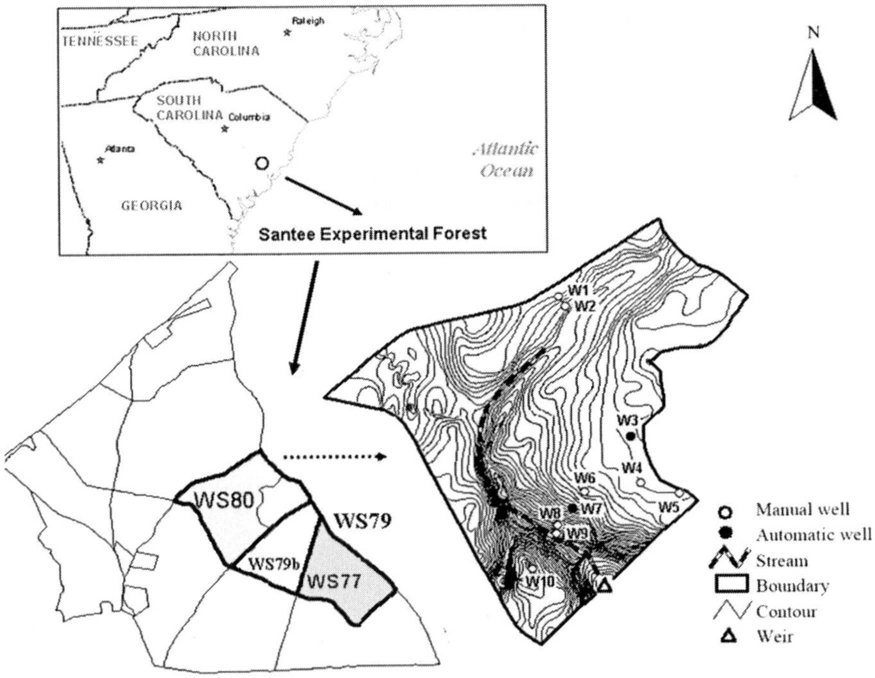

Figure 1. Location of the study watershed (WS80) on the Santee Experimental Forest, South Carolina, USA. The location of the groundwater wells is indicated on the topographic map figure as W1 through W10. WS79 is a paired watershed system; WS80 serves as a control catchment, WS77 as a treatment catchment, WS79b as a mixed watershed with treatment and control.

The soils developed in coastal plain sediments are hydric [44,45], and typified by a loam surface and clayey subsoil, which is moderately well to somewhat poorly drained in the upland and poorly drained in the riparian zone [46] (Figure 2(a)). Clay content is ≤30% in topsoil (within 30 cm), 40% - 60% in subsoil (>30 cm) [46]. Soil reaction is acidic; pH is between 4.5 and 6.5. Soil organic carbon content (SOC) is between 0.1% and 2.5% from uplands to riparian zone [47,48]. Organic matter (litter and duff, exclusive coarse woody debris) on the forest floor is 0.6 - 2.2 $kg \cdot m^{-2}$ (unpublished data).

As a reference forested watershed in the paired watershed system, WS80 has not been managed more than five decades. However, Hurricane Hugo categorized as IV storm passed through the Francis Marion National Forest in September of 1989, over 80% of dominant trees in this catchment were, therefore, broken or uprooted [19,49, 50]. After the hurricane, the watershed remained unmanaged, without biomass removal or salvage logging. The current forest cover is from residuals and natural regeneration, which consists of bottomland hardwoods in the riparian zone and mixed pine-hardwoods elsewhere (Figure 2(b)) [40]. The dominant trees are loblolly (Pinus taeda L.), sweetgum (Liquidambar styraciflua) and a variety of oak species (Queercus spp.) [19,40].

2.2. Data and Field Measurements

There are four weather stations on SEF. Daily precipitation and daily minimum and maximum temperature were measured manually at the weather station at the Santee Experimental Forest Headquarters (SHQ) between 1946 and 1996 and automatically afterwards; relative humidity, evaporation, wind speed, wind direction, vapor pressure, solar and net radiation were measured at

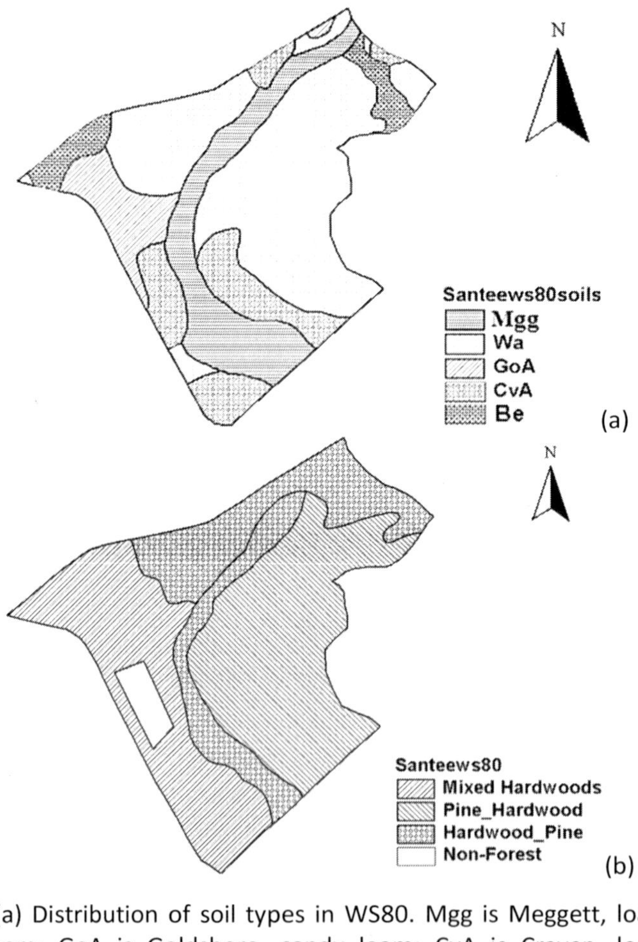

Figure 2. (a) Distribution of soil types in WS80. Mgg is Meggett, loam; Wa is Wahee, loam; GoA is Goldsboro, sandy loam; CvA is Craven, loam; Be is Bethera, loam; (b) Vegetation types on WS80.

30-minute intervals since 2003. The other three onsite weather stations (Met 5 on WS77, Lotti adjacent to WS79b and Met25 on WS80) started to manually record daily precipitation and temperature from 1963, 1964 and 1990, respectively, and automatically collect data at hourly intervals since 2003. Although the climate data from these stations were comparable, there were some differences in summer precipitation among the stations, especially storms.

However, some of precipitation and temperature data were missing at different stations and/or in different time periods before 2003 due to equipment failure for some time and natural disasters such as Hurricane Hugo in 1989. The missing temperature data were at first established by using the regression equations developed from the measurements of daily maximum and minimum temperature between 1972 and 2001 at SHQ and Lotti (about 3 km away from SHQ); and then the remaining missing temperature for SHQ were established by using the regression equations developed from the measurements between 1950 and 2008 at SHQ and Charleston International Airport (CHS) located about 50 km away from the SHQ, and the measurements between 1994 and 2008 at the SHQ and Moncks Corner (MC), about 20 km away. The missing precipitation data for simulating hydrology and C dynamics were filled by the measurements from the nearest weather stations in terms of comparing those precipitation measurements from 15 weather stations around SHQ, based on the assumption that there was no precipitation at SHQ if precipitation was found only at one side of SHQ and/or without an increase in observed steam flow, otherwise the precipitation at SHQ was the same as the precipitation at the nearest weather station or the average from the weather stations at both sides of SHQ.

Stream flow gauge heights above a compound V-notch weir were measured at 10-minute intervals in 1969-2007. The stream flow was calculated using standard rating curve method developed for the weir [51], and the 10- minute values were integrated into daily, monthly and annual flow in cubic meters per second (cms), and then normalized from cms to millimeters (mm) per day by dividing the watershed area to compare with daily precipitation [52]. However, a lot of flow data were missing due to equipment failures, which remained missing. Daily distributed water table depth was measured to calibrate and validate the distributed hydrologic model MIKE SHE [39] providing distributed water table depth for simulating C dynamics using Forest-DNDC. Thirty three manual wells were installed across the watershed with weekly measurements in 1992-1994; eight manual wells were selected from the 33 wells with biweekly measurements in 2003-2005. Two automatic wells were installed in an upland area and a lowland location to record water table depth at four-hour intervals since 2003.

Twenty four plots were set to measure tree height (TH, m) and diameter at breast height (DBH, cm) for a biomass inventory in the paired watershed in

2006 (unpublished data), in which four plots were used to measure TH and DBH in 2007. Eight plots were used to measure soil temperature, moisture, and soil CO_2 in 2006-2008. The plot size was about 1022 m^2 (fourth acre, 105 feet by 105 feet for the plots 1 - 16; 50 feet by 220 feet for the plots 17 - 24). Soil temperature, moisture, and soil CO_2 were measured at 4 - 6 collars in each plot. Soil temperature and moisture were measured by using probes inserted into soils 10 cm deep. Soil CO_2 was measured at monthly intervals using LI-COR 6200 portable photosynthesis system included a LI-6250 infrared gas analyzer. The PVC soil collars were installed permanently 3 cm into the mineral soils [53]. The measured values from the collars for each parameter in a plot were integrated into plot average.

The trees <10 and ≥10 cm in DBH and shrub were measured, respectively, in 2006 and 2007. Four sub-plots (9 m^2 each) were installed in each plot to measure the aboveground biomass for those trees <10 cm in DBH; the trees were harvested, dried and weighted. Shrub was collected in the four sub-plots for every plot, and then dried and weighted. The volume for those trees ≥10 cm in DBH was calculated using the equation developed by Saucier and Clark III [54] for tree species in Southern U.S. based on the DBH and height. The aboveground biomass was estimated by using the equation developed by Smith et al. [55] on the basis of vegetation types in Southeastern U.S. The total wood biomass of a plot was the sum of the biomass of all trees in the plot, and then normalized to unit weight, $Mg \cdot C \cdot ha^{-1}$, by dividing the plot area (1022 m^2). There were four sub-plots in each plot for collecting litter and duff. The area of the basket for litter collection was 1470 cm^2, and the area for collecting duff was 729 cm^2. The litter and duff were dried under 60°C, and weighted.

The impact of Hurricane Hugo on the biomass C storage in the stands in the entire Santee Experimental Forest was estimated by Hook et al. [42]; the aboveground biomass was about 78.0 $Mg \cdot C \cdot ha^{-1}$ before the hurricane, and 13.9 $Mg \cdot C \cdot ha^{-1}$ after the hurricane; the mean aboveground wood biomass loss was about 64.1 $Mg \cdot C \cdot ha^{-1}$ in the entire forest. In this study catchment, all the aboveground biomass C loss to the hurricane remained on forest floor due to without biomass removal and salvage logging; the dead roots produced by the hurricane was about 19.2 $Mg \cdot C \cdot ha^{-1}$, estimated using the coefficient of roots to aboveground biomass [56,57], in which most were in the soils, a few were on the ground due to some trees uprooted by the hurricane.

2.3. Model Validation

Both Forest-DNDC and MIKE SHE were validated. MIKE SHE was validated using six years of daily stream flow and spatial water table for assessing the impact of climate variability on hydrology in this watershed [11, 58]. It was validated against stream flow and water table observed in 2003-2007 for this study. The Forest-DNDC was validated using the biomass inventory for a biomass study in

2006 and measurement in 2007 for this study; it was also validated against the soil CO_2, soil temperature and moisture measured in 2006-2008.

The model performance was evaluated employing widely used quantitative methods, coefficient of determination ($R^2 \leq 1$) (squared correlation coefficient) and model efficiency ($-\infty < E \leq 1$) [5,35,59]. The model performance efficiency is calculated as

$$E = 1 - \frac{\sum (O_i - P_i)^2}{\sum (O_i - \bar{O})^2} \tag{1}$$

where O_i, $\bar{O}$ and P_i are the observed values, mean observation and simulated values, respectively.

2.4. Model Setup and Parameterization

In order to obtain a better understanding the C dynamics in forested wetland ecosystems, Forest-DNDC was configured to simulate C dynamics for forested wetland ecosystems [31] with a spatially explicit modeling approach to utilize spatiotemporal physical and biogeochemical characteristics of the watershed, including vegetation, hydrology, soil, and managements across the forest [35]. Therefore, the catchment (160 ha) was divided into 675 simulating cells with a size of 50 m × 50 m to simulate carbon dynamics using the spatially explicit modeling approach. The main parameters for modeling C dynamics in this catchment were presented in Table 1.

The simulation started from 1950 with the assumption that the mean tree age of the dominant trees at that time (1950) was 17 years old due to clear cutting in 1930's [41,42,60] (also see the assumption 1 below). GIS-based data files for model inputs were created using ArcGIS 9.2 [61] and Microsoft Excel 2003 for the spatial variation in hydrology, soil and vegetation. The spatial hydrologic data were supplied by MIKE SHE [39] calibrated and validated using spatial (water table) and temporal (water table and flow) observations [58,62]. The soil organic carbon (SOC), pH, clay, and hydraulic properties were obtained from soil survey and field samples for previous studies [40,46]. The details of some soil parameters can be found in the publication for simulating hydrology in this watershed [58]. The vegetation species and distribution were obtained from the assessment of Hurricane Hugo damage [19,42] and biomass inventories in 2006 and 2007 (unpublished data). For this study, every file was a dataset with 675 elements (cells), and each represented 0.25 ha (50 m × 50 m) of the watershed.

All assumptions for this study were that 1) the vegetation in this watershed before the hurricane naturally regenerated in 1930's when the government obtained the land used as forestland due to clear cutting before the land acquired by the government [41,42,60]; 2) only 15% of pre-hurricane trees remained for all simulating units based on the assessment of the impact of the Hurricane Hugo on the forest in Francis Marion National Forest [19] and the current forest cover is from residuals and natural regeneration; 3) vegetation type before the hurricane was similar to the current type on the basis of the study of South Carolina forest land research and management related to the storm conducted by Hook et al. [42], Nix et al. [49] and Wilson et al. [50].

Table 1. Key vegetation and soil parameters for Forest-DNDC.

Parameter	Value (hardwood/pine)
Initial leaf N (%)	2/1.3
AmaxA ($mol \cdot g^{-1} \cdot s^{-1}$)	−46/9.3
AmaxB	71.9/21.5
Optimum photosynthetic temperature (°C)	24/24
Minimum photosynthetic temperature (°C)	4/2
Amax fraction	0.76/0.76
Growth respiration fraction	0.25/2.5
Dark respiration fraction	0.1/0.075
Wood maintain respiration fraction	0.07/0.07
Root maintain respiration fraction	1/1
Light half saturation constant	200/200
Respiration Q10	2/2
Canopy light attenuation	0.5/0.58
Water use efficiency	13.9/13.9
DVPD1	0.05/0.05
DVPD2	2/2
Maximum leaf growth rate ($\% \cdot yr^{-1}$)	0.9/0.35
Maximum wood growth rate	0.8/0.9
Leaf start TDD	400/900
Wood start TDD	400/900
Leaf end TDD	1300/1600
Wood end TDD	1300/1600

Leaf N retrainslocation	0.5/0.5
Senescence start day	260/270
Leaf C/N	23/35
Wood C/N	200/200
Leaf retention years	1/2.25
C reserve fraction	0.75/0.75
C fraction of dry matter	0.45/0.45
Specific leaf weight ($g \cdot m^{-2}$)	100/280
Minimum wood/leaf	1.4/1.25
Leaf geometry	2/1
Maximum N storage ($kg \cdot N \cdot ha^{-1}$)	200/200
Spatial soil, climate, vegetation and hydraulic parameters	
Soil organic carbon (%)	Hydraulic conductivities ($cm \cdot hr^{-1}$)
pH	Wilting point (0 - 1)
Clay (%)	Capacity (0 - 1)
Soil depth (cm, ≤150 cm)	Porosity (0 - 1)
Overstory species	Overstory age
Understory species	Understory age
Ground growth (sedge and moss)	Daily minimum temperature (°C)
Daily maximum temperature (°C)	Daily precipitation (mm)

3. RESULTS AND DISCUSSION

Sharing a same set of input databases, MIKE SHE and Forest-DNDC worked in conjunction that passed the MIKE SHE simulated water table data to Forest-DNDC on daily basis. The modeled results are reported as follows.

3.1. Model Validation

MIKE SHE [39] was calibrated and validated using daily flow measurements during the period from 1969- 1971 (before Hurricane Hugo) [11] and another period from 2003-2007 (after the hurricane), and monthly flow in the period from 1969-2007 (only 238 months due to data missing) [11]. The simulated stream flow was in agreement with the observations with model performance efficiency ($-\infty < E \leq 1$) [59] from 0.66 - 0.72 for daily flow, 0.92 - 0.94 for monthly flow in the two periods, 0.88 for monthly flow in the period from 1969-2007 (Table 2). The E was within the range of very good model performance rating [63].

The spatial water table dynamics are important to simulate C dynamics at watershed scale. Daily water table depth was used to calibrate and validate MIKE SHE model. The simulated daily water table depth for the 33 manual wells from 1992-1994 [11], and for two automatic wells and 8 manual wells in 2003-2007 was in good agreement with the observations, with E of 0.81 and 0.79 and the small difference in mean water table depth, less than 10 cm (between −3.5 and 3.9 cm) [11], between the observation and simulation for the two validation periods. These quantitative assessments show that MIKE SHE is applicable to provide proper water table dynamics for this C simulation using spatially explicit modeling approach.

The results from Forest-DNDC validation using soil CO_2, soil temperature and soil moisture, and biomass were presented in Figure 3 and Table 3. Although the simulated average soil moisture was approximately to observed and the slope of regression model between the observations and simulations was close to 1 (b = 1.02), the model efficiency (Table 3) from the validation showed that the simulated soil moisture was in general agreement with the field measurements (E = 0.4 and R^2 = 0.64, n = 216) (Figure 3(a)). However, there was good agreement between the observations and simulations for soil temperature (E = 0.7 and R^2 = 0.78, n = 216; Figure 3(b)), soil CO_2 (E = 0.67 and R^2 = 0.67, n = 200; Figure 3(c)) and aboveground biomass (E = 0.82, R^2 = 0.85, n = 24, for plots p1 - p24 in 2006; and E = 0.86, R^2 = 0.99, n = 8, for plots p21 - p24 in 2006 and 2007; Figure 3(d)).

There were small differences between the field measurements and simulations for soil moisture. The simulated soil moisture trended to be slightly higher than the field measurements during wet periods and slightly lower during dry periods. These differences may be related to the location and distribution of the measurement plots, which include both upland and wetland. The field measurements may also be influenced by the micro-topography within a measurement plot; whereby the 4 - 6 measurement points chosen for a plot were inadequate to provide a representative mean for the plot (0.25 ha). The distribution of measurement points tended to be on the slightly higher (i.e. 10 - 20 cm) micro-topographic positions, which would not be submerged during periods of high-water table, although those events are not uncommon. Since the simulated results represent an average across the plot, it is reasonable that the simulated soil moisture could be higher than the measured values for wet periods, especially in the riparian zone.

The Figure 3(d) showed that the Forest-DNDC slightly over predicted the biomass C accumulated in wood product for plots 5, 13, 14, 19 - 23, and under predicted for the plots 1, 3, 4, 8, 12, 17 and 18. There are many factors that can produce these errors, including vegetation distribution, differences in plot sizes between inventory (fourth acre, about 0.1 ha) and simulation (fourth hectare), and errors from measurements of tree height (m) and diameter at breast height (DBH) (cm). An important error with respect to the over or under estimations of the wood mass can be resulted from estimating

Table 2. The simulated (S) and measured (M) mean water table and mean stream flow in watershed WS80 and rainfall in the 5-year period, and model efficiency (E), and number of samples (n).

Year	Rainfall (mm)	Streamflow (mmd^{-1})					Water table depth from ground surface (m)[a]				
		M	S	R^2	E	*n*	M	S	R^2	E	*n*
2003	1671	2.01	2.00	0.62	0.55	365	0.58	0.63	0.88	0.87	556
2004	962	0.28	0.37	0.63	0.56	366	0.88	0.98	0.79	0.76	477
2005	1540	0.84	0.82	0.66	0.63	365	0.77	0.86	0.77	0.75	699
2006	1255	0.38	0.52	0.83	0.81	365	1.02	1.09	0.66	0.65	736
2007	923	0.16	0.21	0.78	0.78	365	1.06	1.26	0.73	0.56	483

[a]The water table depth is the distance from ground surface to water table level where is below the ground surface, and annual average water table depth from all wells. M is measurement, S is simulation, R^2 is the coefficient of determination, E is Nash-Sutcliffe model efficiency (1970), the unit of stream flow is mm per day, normalized from cms (cubic meters per second).

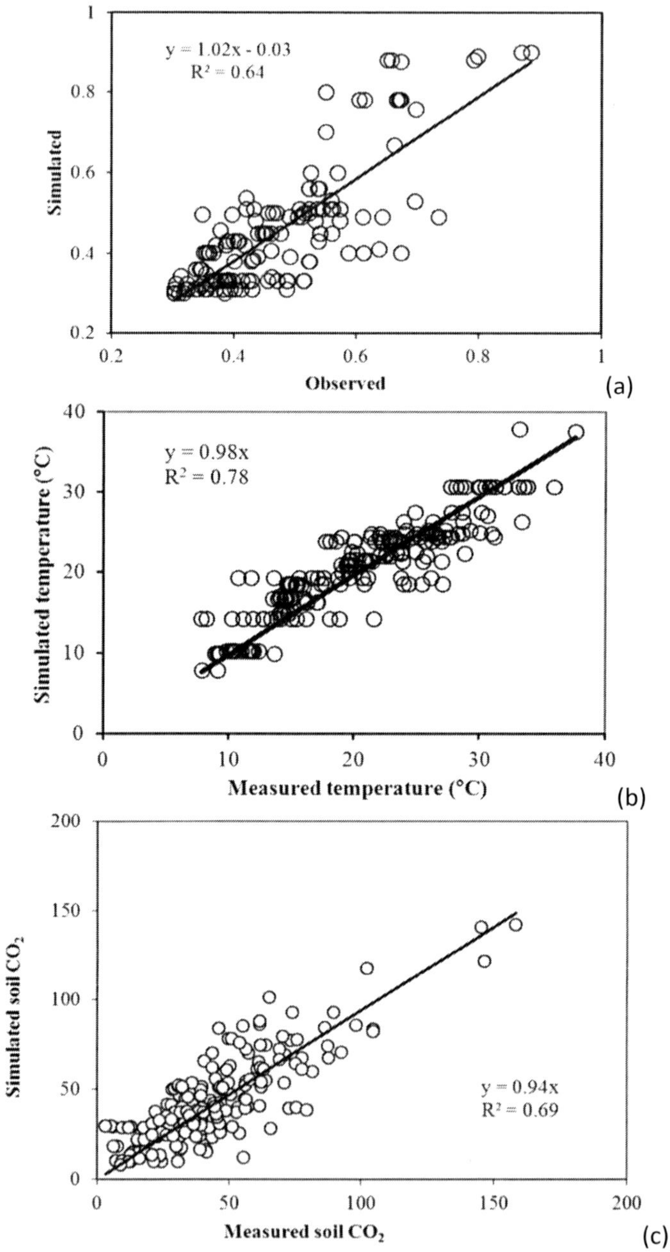
y = 1.02x - 0.03
R² = 0.64
Simulated
Observed
(a)
y = 0.98x
R² = 0.78
Simulated temperature (°C)
Measured temperature (°C)
(b)
y = 0.94x
R² = 0.69
Simulated soil CO2
Measured soil CO2
(c)

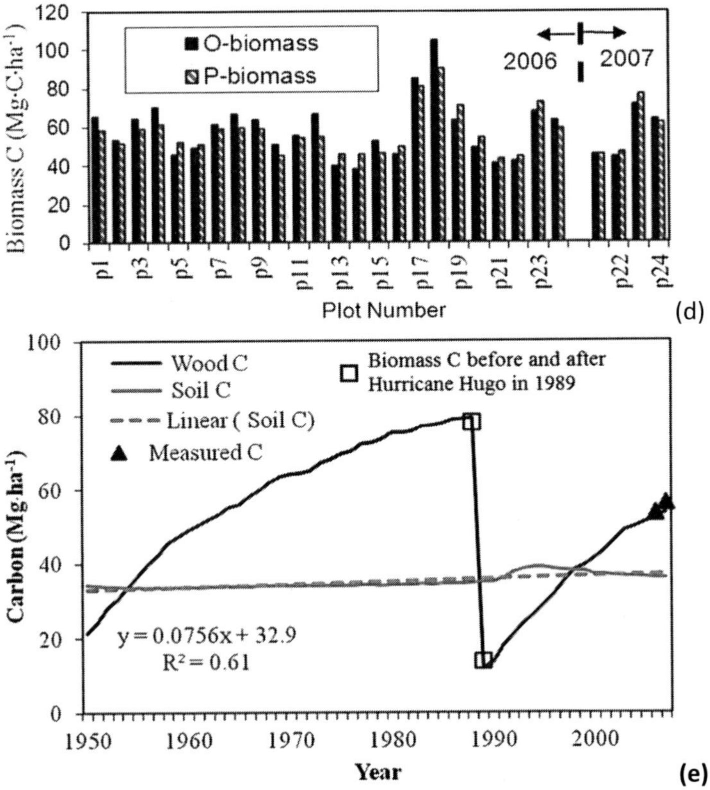

Figure 3. (a) Observed and simulated daily soil moisture (0 - 1) for eight plots (plot average) (n = 216); (b) Observed and simulated daily soil temperature (˚C) for eight plots (plot average) (n = 216); (c) Observed and simulated soil CO_2 fluxes (kg·C·ha^{-1}·d^{-1}) for eight plots (plot average) (n = 200); (d) Observed and simulated aboveground biomass (Mg·C·ha^{-1}) for plots p1 - p24 in 2006 and p21 - p24 in 2007 (O-biomass is the measured biomass C; P-biomass is the simulated); (e) Temporal changes in biomass and mineral soil organic carbon (Mg·C·ha^{-1}) within a depth of 30 cm; ([Wood C] is simulated annual mean aboveground biomass C for the 58-year period from 1950-2007; [Measured C] is the estimated aboveground biomass for 2006 and 2007 using plot average based on the biomass measurements in 2006 and 2007, respectively; [Biomass C before and after Hurricane Hugo in 1989] is the aboveground biomass C estimated by Hook et al. (1996); n = 39 for the linear (Soil C) regression).

Table 3. Observed (M) and simulated (S) averages, model efficiency (E), the slope (b) of regression between observations and simulations, and numer of samples (n)[a].

Parameter	M	S	E	*b*	*n*
Soil CO_2 ($kg{\cdot}C{\cdot}ha^{-1}{\cdot}d^{-1}$)	43.3	42.6	0.67	0.94	200
Soil temperature (°C)	20.4	18.6	0.70	0.91	216
Soil moisture ($m^3{\cdot}m^{-3}$)	0.48	0.46	0.40	1.02	216
Biomass ($Mg{\cdot}C{\cdot}ha^{-1}$)[c]	55.0	58.9	0.86	1.13	8
Biomass ($Mg{\cdot}C{\cdot}ha^{-1}$)[d]	76.1	74.5	0.82	0.96	24

[a]M is observed mean; S is simulated mean; E is model efficiency (Nash and Sutcliffe, 1970); *b* is the slope of regression model between observations and simulations; [c]The biomass was only aboveground part and measured for plots p21-p24 in 2007; [d]The biomass was the sum of aboveground and belowground biomass and observed in twenty four plots in 2006.

biomass loss to Hurricane Hugo in 1989. Because plotbased measurements for the biomass loss to the hurricane were not available, an average wood mass loss was used for all simulating units in this study. The estimated average loss was about 85% based on the damage investigation and studies [19,49,50]. In fact, the wood mass loss to the hurricane in this catchment was heterogeneous.

The Figure 3(e) presents the simulated annual mean aboveground biomass across the study catchment in the 58-year period from 1950-2007, and the biomass before and after the hurricane estimated by Hook et al. [42] (blank square) and the biomass inventoried in 2006 and 2007 (triangle) (unpublished data). The figure shows that the biomass simulated for pre-hurricane (79.3 $Mg\cdot C\cdot ha^{-1}$) is in good agreement with the estimated biomass (78.0 $Mg\cdot C\cdot ha^{-1}$) (blank squares on Figure 3(e)) for assessing the hurricane damage. However, the modeled biomass (12.1 $Mg\cdot C\cdot ha^{-1}$) for post-hurricane was slightly lower than the biomass measurements (13.9 $Mg\cdot C\cdot ha^{-1}$) for the hurricane damage assessment. The under-estimation from this study might be related to an overestimate in the mean biomass loss (85%) to the hurricane for the simulation. The simulated annual mean biomass across the catchment for the two inventories in 2006 and 2007 was slightly lower than the inventoried averages, differentiated about −1.8% and −5.0%, respectively. This difference might be in relation to the overestimation of the biomass loss to the hurricane leading a subsequent under prediction, but the difference should be within the errors produced by biomass measurements and the methods used to computing biomass based on DBH and heights of trees.

The model validation showed that the Forest-DNDC performed properly well for simulating soil CO_2 flux, soil temperature and biomass, with $E = 0.86$ for biomass, $E = 0.7$ for soil temperature, and $E = 0.67$ for soil CO_2, which are within the range of good model performance rating ($0.5 \leq E < 0.75$) [63]. Although the E (0.4) is within the range of general model performance rating for modeling soil moisture, the slope of the regression model between the observations and simulations is close to 1.0 ($b = 1.02$), and the simulated mean soil moisture is approximately to the measured. The differences in the biomass between the simulation and measurements are small (4.9% for biomass loss to the hurricane and 3.4% for biomass inventories). The metrics indicate that the Forest-DNDC can be applicable for modeling biomass and soil CO_2 dynamics for this forest with hurricane disturbance.

3.2. Spatial Differences in Net Ecosystem Exchange (NEE) and Soil CO_2 Flux

The cell-based (simulation unit) annual average soil CO_2 flux in this watershed spatially ranged from 3.2 - 4.8 Mg·C·ha^{-1} during the period from 1950-2007 (Figure 4(a), Table 4), which showed that there was a substantially spatial difference in the flux. The spatial difference in soil CO_2 flux is likely resulted from differentiated soil conditions, including soil moisture and soil organic content, and vegetation distribution [64,65]. The arithmetic mean, median and geometric mean of the 58-year soil CO_2 flux on WS80 were same, 3.8 Mg·C·ha^{-1}, which indicates a spatially normal distribution in this watershed. High CO_2 flux occurred in the uplands, and low flux in wetlands.

The spatial variation of NEE in the catchment was large, but the annual mean C storage was low (high NEE). The cell-based annual average NEE ranged from −0.4 to −2.5 Mg·C·ha^{-1} (Figure 4(b)) with an arithmetic mean of −1.1 and geometric value of −1.0 Mg·C·ha^{-1} during the time period from 1950-2007. The high NEE at this site was primarily resulted from the decomposition of large amount of debris left by Hurricane Hugo in 1989, including fallen trees (178 m^3·ha^{-1} - 24 m^3·ha^{-1} = 154 m^3·ha^{-1} woody debris) due to without salvage logging and biomass removal after the hurricane and dead roots (about 19.2 Mg·C·ha^{-1}) produced by the destruction of - the forest (see more discussion on the hurricane impact below). This is because CO_2 released from the decomposition of the residues remained in the forest was much higher than the amount consumed by the regenerating forest within several years after the storm.

Compared to the NEE (−1.85 Mg·C·ha^{-1}·yr^{-1}) in the same period from 1950-2008 at Savannah River Site in South Carolina where the forest was not disturbed by any hurricane [62], the high NEE in this catchment showed that the hurricane was highly impacted on the C sequestration to this forest.

3.3. Temporal Changes in Soil CO_2 Flux

The hurricane did not only destroy the forest in this catchment, but also it added a large amount of coarse wood debris (CWD) to forest floor and dead roots to mineral soil. Therefore, CO_2 flux was substantially higher within a short time period after the hurricane than before the disastrous wind due to the contribution from the decomposition of organic C added to the soils (SOC) by the hurricane, excluding CWD, litter and duff, (Figure 5), which was primarily as the result from the dead roots. However, the Figure 5 also shows that the impact of dead roots added by the hurricane has likely been small at present, about 15 years after.

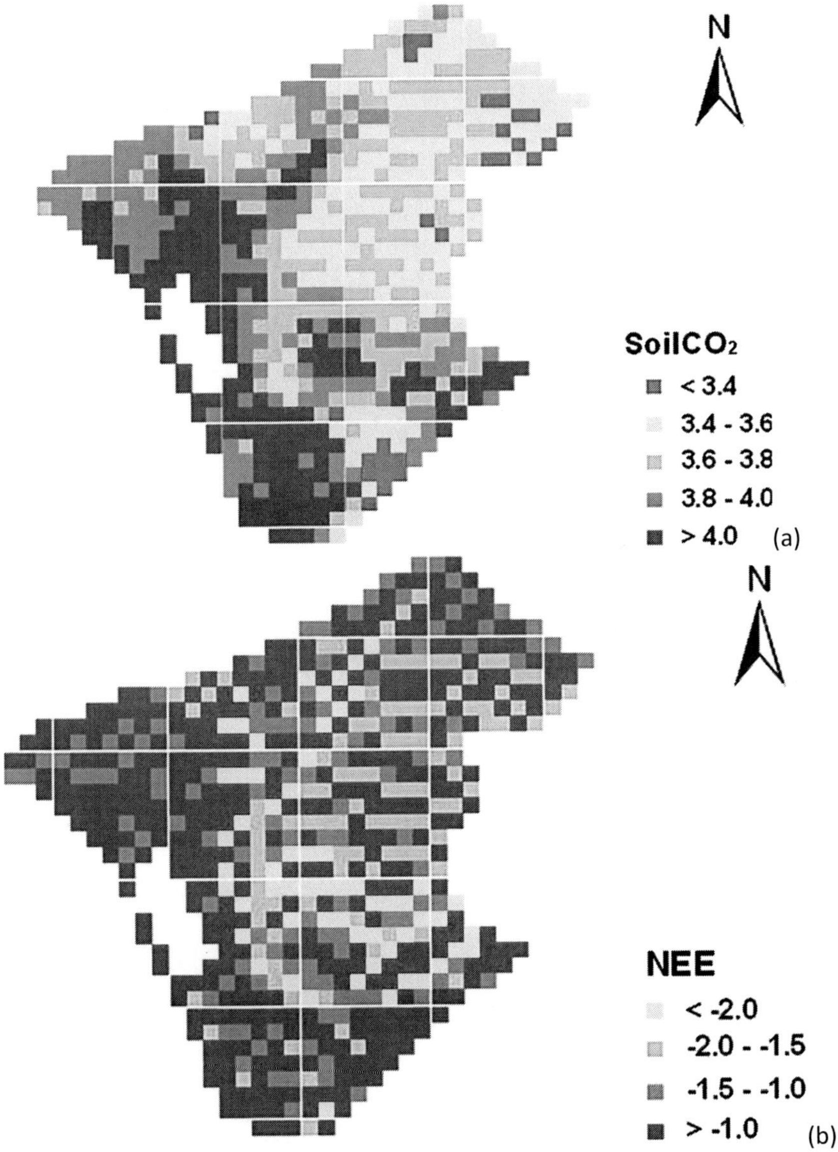

Figure 4. (a) Spatial differences in annual average soil CO_2 flux ($Mg \cdot C \cdot ha^{-1} \cdot yr^{-1}$) in the period from 1950-2007; (b) Spatial differences in annual average NEE ($Mg \cdot C \cdot ha^{-1} \cdot yr^{-1}$) in the period from 1950-2007.

Table 4. Simulated spatial differences in annual averages of GPP, NPP NEE, biomass C and soil CO_2-C[a].

Parameter	Min	Max	AM	Median	GM
GPP ($Mg·C·ha^{-1}·yr^{-1}$)	12.3	28.4	16.5	16.9	16.4
NPP ($Mg·C·ha^{-1}·yr^{-1}$)	2.1	6.1	4.3	4.2	4.3
NEE ($Mg·C·ha^{-1}·yr^{-1}$)[b]	−0.4	−2.5	−1.1	−0.9	−1.0
Woody C ($Mg·C·ha^{-1}·yr^{-1}$)[c]	2.1	6.0	3.8	3.6	3.7
Soil CO_2 ($Mg·C·ha^{-1}·yr^{-1}$)	3.2	4.8	3.8	3.8	3.8

[a]All values were annual average; Min is minimum; Max is maximum; AM is arithmetic mean; GM is geometric mean. [b]Negative values here for NEE represents C sink; normally −0.4 should be larger than −2.5; however, from the C sink, the value of −0.4 represents a smaller C sink than −2.5; [c]The woody C is above-ground biomass C that is the sum of overstory and understory biomass; number of samples (n) is equal to the number of cells, 675.

The annual soil CO_2 flux from the study catchment between 1950 and 2007 was presented in Figure 5. The figure showed an abnormally high CO_2 flux in a short time period after the hurricane. The abnormal flux was mainly resulted from decomposition of the dead woods and debris left by the hurricane, added over 64 Mg·C·ha^{-1} to forest floor and about 19 Mg·C·ha^{-1} to soils (see the Impact of Hurricane Hugo on C sequestration below), because this watershed serves as control in the paired watershed system. However, the CO_2 flux is not constant even if there is no a hurricane disturbance. Figure 5 indi cated that the annual flux fluctuated and linearly increased at an average rate of about 27 kg·C·ha^{-1}·yr^{-1} before the hurricane. The slow and linear increase in soil CO_2 before the hurricane was likely resulted from two factors, an increase in biomass, including roots and leaf that can be the main factor leading to an increase soil CO_2 flux, and the increase in air temperature which can cause a small increase in SOC decomposition [64] because the temperature increased at an average rate of 0.21°C per decade during the time period from 19501988 (the pre-hurricane period) [11]. However, the inter-annual soil CO_2 flux fluctuation in the same period was principally related to the changes in soil moisture regime (see more discussion below). Figure 5 also showed that the hurricane significantly influenced soil CO_2 flux. This was because the hurricane left a large amount of debris on the forest floor and dead roots in the soils, and those fine and small residues can decompose fast under the subtropical climate, producing a high CO_2 flux within a short period after the hurricane.

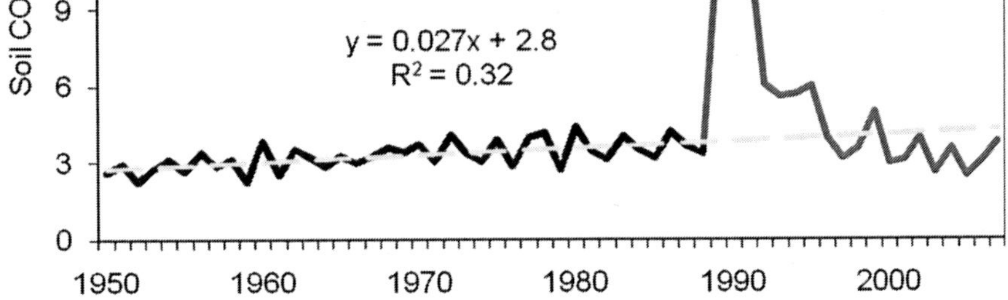

Figure 5. Temporal changes in soil CO_2 flux (Mg·C·ha^{-1}·yr^{-1}) in the period from 1950-2007 (n = 39 for the pre-hurricane period, n = 19 for post-hurricane period).

3.4. Impact of Climatic Variability on CO_2 Sequestration

We only analyzed the C dynamics in this forest in the period from 1950-1988 in order to understand well the impact of climate variability on C sequestration due to without a hurricane perturbation in the period. The results from simulation showed that climatic variability hardly influenced C sequestration to the forest product in this study catchment during the period from 1950-1988 (before the hurricane), including without substantial impacts of the variation in temperature (R^2 = 0.0004) and precipitation (R^2 = 0.015). However, the climate variability influenced C sequestration to the forested ecosystems, even if there were no impacts of hurricanes. The NEE increase significantly with a decrease in precipitation before the hurricane in this forested watershed ($P < 0.02$) (Figure 6), but insignificantly with an increase or decrease in temperature (R^2 = 0.01, $P > 0.1$). The increase in NEE was principally resulted from an increase in soil CO_2 flux caused by the changes in soil moisture regime regulating C dynamics in this watershed. Soil CO_2 flux from this forest was lower in the wet years than in dry years because the soil was saturated for a long period in wet years [62]. For example, mean soil CO_2 flux from this watershed in 2004 was over 1 $Mg{\cdot}C{\cdot}ha^{-1}$ higher than the flux in the two adjacent years (2003 and 2005) because 2003 and 2005 were much wetter than 2004, about 710 and 580 mm precipitation higher than 2004, respectively, but the temperature in the three years was approximate. High soil CO_2 flux in dry years is primarily as wetlands loss to a decrease in the water table level regulated by precipitation and evapotranspiration in this forest. This result was similar to the finding of Pietsch et al. [7] that water table level decrease in their sites led to an increase in soil carbon loss. This result implies that global warming can lead more soil CO_2 release from this type of forested wetlands because the warming does not bring more rain to this subtropical area [10,11].

3.5. Impact of Hurricane Hugo on Carbon Sequestration

Carbon sequestration in this forest did not monoto nously change in time, although the forest has not been managed near six decades, because there was a large fault (Figure 3(e)). The fault of C sequestration in the forest is primarily resulted from the disastrous storm, Hurricane Hugo in 1989, which blew down over 80% of trees in this watershed [19]. Both the simulation and measurements showed the hurricane substantially influenced C storage in the stands in this forest. The simulated mean aboveground biomass altered from 79.3 $Mg{\cdot}C{\cdot}ha^{-1}$ before the hurricane to 12.1 $Mg{\cdot}C{\cdot}ha^{-1}$ after the hurricane; the measured biomass reduced from78.0 $Mg{\cdot}C{\cdot}ha^{-1}$ to 13.9 $Mg{\cdot}C{\cdot}ha^{-1}$. These results indicated that the hurricane reduced the wood biomass C storage in the stands by over 82% in the forest in 1989.

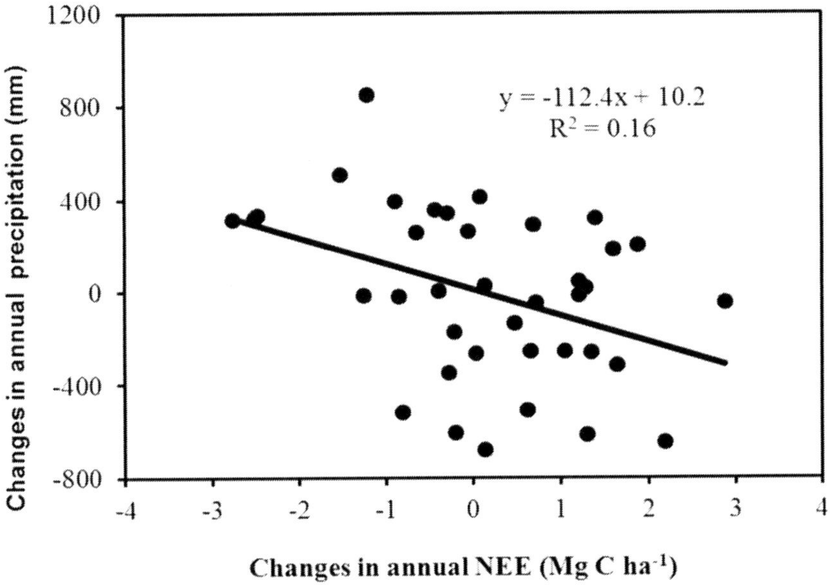

Figure 6. Impact of precipitation on NEE during the period from 1950-1988 (n = 39, the pre-hurricane period).

Hurricane Hugo did not only cut the C storage in the stands of the forest, but also it impacted the soil CO_2 flux due to the decomposition of the residues left by the hurricane (Figure 5). A substantial increase in CO_2 flux (the abnormal high flux) occurred in this forest within several years after the hurricane. The abnormal CO_2 flux after the hurricane is primarily as the result of this catchment (WS80) serving as a control watershed within the paired watershed system (see Figure 1) that is used to study forest and hydrology (see Section of Materials and Methods). Accordingly, all fallen trees and debris (about 64 $Mg \cdot C \cdot ha^{-1}$) left by the hurricane remained on the forest floor, led to a subsequent increase in CO_2 flux due to the decomposition of those residues. Another factor with respect to an increase in soil CO_2 flux is the dead roots, about 19.2 $Mg \cdot C \cdot ha^{-1}$ estimated on the basis of the radio of aboveground biomass to the roots [59,60], left in the soils due to the destruction of the forest. However, the impact of the dead roots on soil CO_2 flux has become very small about 10 years late after the hurricane due to decomposing year-by-year with an exponential decrease in the mass [66].

4. CONCLUSIONS

This study indicates that Forest-DNDC with a spatially explicit modeling approach is applicable for estimating C sequestration for the coastal forested watershed with a mosaic landscape consisting of wetlands and uplands. The agreement result between observation and simulation for biomass with a disastrous hurricane disturbance shows that this model can simulate C sequestration not only for the forested ecosystems without impacts of hurricanes but also for the similar forested ecosystems with hurricane perturbations.

This study indicates that hurricane can substantially influence CO_2 sequestration in the coastal forest ecosystems, including reducing C storage in the stands and increasing soil CO_2 flux within a short period. Climate variability and change can alter soil CO_2 flux from this forested watershed. CO_2 fluxes can increase with a decrease in water table level regulated by climate. Projecting those impacts to future climate change, soil CO_2 flux can increase due to temperature rise without bringing more precipitation to this forest, leading to an increase in evapotranspiration demands that incurs a decrease in water table level.

REFERENCES

1. Hansen, J., Sato, M., Ruedy, R., Lo, K., Lea, D.W. and Medina-Elizade, M. (2006) Global temperature change. Proceedings of the National Academy of Sciences of the United States of America, 103, 14288-14293.
2. Trenberth, K.E., Jones, P.D., Ambenje, P., Bojariu, R., Easterling, D., Klein Tank, A., Parker, D., Rahimzadeh, F., Renwick, J.A., Rusticucci, M., Soden, B. and Zhai, P. (2007) Observations: Surface and atmospheric climate change. In: Solomon, S., Qin, D., Manning, M., Chen, Z., Marquis, M., Averyt, K.B., Tignor, M. and Miller, H.L., Eds., Climate Change 2007: The Physical Science Basis, Contribution of Working Group I to the Fourth Assessment Report of the Intergovernmental Panel on Climate Change, Cambridge University Press, Cambridge and New York.
3. Trettin, C.C. and Jurgensen, M.F. (2003) Carbon cycling in wetland forest soils. In: Kimble, J., Birdsie, R. and Lal, R., Eds., Carbon Sequestration in US Forests, CRC Press LLC, Boca Raton.
4. Trettin, C.C., Laiho, R., Minkkinen, K. and Laine, J. (2006) Influence of climate change factors on carbon dynamics in northern forested peatlands. Canadian Journal of Soil Science, 86, 269-280. doi:10.4141/S05-089
5. Miehle, P., Livesley, S.J., Feikema, P.M., Li, C. and Arndt, S.K. (2006) Assessing productivity and carbon sequestration capacity of Eucalyptus globules plantation using the process model Forest-DNDC: Calibration and validation. Ecological Modeling, 192, 83-94. doi:10.1016/j.ecolmodel.2005.07.021

6. Pan, Y., Birdsey, R.A., Fang, J., Houghton, R., kauppi, P.E., Kurz, W.A., Phillips, O.L., Shvidenko, A., Lewis, S.L., Canadell, J.G., Ciais, P., Jackson, R.B., Pacala, S.W., McGuire, D.A., Piao, S., Rautiainen, A., Sitch, S. and Hayes, D. (2011) A large and persistent carbon sink in the world's forests. Science, 333, 988-993.
7. Pietsch, S.A., Hasenauer, H., Kucera, J. and Cermak, J. (2003) Modeling effects of hydrological changes on the carbon and nitrogen balance of oak in floodplains. Tree Physiology, 23, 735-746.
8. Riveros-Iregui, D.A. and McGlynn, B.L. (2009) Landscape structure control on soil CO_2 efflux variability in complex terrain: Scaling from point observation to watershed scale fluxes. Journal of Geophysical Research: Biogeosciences, 114, G02010.
9. Pacific, V.J., McGlynn, B.L., Riveros-Iregui, D.A., Epstein, H.E. and Welsch, D.L. (2009) Differential soil respiration response to changing hydrologic regimes. Water Resources Research, 45, W07201.
10. Zhang, X., Zwiers, F.W., Hegerl, G.C., Hugo Lambert, F., Gillett, N.P., Solomon, S., Stott, P.A. and Nozawa, T. (2007) Detection of human influence on twentieth-century precipitation trends. Nature, 448, 461-465.
11. Dai, Z., Amatya, D.M., Sun, G., Trettin, C.C., Li, C. and Li, H. (2011) Climate variability and its impact on forest hydrology on South Carolina Coastal Plain, USA. Atmosphere, 2, 330-357.
12. Lu, J., Sun, G., McNulty, S.G. and Comerford, N. (2009) Sensitivity of pine flatwoods hydrology to climate change and forest management in Florida, USA. Wetlands, 29, 826-836.
13. Mulhouse, J.M., Steven, D.D., Lide, R.F. and Sharitz, R.R. (2005) Effects of dominant species on vegetation change in Carolina bay wetlands following a multi-year drought. The Journal of the Torrey Botanical Society, 132, 411-420.
14. Pielke, R.A., Landsea, C., Mayfield, M., Laver, J. and Pasch, R. (2005) Hurricanes and global warming. Bulletin of the American Meteorological Society, 86, 1571-1575.
15. Pezza, A.B. and Simmonds, I. (2005) The first South Atlantic hurricane: Unprecedented blocking, low shear and climate change. Geophysical Research Letters, 32, L15712. doi:10.1029/2005GL023390
16. Trenberth, K. (2005) Uncertainty in hurricanes and global warming. Science, 308, 1753-1754.
17. Elsner, J.B. (2006) Evidence in support of the climate change—Atlantic hurricane hypothesis. Geophysical Research Letters, 33, L16705.
18. Landsea, C.W., Harper, B.A., Hoarau, K. and Knaff, J.A. (2006) Can we detect trends in extreme tropical cyclones? Science, 313, 452-454.
19. Hook, D.D., Buford, M.A. and Williams, T.M. (1991) Impact of Hurricane Hugo on the South Carolina coastal plain forest. Journal of Coastal Research, 8, 291-300.

20. Sheikh, P.A. (2006) The impact of Hurricane Katrina on biological resources. http://assets.opencrs.com/rpts/RL33117_20051018.pdf
21. Chapman, E.L., Chambers, J.Q., Ribbeck, K.F. and Baker, D.B. (2008) Hurricane Katrina impacts on forest of Louisiana's Pearl River basin. Forest Ecology and Management, 256, 883-889.
22. Oswalt, S.N. and Oswalt, C.M. (2008) Relationships between common forest metrics and realized impacts of Hurricane Katrina on forest resources in Mississippi. Forest Ecology and Management, 255, 1692-1700.
23. Middleton, B.A. (2009) Effects of Hurricane Katrina on the forest structure of Taxodium Distichum swamps of the Gulf coast, USA. Wetlands, 29, 80-87.
24. Wang, Y.P. and Jarvis, P.G. (1990) Description and validation of an array model—MAESTRO. Agricultural and Forest Meteorology, 51, 257-280.
25. Li, C., Aber, J., Stang, F., Butter-Bahl, K. and Papen, H. (2000) A process-oriented model of N_2O and NO emissions from forest soils. 1. Model development. Journal of Geophysical Research: Atmospheres, 105, 4369-4384.
26. Thornton, P.E., Law, B.E., Gholz, H.L., Clark, K.L., Falge, E., Ellsworth, D.S., Goldstein, D.S., Monson, R.K., Hollinger, D., Falk, M., Chen, J. and Sparks, J.P. (2002) Modeling and measuring the effects of disturbance history and climate on carbon and water budgets in evergreen needleleaf forests. Agricultural and Forest Meteorology, 113, 185-222.
27. Battaglia, M., Sands, P., White, D. and Mummery, D. (2004) CABALA: A linked carbon, water and nitrogen model of forest growth for silvicultural decision support. Forest Ecology and Management, 193, 251-282.
28. Hanson, P.J., Amthor, J.S., Wullschleger, S.D., Wilson, K.B., Grant, R.F., Hartley, A., Hui, D., Hunt, E.R., Johnson Jr., D.W., Kimball, J.S., King, A.W., Lou, Y., McNulty, S.G., Sun, G., Thornton, P.E., Wang, S., Williams, M., Baldocchi, D.D. and Cushman, R.M. (2004) Oak forest carbon and water simulations: Model intercomparisons and evaluations against independent data. Ecological Monographs, 74, 443-489.
29. Stange, F., Butterbachl, K., Papen, H., Zechmeister-Boltenstern, S., Li, C. and Aber, J. (2000) A process-oriented model of N_2O and NO emissions from forest soils. Journal of Geophysical Research: Atmospheres, 105, 4385- 4398.
30. Zhand, Y., Li, C., Trettin, C.C., Li, H. and Sun, G. (2002) An integrated model of soil, hydrology, and vegetation for carbon dynamics in wetland ecosystems. Global Biogeochemical Cycles, 16, 1-17.
31. Li, C., Cui, J., Sun, G. and Trettin, C.C. (2004) Modeling impacts of management on carbon sequestration and trace gas emissions in

forested wetland ecosystems. Environmental Management, 33, S176-S186.

32. Kesik, M., Ambus, P., Baritz, R., Bruggemann, N., Butterbach-Bahl, K., Damm, M., Duyzer, J., Horvath, L., Kiese, R., Kitzler, B., Leip, A., Li, C., Pihlate, M., Pilegaard, K., Seufert, G., Simpson, D., Skiba, U., Smiatek, G., Vesala, T. and Zechmeister-Boltenstern, S. (2005) Inventories of N_2O and NO emissions from European forest soils. Biogeosciences, 2, 353-375.
33. Kesik, M., Bruggemann, N., Forkel, R., Kiese, R., Knoche, R., Li, C., Seufert, G., Simpson, D. and Butterbach-Bahl, K. (2006) Future scenarios of N_2O and NO emissions from European forest soils. Journal of Geophysical Research: Atmospheres, 111, G02018.
34. Kurbatova, J., Li, C., Varlagin, A., Xiao, X. and Vygodskaya, N. (2008) Modeling carbon dynamics in two adjacent spruce forests with different soil conditions in Russia. Biogeosciences, 5, 969-980.
35. Dai, Z., Trettin, C.C., Li, C., Li, H., Sun, G. and Amatya, D.M. (2012) Effect of assessment scale on spatial and temporal variations in CH_4, CO_2 and N_2O fluxes in a Forested Watershed. Water, Air, and Soil Pollution, 223, 253-265.
36. Butterbach-Bahl, K., Stange, F. and Papen, H. (2001) Regional inventory of nitric oxide and nitrous oxide emissions for forest soils of southeast Germany using the biogeochemical model PnET-N-DNDC. Journal of Geophysical Research: Atmospheres, 106, 34155-34166. doi:10.1029/2000JD000173
37. Cui, J., Li, C. and Trettin, C.C. (2005) Analyzing the ecosystem carbon and hydrologic characteristics of forested wetland using a biogeochemical process model. Global Change Biology, 11, 278-289.
38. Li, C. (2007) Quantifying greenhouse gas emissions from soils: Scientific basis and modeling approach. Soil Science and Plant Nutrition, 53, 344-352.
39. DHI (2005) DHI water and environment. Danish Hydraulic Institute, Denmark.
40. Harder, S.V., Amayta, D.M., Callahan, T.J., Trettin, C.C. and Hakkila, J. (2007) Hydrology and Water Budget for a Forested Atlantic Coastal Plain Watershed, South Carolina. Journal of American Water Resources Association, 43, 563-575.
41. Trettin, C.C., Amatya, D.M., Kaufman, C., Levine, N. and Morgan, R.T. (2008) Recognizing change in hydrologic functions and pathways due to historical agricultural use—Implications to hydrologic assessments and modeling. The 3rd Interagency Conference Research in the Watersheds, Estes Park, 8-11 September 2008, 1-5.
42. Hook, D.D., Buford, M.A. and Williams, T.M. (1996) Impact of Hurricane Hugo on the South Carolina coastal plain forest. In: Haymond, J.L., Hook, D.D. and Harms W.R., Eds., Hurrican Hugo: South

Carolina Forest Land Research and Management Related to the Storm, Southern Research Station, USDA Forest Service, Ashville, 540.
43. Sun, G., Lu, J., Gartner, D., Miwa, M. and Trettin, C.C. (2000) Water bugets of two forested watersheds in South Carolina. In: Higgins, R.W., Ed., Proceedings of the Spring Special Conference, American Water Resources Association, Miami, 199-202.
44. Federal Register (1994) Changes in hydric soils of the United States, July 13, 1994. https://www.federalregister.gov/articles/
45. Federal Register (2002) Hydric soils of the United States, September 18, 2002. http://soildatamart.nrcs.usda.gov/
46. Long, B.M. (1980) Soil survey of Berkeley County, South Carolina. United States Department of Agriculture, Washington DC.
47. Fissore, C., Giardian, C.P., Kolka, R.K. and Trettin, C.C. (2009) Soil organic carbon quality in forested mineral wetlands at different mean annual temperature. Soil Biology & Biochemistry, 41, 458-466.
48. Soil Survey Staff (2011) Natural Resources Conservation Service, United States Department of Agriculture. Soil Survey Geographic (SSURGO) Database for Berkeley County, South Carolina. http://soildatamart.nrcs.usda.gov
49. Nix, L.E., Hook, D.D., Williams, J.G. and Blaircom, D.V. (1996) Assessment of hurricane damage to the santee experimental forest and the Francis Marion National Forest with a geographic information system. In: Haymond, J.L., Hook, D.D. and Harms, W.R., Eds., Hurricane Hugo: South Carolina Forest Land Research and Management Related to the Storm, US Department of Agriculture, Asheville.
50. Wilson, L., Amatya, D., Callahan, T. and Trettin C. (2006) Hurricane impact on stream flow and nutrient exports for a first-order forested watershed of the lower coastal plain, South Carolina. 2nd Interagency Conference on Research in the Watersheds, Otto, 16-18 May 2006, 169-179.
51. Amatya, D.M. and Trettin, C.C. (2007) Development of watershed hydrologic research at Santee Experimental Forest, Coastal South Carolina. In: Furniss, M.J., Clifton, C.F. and Ronnenberg, K.L., Eds., Advancing the Fundamental Sciences: Proceedings of the Forest Service National Earth Sciences Conference, US Department of Agriculture, Forest Service, Pacific Northwest Research Station, San Diego.
52. Amatya, D.M., Miwa, M., Harrison, C.A., Trettin, C.C. and Sun, G. (2006) Hydrology and water quality of two order forested watersheds in coastal South Carolina. In: Williams, T., Ed., Proceedings of the 2006 International Conference on Hydrology and Management of Forested Wetlands Conference, St. Joseph, 8-12 April 2006, 15-25.
53. Sulzman, E.W., Brant, J.B., Bowden, R.D. and Lajtha, K. (2005) Contribution of aboveground litter, belowground litter, and

rhizosphere respiration to total soil CO_2 efflux in an old growth coniferous forest. Biogeochemistry, 73, 231-256.

54. Saucier, J.R. and Clark III, A. (1985) Tables for estimating total tree and product weight and volume of major southern tree species groups. Southwide Energy Committee, American Pulpwood Association Inc., Washington DC.
55. Smith, J.E., Heath, L.S. and Jenkins, J.C. (2003) Forest volume-to-biomass models and estimates of mass for live and standing dead trees of US forests. US Department of Agriculture, Forest Service, Northeastern Research Station, Newtown Square.
56. Schroth, G. (1995) Tree root characteristics as criteria for species selection and systems design in agroforestry. Agroforestry Systems, 30, 125-143.
57. IPCC (2003) Chapter 3. In: Penman, J., Gytarsky, M., Hiraishi, T., Krug, T., Kruger, D., Pipatti, R., Buendia, L., Miwa, K., Ngara, T., Tanabe, K. and Wagner, F., Eds., Good Practice Guidance for Land Use, Land-Use Change and Forestry, IPCC, Kanagawa, 3.1-3.150.
58. Dai, Z., Li, C., Trettin, C.C., Sun, G., Amatya, D.D. and Li, H. (2010). Bi-criteria evaluation of the MIKE SHE model for a forested watershed on the South Carolina coastal plain. Hydrology and Earth System Sciences, 14, 1033-1046.
59. Nash, J.E. and Sutcliffe, J.V. (1970) River flow forecasting through conceptual models part I: A discussion of principles. Journal of Hydrology, 10, 282-290.
60. Hawley, N.R. (1949) The old rice plantation in and around the Santee Experimental Forest. Agricultural History, 23, 86-91.
61. Omsby, T., Napoleon, E., Burke, R., Groessl, C. and Feaster, L. (2004) Getting to know ArcGIS desktop: The basics of ArcView, ArcEditor, and ArcInfo Updated for ArcGIS 9. ESRI Press, Redlands, 569.
62. Dai, Z., Trettin, C.C., Li, C., Amatya, D.M., Sun, G. and Li, H. (2010) Sensitivity of stream flow and water table depth to potential climatic variability in a coastal forested watershed. Journal of American Water Resources Association, 46, 1036-1048.
63. Moriasi, D., Arnold, J., Liew, M.W.V., Bingner, R., Harmmel, R. and Veith, T.L. (2007) Model evaluation guidelines for systematic quantification of accuracy in watershed simulations. American Society of Agricultural and Biological Engineers, 50, 885-899.
64. Xu, M. and Qi, Y. (2001) Soil-surface CO_2 eflux and its spatial and temporal variations in a young ponderosa pine plantation in northern California. Global Change Biology, 7, 667-677.
65. Miller, R.D. and Johnson, D.D. (2002) The effect of soil moisture tension on carbon dioxide evolution, nitrification, and nitrogen mineralization. Soil Science Society of America, 28, 644-647.

66. Ludovici, K.H., Zarnoch, S.J. and Richter, D.D. (2002) Modeling in-situ pine root decomposition using data from a 60-year chronosequence. Canadian Journal of Forest Research, 32, 1675-1684.

Chapter 10

Ocean Carbon Sequestration by Direct Injection

Aaron Chow[1]

[1] Berkeley Research Group, Waltham, Massachusetts, USA

1. INTRODUCTION

The ocean is the largest sink of atmospheric CO_2 (about 7 petagrams (Pg) per year) (1 Pg = 1 gigaton = 10^{15} g). Dissolved CO_2 (passively entering the ocean via diffusion from the atmosphere) has already acidified the surface ocean, the most productive region of the ocean. Ocean carbon sequestration (OCS) is a method to distribute CO_2 more evenly throughout ocean depth and minimize surface ocean impacts. There are two major methods of OCS – direct injection and ocean fertilization (promoting photosynthetic fixation of CO_2 by ocean organisms). This chapter focuses only on the direct injection as a method of OCS. This chapter will first describe the physical mechanism by which CO_2 can be stored in the ocean water column at depth. It will summarize past ocean direct injection studies, and outline the effects of increased dissolved CO_2 and locally increased CO_2 partial pressure on marine organisms. It will also include a discussion of the engineering challenges of delivering CO_2 to the water column, including the selection of injection sites to minimize CO_2 outgassing to the atmosphere as well as minimizing marine life impacts. Finally, this chapter will address the legal, policy and public outreach issues that have ultimately precluded implementation of OCS using direct injection.

1.1. Motivation of Ocean Carbon Sequestration (OCS)

The ocean is presently the largest sink of atmospheric CO_2 (about 7 Pg per year) [1]. The Earth's oceans cover over 70% of the Earth's surface, and have an average depth of 3,800 m. However, dissolved CO_2 is already causing surface ocean acidification (most productive region of ocean) as it equilibrates with the atmospheric CO_2[2]. By 1994, the total atmospheric release of anthropogenic (i.e., man-made) carbon was about 244 Pg of carbon (PgC) from fossil fuel combustion, and about 140 PgC from land use change (e.g., deforestation) [3]. The oceans have absorbed about one-third of anthropogenic CO_2 (the atmosphere retained about 43%, while the oceans absorbed about 30%), leading to a decrease of surface-ocean total pH by about 0.1 units from about 8.2 to 8.1. If CO_2 emissions continue unabated the subsurface ocean total could decline by 0.7 units by 2300. To place in a geological perspective, the surface ocean pH (on a total scale) has not been below 8.1 during the past 2.1 million years. The total addition of carbon into the atmosphere is expected to be about 5000 PgC – the estimated fossil fuel reserves excluding hydrates -- in the next 500 years. This is a higher rate of carbon addition than ever experienced by the earth over a short geological time scale [3].

At the same time the ocean pH in the deep ocean has been decreasing at a relatively slower rate compared with the surface ocean [1, 4]. Rising concentrations of greenhouse gases in the atmosphere are implicated in adverse climate changes and two-thirds of the change is attributed to CO_2 [1]. Ocean carbon sequestration was conceived as a method to distribute CO_2 more evenly throughout the ocean column, especially into deep ocean waters, and minimize surface ocean impacts while the ocean CO_2 levels equilibrate with the atmosphere.

There are two major methods of OCS – direct injection and ocean fertilization (promoting photosynthetic fixation of CO_2 by ocean organisms). This chapter focuses only on the direct injection of CO_2.

2. PHYSICAL DESCRIPTION OF DIRECT INJECTION

2.1. Physical properties of ocean/CO_2 system

The conditions under which CO_2 can exist in a gas, liquid, solid or hydrate, and aqueous phases are depicted in the phase diagram (see Figure 1) [5] At typical ocean temperatures and pressures, CO_2 exists as a gas above 500 m depth, and a liquid below this depth. Between 500 and 2700 m depth, liquid CO_2 is less dense than seawater and would float, while below 2700 m CO_2 is denser than seawater and would sink. A solid CO_2 hydrate phase is thermodynamically stable in the ocean at low temperatures; CO_2 hydrates are discussed in Section 2.3.

2.1.1. Solubility of CO_2 in the ocean

CO_2 dissolves in ambient seawater that is not saturated with CO_2. Once dissolved, aqueous CO_2 exists in various charged forms in water according to these main reactions, known as the carbonate system [6]:

$$CO_2(aq) + H_2O = H_2CO_3(aq) \tag{1}$$

$$H_2CO_3(aq) = H^+ + HCO_3^- \tag{2}$$

$$HCO_3^- = H^+ + CO_3^{2-} \tag{3}$$

The total dissolved inorganic carbon (C_T) is defined as:

$$C_T = [H_2CO_3(aq)] + [HCO_3^-] + [CO_3^{2-}] \tag{4}$$

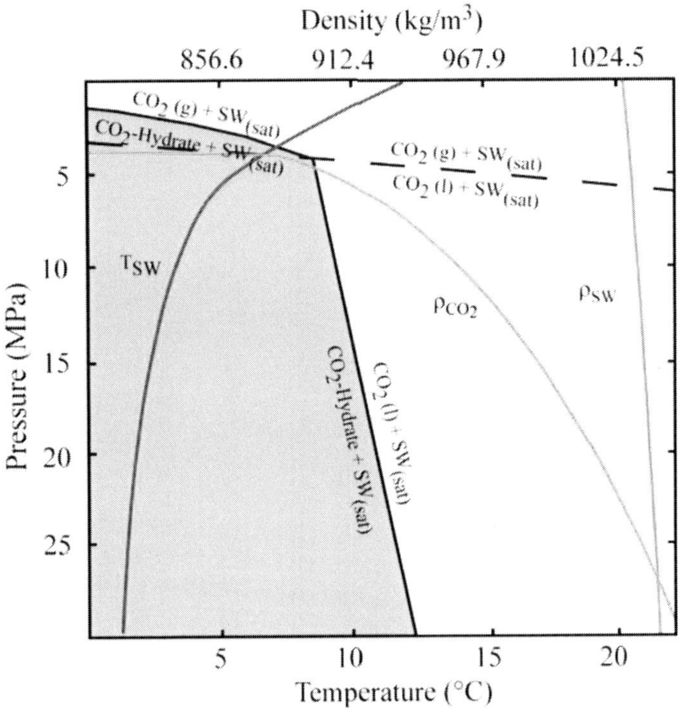

Figure 1. Phase Diagram for CO_2 in the ocean [5]. Gray area: region of stability for CO_2 hydrates; dashed line – gas/liquid phase transition for pure CO_2; red line – depth - temperature profile of the Pacific Ocean at 152°W, 40°N; green line – CO_2 density; blue line – seawater density (35.0 PSU).

The result of this is that increasing dissolved CO_2 will shift the equilibria (1) and (2) above to the right, and lower the local pH of the ambient seawater. [7]. At a typical surface seawater water pH of 8.2, the equilibrium (3) is shifted to the left with the addition of CO_2. The net result of an increase in dissolved CO_2 is the decrease of pH, an increase in HCO_3^- and a decrease of CO_3^{2-} (Figure 2).

In turn, the dissolved CO_2 causes an increase of the density of the seawater via the solute density effect [9] that has implications for the physical design of a direct injection of CO_2 into the ocean.

2.1.2. CO_2 partial pressure (pCO_2)

The partial pressure of CO_2 of a sample of water, denoted by pCO_2, is the pressure of gaseous CO_2 which, if allowed to equilibrate with water, will result in the same amount of dissolved CO_2 as observed in the sample. It is related to the solubility of CO_2, C_s, and the concentration of $H_2CO_3(aq)$ by the following:

$$pCO_2 = \frac{[H_2CO_3(aq)]}{C_s(T,S)} \tag{5}$$

where $C_s(T,S)$ (usually expressed in µatm) is dependent on the local temperature T and salinity [10] provide empirical relations to obtain C_s.

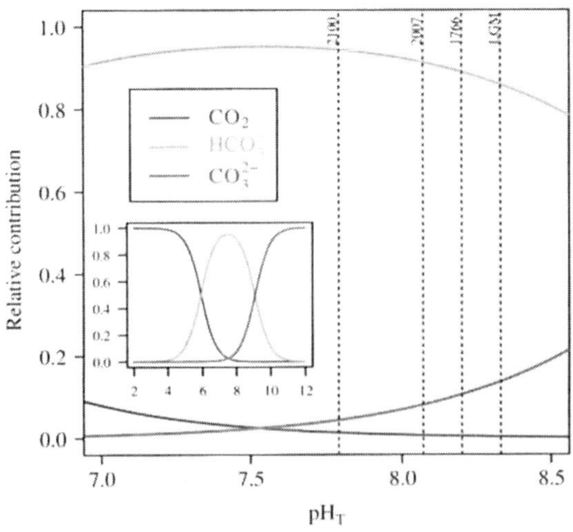

Figure 2. Bjerrum (pH_T (total scale pH) – relative speciation) plot showing the relative contributions of CO_2, HCO^{3-} and CO_3^{2-} to the dissolved inorganic carbon as a function of pH, at 15 deg C and a salinity of 35 PSU. The dashed vertical lines indicate the average open ocean surface pH_T during the Last Glacial Maximum (LGM), 1766, 2007 and 2100 (projected) [8].

As the atmospheric CO_2 concentration increases, pCO_2 levels increase in the surface ocean as it equilibrates with the atmosphere. The ratio of the relative change in pCO_2 to the relative change in C_T is known as the Revelle factor, and is inversely proportional to $[CO_3^{2-}]$. The inverse of the Revelle factor is also often termed as the buffering capacity of the ocean. According to this relation, a doubling in atmospheric CO_2 would only increase the total dissolved CO_2 by about 10%. [11]

While sea-air equilibria for most gases like oxygen occur over a time scale of days, it can take ~8 months for CO_2 to reach equilibrium at the surface, because the dissolved CO_2 in the carbonate system does not remain a dissolved gas but instead causes an increase in HCO_3^-. [4].

2.1.3. Calcium carbonate saturation state of seawater

The $CaCO_3$ saturation state of seawater is defined as Ω:

$$\Omega = \frac{\left[Ca^{2+}\right]\left[CO_3^{2-}\right]}{K_{sp}} \tag{6}$$

where $[Ca^{2+}]$ and $[CO_3^{2-}]$ are the seawater concentrations of Ca^{2+} and CO_3^{2-}, and K_{sp} is the solubility product of either calcite or aragonite (the two major forms of $CaCO_3$). If Ω for aragonite (Ω_a) for instance is greater than 1, then aragonite is supersaturated and solid aragonite would begin to precipitate; if Ω_a drops to below 1 then aragonite is undersaturated with respect to the ambient ocean and solid aragonite would begin to dissolve. Because K_{sp} increases with pressure, for both aragonite and calcite there is a transition of the saturation state from $\Omega > 1$ to $\Omega < 1$ sediments with depth [3]. The depth at which $\Omega = 1$ for a mineral is known as its saturation horizon.

2.2. Methods of direct injection

CO_2 sequestration first involves capture from their sources, of which one major type of the coal-fired power plant. The CO_2 emissions are relatively pure from coal-fired power plants and could be isolated and injected into the ocean. A typical 500 MW power plant produces about 130 kg/s of CO_2. [12]. After CO_2

capture, the CO_2 would be transported to the ocean via a pipe or ship to the ocean for direct injection. Technologies for CO_2 direct injection include: Liquid CO_2 droplets [13]; CO_2 laden seawater [9,14,15]; Solid CO_2 (dry ice) [16,17]; and CO_2 lake formation (See Figure 3)

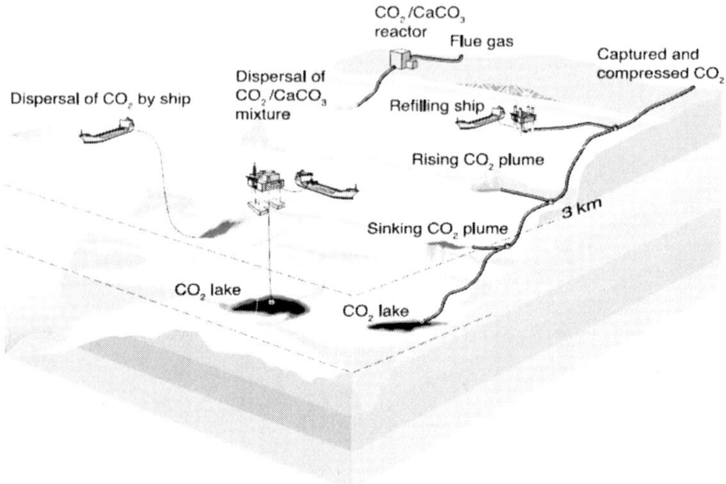

Figure 3. Ocean storage strategies (From Goddard, in [1]).

2.3. CO_2 hydrates

As shown in Figure 1, at lower temperatures (below about 5 - 10 degrees C) and at high pressure (corresponding to an ocean depth of about 400 m or greater) solid CO_2 hydrates are thermodynamically stable. CO_2 hydrates consist of molecules of CO_2 inside a cage-like structure of hydrogen-bonded water molecules [18]. They are of interest as a possible vehicle for deeper ocean carbon sequestration, because they are denser than seawater, and will sink unaided while dissolving to promote dispersion in the ocean. Pure hydrate particles are difficult to produce, but the Oak Ridge National Laboratory (ORNL) has designed a continuous CO_2-seawater co-flowing injector to create cylindrical composite particles comprised of CO_2 hydrate (negatively buoyant), liquid CO_2 (slightly positively buoyant at 1000-1500 m depths) and seawater [19].

Although CO_2 hydrates are thermodynamically stable, they will dissolve in ambient seawater upon release, because CO_2 is under-saturated in the ambient water. Field and laboratory observations confirmed that both pure hydrates and partially reacted cylindrical composite particles dissolved in the ambient seawater [20,21,23].

2.4. CO_2 droplet and hydrate studies

Numerical efforts to simulate the behaviour of CO_2 droplet plumes have included solving the full three dimensional Navier-Stokes equations in quiescent ambient sea conditions [7,22,24]. Bubble plume models calibrated using laboratory observations have also been applied to CO_2 droplet releases [25-27].

Field tests were conducted using CO_2 hydrate composite injectors [23,28]. The latest survey, with a hydrate reactor located at an ocean depth of ~1500 m, produced curved negatively buoyant cylindrical particles with diameters ~2.2 cm and lengths up to ~1 m. Applying a drag coefficient model to observed initial settling velocities and dissolution rates during the most recent survey [29,30], the hydrate conversion efficiency (percentage of liquid CO_2 converted to hydrate) in the field was ~ 15-20% resulting in particles with specific gravity 1-2% greater than seawater, which lead them to sink to a depth below discharge of roughly 100 m. Greater sinking could be achieved using larger particles. Discharging particles with a range of sizes and densities (reflecting different conversion rates) would cause differential settling resulting in spreading in the down-current and vertical directions. Furthermore, towing the source from a moving ship would contribute additional dispersion [29].

An alternative approach to enhancing mixing and vertical descent is to release a continuous stream of particles, forming a dense plume which would sink both due to the density of the particles as well as the increased density of seawater containing dissolved CO_2. An integral double plume model [25,29,31] was used to simulate the behavior of continuous streams of composite particles released to a quiescent ocean, with typical ambient stratification, at CO_2 loadings of 0.01 to 1000 kg/s. Results showed that, for a CO_2 release of 100 kg/s (roughly the emission from a 500 MW coal-fired power plant), a plume composed of 2.2 cm diameter composite particles with 16% reaction efficiency would sink about 1000 m, approximately 10 times the individual particle sinking depth. A plume composed of similar particles, but with a diameter of 5 cm, would sink about 2000 m (~5 times the individual particle depth), while plumes composed of larger particles, or particles exhibiting higher reaction efficiency, would reach the seafloor (as would the individual particles).

Two ambient effects reduce the performance of a plume: stratification and ocean currents. Plume sinking is hampered by strong ambient stratification which causes trapping of entrained seawater at intermediate depths below release. Density stratification weakens at depths below 1500 m [32], so from the perspective of reduction of plume trapping, regions of the ocean deeper than 1500 m are potentially favourable for depositing CO_2 [33].

3. ENVIRONMENTAL IMPACTS/CHALLENGES

Some of the concepts relevant to the impacts of OCS by direct injection (e.g. ocean acidification) are presented in this section. The reader is directed to [1,11] for a more detailed and comprehensive summary of the causes and effects of ocean acidification.

3.1. Long term stability of dissolved CO_2 in the ocean

Investigations and estimation of the long term stability is described in greater detail in [1]. Numerical ocean models indicate that placing CO_2 in the deep ocean would isolate most of the CO_2 from the atmosphere for several centuries, but over longer times the ocean and atmosphere would equilibrate.

Relative to direct atmospheric release, direct injection of CO_2 into the ocean could reduce the rise and peak of atmospheric CO_2 levels over the next several centuries. After several centuries, the CO_2 released in the ocean would be transported back to the ocean surface and interact with the atmosphere again. However, in the new equilibrium, most (66% to 85%) of the injected CO_2 would still remain in the ocean despite contacting the atmosphere [1].

Generally, carbon injected in the deep ocean would equilibrate with the atmosphere over a time scale of 300 to 1000 years, based on radiocarbon and other tracer dating to estimate the age of the deep seawater. The estimated age of the North Pacific deep water is 700 – 1000 years, while the North Atlantic deep water is estimated to be only about 300 years old. A large number of numerical three dimensional ocean general circulation models were used to study CO_2 retention. The models generally predict a higher retention time with a deeper injection depth (isolation of CO_2 from the atmosphere is nearly complete for 100 years with an injection depth of 3000 m). Consistent with the radioactive tracer dating, many of the models suggest that the Pacific Ocean would retain a larger fraction than the Atlantic Ocean. However, the models vary greatly in their predictions on the actual time taken for CO_2 injected at a particular site to once again make contact with the atmosphere [1, 34].

Additionally, other geochemical factors may affect these predictions. For example, a higher ocean temperature, as well as a higher dissolved inorganic carbon concentration may lead to a lower efficiency for the ocean to absorb additional CO_2. (See [11]).

3.2. Potential pH and carbonate system changes from added CO_2

As described in Section 2.1, ocean acidification has been occurring since the Industrial Revolution. This section describes the effect of continued ocean acidification on the ocean's carbon cycle and marine ecosystems. Between

1991 and 2006, North Pacific ocean pHT showed a decrease of 0.06 units over the upper 500 m of ocean. In the Iceland Sea, the trend of pH_T decrease between 1985 and 2008 in the surface ocean was 0.0024 units per year, with a corresponding decrease in Ω_a of 0.0117 units per year. The decline in pH_T below 1,500 m in the Iceland Sea was one-quarter of that on the surface, with a corresponding decrease in Ω_a at 0.0009 units per year [4].

Another consequence of the increased dissolved CO_2 in the ocean, as described in Section 2.1.1, is the increase of HCO_3^- and a decrease of CO_3^{2-} in the ocean. The decreased CO_3^{2-}in turn leads to the decrease of the local value of Ω in the ocean. As there is a transition from saturation to undersaturation from $\Omega = 1$, this means that the saturation horizons for both aragonite and calcite would both become less deep with time [3]. The decrease in Ω_a caused the aragonite saturation horizon (ASH), the interface between supersaturated waters above and undersaturated waters below, to rise (shoal) at a rate of 4 m per year. The decrease in Ω, and therefore the shoaling rate for the ASH, is predicted to be more pronounced near the poles, and more severe in the Arctic Ocean than the Southern Ocean, partly because the polar oceans have lower initial concentrations of CO_3^{2-} [4].

It was proposed [3] that the addition of CO_2 followed by global increase in surface temperature can be compared to that which occurred during the Paleocene-Eocene Thermal Maximum (PETM, ~55 million years ago). During PETM, about 3000 PgC was added to the over an estimated 6000 years. However, the current estimate for expected total anthropogenic carbon addition is a larger rate of carbon input over a shorter period of time, about 5000 PgC over about ~500 years. The next highest global carbon addition was experienced by the earth during the Paleocene-Eocene Thermal Maximum, (~55 million years ago) where about 3000 PgC was added over ~6000 years. During the PETM, the effects of ocean acidification on surface calcifying organisms was limited, but the conditions of the PETM were not identical to the predicted future scenario, notably in that the carbon input rate was still much slower than the modern anthropogenic carbon addition. Nevertheless, studies of the PETM may inform future predictions of the behavior of ocean marine life with a large increase of atmospheric CO_2. [3,4].

3.3. Effect of pCO_2 increase on organisms

Effects of elevated CO_2 levels and acidified seawater on marine organisms are explained in in more detail in [1,11,35,36].

At acute levels CO_2 has a narcotic effect on animals and causes respiratory distress and death. The work of [37 – 41] that model the lowered pH on passive marine organisms such as zooplankton that spend varying times in and out of a

CO_2 plume, and found that minimizing the local dissolved CO_2 and pH drops will reduce the mortality rate.

Non-lethal effects have also been observed due to hypercapnia (elevated CO_2 exposure) [42-44]. Tamburri et al. [42] have observed the narcotic effects of increased CO_2 levels on mobile deep sea animals in the field; they also observe that while many tend to avoid CO_2 plumes, some may risk the narcotic effects to obtain food. They note [42] that increased partial pressure of carbon dioxide will also have a detrimental effect on marine organisms, such as causing slow respiratory distress and inducing a narcotic effect on fish. Passive marine animals may experience depressed ion exchange capability and metabolism when exposed to lower, chronic levels CO_2. Some studies also show slowed growth in mussels and corals, as well as developmental effects on some marine larvae and eggs (brittle stars and bivalves) [35].

The primary effect of acidified seawater exposure by organisms is acidosis, the decrease of pH in body fluids. Intracellular and extracellular processes have been shown to be disrupted when seawater pH drops to a range of about 6.0 – 7.8. Many marine animals counter acidosis by increasing bicarbonate ion production (e.g. in the gills) [35]. Barry et al. report that organisms that have weaker control of their internal fluid chemistry, and that rely on passive molecular diffusion for gas exchange such as sponges, echinoderms, may have greater sensitivity to ocean acidification [45].

Some organisms may adapt to hypercapnia (elevated CO_2) better than others [46]. For example, tropical fishes, as they live closer to the edge of oxygen limitation than temperate fishes, may make them more sensitive to the combined effects of ocean temperature and ocean acidification than their temperate counterparts. For example, studies on acutely exposed tropical cardinal fishes to 1 week of pCO_2 of 1000 μatm resulted in decreases of aerobic scope and critical swimming speeds by about 40 – 50%, but a similar study conducted for Atlantic cod after 12 months of exposure to both 3000 and 6000 μatm did not result in any significant change in swimming capacity.

High CO_2 levels (up to a pCO_2 of 16,000 ppm [47] have also been observed in ocean bottom waters and marine sediments where there are high rates organic matter oxidation and low rates of mixing with the overlying seawater. Under these conditions, high CO_2 concentrations are often accompanied by low O_2 concentrations. Near the surface at night, respiratory fluxes in some relatively confined rock pools of the intertidal zone can produce high CO_2 levels. [1]. Portner et al. [46] report that high pCO_2 is found in oxygen minimum layers. They report that elevated pCO_2 is linked to acid-base regulation and respiration in fish. However, they also report that coastal and mid-water animals (both pelagic and benthic) regularly experience a large range of pCO_2 values (500 to 9400 μatm) in estuaries [46]. "These patterns suggest that in some environments, organisms have evolved to tolerate relatively wide pH oscillations and/or low pH values." [1]

Organisms such as the Humboldt squid, although thought not to be able to adapt physiologically to future changes to the oceans oxygen balance, have been observed to thrive in oxygen minimum layers which tend to have low pH and are undersaturated with respect to calcium carbonates [46].

Deep sea ecosystems depend on sinking particles of organic carbon, made by photosynthesis near the ocean, settling down through the water. Most species living in the deep sea display very low metabolic rates [48, 49], especially in oxygen minimum layers [51]. Organisms living in the deep seawaters have adapted to the energy limited environment by conserving energy stores and minimizing energy turnover. Turley et al. also suggest the depletion of oxygen as a contributing factor to the increased prevalence of harmful algal blooms, though the link between anthropogenic CO_2 and algal blooms remains controversial [36].

Finally, as many marine organisms synthesize and depend on calcium carbonate structures (e.g. shells), the implication of a lowered CO_3^{2-} and Ω in the ocean is the potential for reduction of their habitats.

As a guide, [1] uses a pH drop of 0.1 units as the threshold pH drop for insignificant marine life impact; it is also within the observed natural variability in the ocean. The US Environmental Protection Agency proposed that the threshold for open waters at depths greater than the euphotic zone, the pH value should not drop more than 0.2 pH units outside the range of natural variation [11]. [39] shows that some theoretically modeled scenarios of carbon dioxide releases (for example, releasing sinking CO2 hydrates from a fixed or moving source at 1,500 m, injecting 10 to 1000 kg/s) would result in local pH drops within this guideline threshold in the vicinity of the release point. Others (e.g. Rockstrom et al.) have introduced the concept of planetary boundaries, and for CO_2 they have proposed a threshold carbonate ion concentration. As a first estimate, they proposed that the oceanic aragonite saturation state Ω_a be maintained at 80% or higher of the average global pre-industrial surface seawater level of 3.44 [50]. As with [1], these planetary boundaries are guides for a sustainable global environment, and (with the exception of the US Environmental Protection Agency for pH) have not been implemented as a regulatory threshold.

3.3.1. Comparison with naturally occurring ocean CO_2 vents

In the ocean, hydrothermal vents are submarine volcanic structures that act as natural sources of CO_2 in the ocean. These have been observed as potential natural analogues of OCS direct injection points. Field observations of hydrothermal vents have shown large fluctuations of pCO_2 (up to 80,000 ppm), over 100 times that observed in typical deep seawater). Over time, the vents have sustained organisms that are specially adapted to living in elevated pCO_2 conditions [52].

Observations near hydrothermal vents have shown that ocean acidification reduced biodiversity below a mean pH_T of 7.8 [53]. While Echinoderms are notably absent from habitats with naturally high CO_2 levels such as hydrothermal vents and shallow CO_2 vents off the coast of Italy [53], sponges appeared to tolerate these same sites.

As observed in [45], "[h]owever, while commonly the literature contains results of short term studies of organism physiology and survival, they may not be indicative of eventual long term consequences of ocean acidification."

4. ENGINEERING FEASIBILITY/CHALLENGES

4.1. Site selection for injection

As described in [1,54], to date there are no publications dedicated to site selection for direct ocean injection. Although numerical models have predicted CO_2 retention time as a function of the injection location, they have not consistently agreed on any individual location for direct injection. The only agreement appeared to be that a larger depth of injection would result in a longer isolation of CO_2 from the atmosphere [34]. In contrast, [55] presented a study of site selection for deep sea geological storage, highlighting the potential of storage in basalt aquifers along particular seismic and aseismic oceanic ridges. This section therefore discusses factors that should be considered site selection criteria based on to be considered when selecting a site for OCS. Environmental goals of site selection include reducing the likelihood of outgassing, and minimizing acute impacts to ocean organisms, as described in Section 3. Additional considerations include the costs of OCS, applicable international policies (such as regulations regarding disposal and cross border transport) – these factors are presented in Sections 4.2 and 4.3.

4.2. Cost of OCS

Costs were estimated for ship transport of liquid CO_2 to an injection platform, with CO_2 injection from a vertical pipe, or a ship trailing an injection pipe, to water at 3000m [1]. The cost estimate of ocean storage is the sum of three major components: tank storage of CO_2 onshore awaiting shipping; the shipping of CO_2; and direct injection of CO_2 into the ocean (either via an ocean platform, a moving ship, or a pipeline). The estimated sum of the three components (including an assumption of 3% CO_2 emissions from boil off and fuel consumption) is 11.9 and 13.2 US$/ton CO_2 net stored from shipping to

100 km and 500 km offshore, respectively [56]. Cost estimates presented do not include transport of CO_2 onshore.

The cost for transporting CO_2 from a power plant located at the shore through a pipeline running on the sea floor to an injection nozzle was also estimated in [56]. CO_2 captured from a pulverized coal fired power plant with a net generation capacity of 600 MWe is transported either 100 or 500 km by a CO_2 pipeline for injection at a depth of 3000 m at a cost of 6.2 US$/ton CO_2 net stored (100 km case) to 31.1 US$/ton CO_2 net stored (500 km case). Other technical challenges that may not be accounted for include: residual chemicals, metals, minerals and oils that may be released during drilling activities; and the fact that liquefied CO_2 is highly corrosive, requiring that piping for CO_2 delivery would require anti-corrosion coatings, which themselves may pose contamination issues [35].

There are no published cost estimates specific to the production of a CO_2 lake on the sea floor; however, given the dominance of pipeline costs, it is reasonable to assume it to be similar to deep water injection. [1,56].

5. POLICY ISSUES/CHALLENGES

Since offshore OCS is likely to take place in international waters, several international environmental agreements may apply, mainly those that aim to minimize potential risk s to the marine environment. The main international treaties are the Law of the Sea, the London Convention, London Protocol, and the OSPAR Convention. A succinct background of these treaties is taken directly from [57]:

"International marine environment protection was established in 1972 with the London Convention to regulate the dumping of wastes and other matter at sea. In 1982, this field was extended through the adoption of the United Nations Convention on the Law of the Seas (UNCLOS). Being an overarching construction, UNCLOS does not contain detailed operative provisions on most maritime issues; rather, it provides a framework for all areas, including marine protection, and allows other, more targeted treaties to fill in the gaps...With regard to marine pollution, global standards are set by the Convention on the Prevention of Marine Pollution by Dumping of Wastes and other Matter, signed in London in 1972 (London Convention). Beneath the London Convention exist several regional agreements that cover specific areas of the ocean [Also listed in [57]]. The most widely known of these is OSPAR, the Convention for the Protection of the Marine Environment of the North-East Atlantic. OSPAR is also notable as its regulations on marine pollution are markedly stricter than those of the London Convention, and its decisions are legally as opposed to politically binding on its Contracting Parties."

5.1. 1996 London protocol

UN Convention on Climate Change encouraged the use of the oceans as a reservoir for CO_2, but the UNCLOS (in force since 1994) did not give clear guidance on OCS [1]. With respect to CO_2 storage, the original London Convention (with 80 contracting parties, and in force since 1975) only applied to storage by aircraft and vessels and platforms in the water column. As a result, the London Convention did not apply to storage of CO_2 in the seabed or the water column itself [57].

In November 1996, the London Protocol was established that prohibited the disposal of "industrial waste" into international waters. The list of prohibited substances that were categorized as "industrial waste" were contained in Annex I of the London Protocol. However, in 1996 the London Protocol did not give an opinion whether CO_2 was categorized as a "waste material generated by manufacturing or processing operations" [1]. The London Protocol entered into force March 2006 [57].

5.2. OSPAR convention

In 1992, the OSPAR Commission for the Protection of the Marine Environmental of the North-East Atlantic, was formed which unified the 1972 Oslo and 1974 Paris Conventions. It brought together the governments of Belgium, Denmark, Finland, France, Germany, Iceland, Ireland, Luxembourg, the Netherlands, Norway, Portugal, Spain, Sweden, Switzerland and the United Kingdom, together with the European Community (EC). It is considered the most comprehensive and strict legal framework governing the marine environment. As mentioned above, the contracting parties are legally bound to OSPAR's decisions [57,58].

5.3. 2007 OSPAR amendments for CO_2 storage and implications to OCS

In June 2007, the OSPAR issued two amendments to the OSPAR Convention: the first, a decision to prohibit the storage of CO_2 streams in the water column or on the sea bed in the Northeast Atlantic; and the second, a decision to allow the storage of CO_2 in subsea sediments. [59,60]

In the first amendment, OSPAR stated that CO_2 storage in the water column or on the sea bed "is not a sustainable storage option, is likely to result in harm to living resources and marine ecosystems and is thus neither a viable solution with regard to mitigating climate change nor compatible with the aims of the

[OSPAR] Convention." However, in the first amendment, OSPAR indicated that ocean storage of CO_2 in the water column or on the seabed is nevertheless still under consideration in international forums. [59]

The OSPAR amendments provided a framework for its contracting national governments to develop permitting programs for CO_2 storage. For example, a list of the minimum items required in an offshore CO_2 storage permit included: a description of the project, including injection rates; types, amounts and sources of CO_2; the location of the facility; characteristics of the geological formation; methods of transport; and a risk management plan, with monitoring and verification measures, mitigation steps and a site closure plan [60].

In July 2011 the contracting parties of the OSPAR Convention ratified the 2007 Amendments to allow for CO_2 storage in subsea geological formations [58].

5.4. 2007 Amendment to the London Protocol

In 2007, an amendment to the London Protocol (Annex 1) allowed for storage of CO_2, if the disposal is into a sub-seabed geological formation, if CO_2 streams are "overwhelmingly" carbon dioxide, and as long as no wastes are added. This amendment provided that CO_2 streams may only be considered if [60,61]:

1. disposal is into a sub-seabed geological formation;
2. they consist overwhelmingly of CO_2. They may contain incidental associated substancesderived from the source material and the capture and sequestration processes used; and
3. no wastes or other matter are added for the purpose of disposing of those wastes or other matter.

The amendments to Annex 1 entered into force on 10 February 2007. In contrast to the OSPAR Convention Amendment that only covered the Northeast Atlantic, the 2007 London Protocol Amendment specifically prohibited direct injection of CO_2 for OCS for all London Protocol contracting parties.

5.5. 2009 London Protocol amendment for transboundary transport of CO_2

Article 6 of the London Protocol (on the export of wastes or other material) was largely interpreted by the contracting parties as prohibiting the export of CO_2 from a contracting party for injection into sub-seabed geological formations. In 2009 Article 6 was amended to allow for cross-border transportation of CO_2. [61]

As of 2011, there were 40 contracting parties to the London Protocol. Of these parties, 27 must also accept the 2009 amendment to Article 6 for it to enter into force. However, not all of the parties have been interested in offshore CO_2 storage or cross-border movement of CO_2, and have placed the ratification of Article 6 as a low priority. Cross-government cooperation will probably be required for ratification to occur. In some countries, the ratification may also be pending other laws and regulations that need to be changed for carbon storage and sequestration in general [61].

Therefore, although geological carbon sequestration in the ocean has been approved in principle the OSPAR Convention and even the London Protocol, the Article 6 amendment may continue to pose a policy barrier to OCS deployment in the foreseeable future.

5.6. Public outreach: Lessons from Hawaii

It is noteworthy that no field studies demonstrating OCS at a significant scale have been conducted so far prior to its prohibition through the 2007 London Protocol and OSPAR Convention amendments. The largest scale attempt at demonstrating OCS was the Hawaii CO_2 direct injection experiment This section outlines the failure of the Hawaii experiment mainly fuelled by a lack of early public outreach, and outlines some lessons learnt from the Hawaii project. [62,63]

In 1997 the US Department of Energy, the New Energy and Industrial Technology Development Organization of Japan (NEDO) and the Norwegian Research Council (NRC) signed an agreement to conduct experiments to evaluate the behaviour of liquid CO_2 releases in to the ocean. While the project was announced in Kyoto in 1997, with a high profile to demonstrate the sponsors' commitment to CO_2 mitigation, few resources were subsequently invested in public outreach.

The project scientists and sponsors selected an area off the coast of the Big Island of Hawaii to conduct the pilot CO_2 study, based on technical feasibility and existing research infrastructure. However, they did not gauge the public perception prior to site selection. The local population only learnt of the injection project planned in their waters when it was first published in a newspaper article. In an area where the ocean is viewed as a major natural resource, the public perceived of the "dumping" of CO_2 as a violation, and strongly opposed its continuation.

Eventually the pilot injection project was abandoned in Hawaii. In order to salvage the project, scientists attempted to instead conduct an injection study in Norway. However, here the actions of Greenpeace stopped any further testing, thus precluding completely any chance of field scale testing of direct-injection OCS.

Although the introduction of OCS was initially high profile the sponsors' commitment to CO_2 mitigation, few resources were subsequently invested in public outreach. Moreover, the sponsors largely did not include the public in their decision to site the pilot injection experiment in Hawaii, nor did they factor public perception of potentially conducting a CO_2 injection experiment in an area where the ocean is viewed as a major environmental resource.

Reiner (2008) cited the US National Institute of Standards and Technology (NIST) 2002 workshop for "Best Practices for Communication of Science and Technology to the Public" as a resource that offered key recommendations for public outreach, including:

- Illustrating both the scientific process and unresolved scientific questions, rather than showing a result based on a black box model
- Using scientists in a public education role for a range of audiences (children and adult)
- Incorporating a wide source of knowledge from others so as to avoid parochialism
- Presenting the issues from the audiences' point of view
- Using face-to-face communications whenever possible to foster trust
- Reaching out beyond the science-attentive public, e.g. presenting at shopping malls, to disadvantaged youth
- Using multimedia and illustrations
- Providing press releases in forms easily usable to time-strapped journalists
- Avoiding perceptions of environmental injustice, bias or conspiracy
- Including the public from the start to avoid charges of "sneaking up on us" or less than full disclosure
- Avoiding letting the project become a "political football" by creating vocal supporters within the community

In addition, Reiner suggested that early outreach to the public via the internet during developmental stages of a project is important before the project becomes newsworthy and receives attention from mainstream media outlets. [63].

The lack of outreach is reflected in the low level of understanding that has remained among the public, as well as relatively low public acceptance of carbon capture and sequestration (both geologic and ocean). Reiner summarized the European Commission's survey of the public from 25 countries of the European Union (the Eurobarometer) that showed that, in 2007 (at the same time as the OSPAR convention and London Protocol amendments), only 21% of those surveyed have heard of carbon capture and storage (geologic or ocean), compared with 53% for hydrogen energy and cars, 41% for fuel cells, and 44% for geothermal energy [63]. In the US in 2004, only 2.5% of 1200 respondents in a web-based survey had previously heard of carbon sequestration. In 2007, Palmgren et al. surveyed 126 community respondents, who ranked OCS less favourable than geological carbon sequestration. Both carbon sequestration options were less favourable to the respondents than nuclear power [64].

6. CONCLUSIONS

Whether CO_2 is introduced intentionally, or passively diffusing from the atmosphere to the ocean, the ocean is and will remain the largest sink of anthropogenic CO_2. In addition to climate change implications of elevated atmospheric CO_2, a further impact is the acidification of the ocean. Effects of increased acidity and pCO_2 in organisms include respiratory distress (but some deep sea organisms take advantage of the CO_2/O_2 balance). There is also a risk of a reduced habitat as calcium carbonate stability zones decrease. However, further study is required to determine the variability of responses among marine species.

This chapter presented several methods by which direct injection of CO_2 into the ocean could be introduced. Some injection technologies were developed that would theoretically, combined with proper siting of injection points, cause a relatively minor impact to marine ecosystems. Some pilot scale field studies began that would have provided more information about environmental impacts, but they were nixed due to public opposition stemming from a lack of extensive and continuous public outreach from the onset. Since 2007, international policies began to prohibit direct discharge of CO_2 into the ocean, while favouring deep sea geological sequestration. CO_2 leaks (e.g. in the form of droplets [65]) from geological structures to the ocean water column are however still possible [35], so continued research and studies about the mechanisms of CO_2 leakage and the effects of increased dissolved carbon in the ocean continues to be an important topic of study for carbon sequestration.

7. ACKNOWLEDGEMENTS

This chapter contains work funded by Ocean Carbon Sequestration Program, Biological and Environmental Research (BER), U.S. Dept. of Energy (grant number DE-FG02-01ER63078), the National Energy Technology Laboratory, U.S. Dept. of Energy (grant number DE-FG26-98FT40334) and the Martin Family Fellows for Sustainability.

REFERENCES

1. Intergovernmental Panel on Climate Change (IPCC) (B. Metz B., Davidson O. eds. Carbon Dioxide Capture and Storage: A Special Report of IPCC Working Group III, Cambridge University Press, Cambridge UK; 2005
2. Caldeira, K., Wickett, ME. Anthropogenic carbon and ocean pH. Nature 2003;425, 365-365.
3. R. Zeebe R., Rigwell A. Past changes in ocean carbonate history. In: Gattuso J-P., Hansson L., eds; Ocean acidification. Oxford; 2010 p.21-40.
4. Orr JC. Past and recent changes in the ocean carbonate system. In: Gattuso J-P., Hansson L., eds; Ocean acidification. Oxford; 2010 p.41-66.
5. Bigalke N., RehderG., Gust G. Experimental investigation of the rising behavioe of CO2 droplets in seawater under hydrate-forming conditions. Environmental Science and Technology 2008;(42)5241 – 5246.
6. Morel FMM., Hering JG. Principles and Applications of Aquatic Chemistry. Wiley-Interscience, 1993.
7. Alendal G., Drange H. Two-phase, near field modeling of purposefully released CO2 in the ocean. Journal of Geophysical Research, 2001;106(C1):1085–1096.
8. Lavigne and Gattuso. In: Gattuso J-P., Hansson L., eds; Ocean acidification. Oxford; 2010.
9. Haugen P., Drange H. Sequestration of CO2 in the deep ocean by shallow injection. Nature 1992;357(28):1065–1072.
10. Dickson G., Goyet C., editors. Handbook of Methods for the Analysis of the Various Parameters of the Carbon Dioxide System in Sea Water. ORNL/CDIAC-74. 1994.

11. Gattuso J-P., Hansson L., eds; Ocean acidification. Oxford; 2010.

12. Herzog HJ., Adams EE., Auerbach D., Caulfield J. Environmental impacts of ocean disposal of CO2. Energy Conversion and Management, 1996;37(6-8):999–1005.

13. Brewer PG., Peltzer ET., Walz P., Aya I., Yamane K., Kojima R., Nakajima Y., Nakayama N., Haugan P., Johannessen T. Deep ocean experiments with fossil fuel carbon dioxide: creation and sensing of a controlled plume at 4 km depth. Journal of Marine Research, 2005;63(1):9–33.

14. Adams EE., Golomb D., Zhang XY., Herzog HJ. Confined release of CO2 into shallow sea water. In Direct Ocean Disposal of Carbon Dioxide, TERRAPUB 1995; Tokyo p.153–164.

15. Sato T., Kajishima T., Naguosa R. CO2 sequestration at sea by gas-lift system of shallow injection and deep releasing. Environmental Science and Technology, 2000:34:4140–4145.

16. Nakanishi N., Ohsumi T., Shitashima K. Sequestering of CO2 in a deep ocean. Technical Report, Central Research Institute of Electric Power Industry, 1994; 1646 Abiko, Abiko-City, Chiba 270-11, Japan.

17. Caulfield JA., Adams EE. Auerbach DI., Herzog HJ. Impacts of ocean CO2 disposal on marine life: II. probablistic plume exposure model used with a time-varying dose-response analysis. Environmental Modeling and Assessment 1997; 2:345–353.

18. Sloan ED. Gas hydrates: Review of physical/chemical properties. Energy & Fuels, 1998;12(2):191–196.

19. Lee S., Liang L., Riestenberg D., West OR., Tsouris C., Adams E. CO2 hydrate composite for ocean carbon sequestration. Environmental Science and Technology, 2003; 37:3701–3708.

20. Rehder G, Kirby SH., Durham WB, Stern LA., Peltzer ET., Brewer PG. Dissolution rates of pure methane hydrate and carbon dioxide hydrate in undersaturated seawater at 1000m depth. Geochimica et Cosmochimica Acta 2004;68:285–292.

21. Holder G., Mokka L., Warzinski R. Formation of hydrates from singlephase aqueous solutions and implications for ocean sequestration of CO2. In: Preprints of Spring 2000 National Meeting in San Diego, CA. ACS Division of Fuel Chemistry, 2001.

22. Chen B. A numerical prediction on plume structure of liquid CO2 in the ocean - a near field model. In: Proceeding of the 5th International Conference on Greenhouse Gas Control Technologies. 2001; p. 417–422, 2001.

23. Tsouris C., McCallum S., Aaron D., Reistenberg D., Gabitto J., Chow A., Adams E. Scale-up of a continuous-jet hydrate reactor for CO2 ocean sequestration. AiChE Journal 2007; 53(4):1017–1027.

24. Sato T., Hama T. Numerical simulation of dilution process in CO2 ocean sequestration. In: Proceedings of the 5th International Conference on Greenhouse Gas Control Technologies. 2001; p. 475–480.

25. Crounse B., Wannamaker EJ., Adams EE.. Integral model of a multiphase plume in quiescent stratification. Journal of Hydraulic Engineering, 2007;133(1):70–76.

26. Socolofsky SA., Adams EE. Multi-phase plumes in uniform and stratified crossflow. Journal of Hydraulic Research, 2002;40(6):661–672.

27. Socolofsky SA., Adams EE. Liquid volume fluxes in stratified multiphase plumes. Journal of Hydraulic Engineering, 2003;129(11):905 – 914.

28. Riestenberg, D., Tsouris C., Brewer PG., Peltzer ET., Walz P., Chow A., Adams E. Field Studies on the Formation of Sinking CO2 Particles for Ocean Carbon Sequestration: Effects of Injector Geometry on Particle Density and Dissolution Rate and Model Simulation of Plume Behavior. Environmental Science and Technology, 2005:39, 7287-7293.

29. Chow A., Adams EE., Israelsson PH., Tsouris, C. Carbon dioxide hydrate particles for ocean carbon sequestration. Energy Proceedia, 2009;1(1), 4937-4944.

30. Chow AC., Adams, EE. Prediction of Drag Coefficient and Secondary Motion of Free-Falling Rigid Cylindrical Particles with and without Curvature at Moderate Reynolds Number. Journal of Hydraulic Engineering 2011; 137(11)1406 – 1414.

31. Wannamaker EJ., Adams EE. Modeling descending carbon dioxide injections in the ocean. Journal of Hydraulic Research, 2006; 44(3):324–337.

32. R. Schlitzer. Electronic atlas of WOCE hydrographic and tracer data now available http://www.ewoce.org. Eos Trans. AGU, 2000; 81(5):45.

33. Chow A., Adams EE. Particle laden flows through an inverted chimney with applications to ocean carbon sequestration. Environmental Fluid Mechanics 2012; 12(1) 3 – 21.

34. Orr, JC. Modelling of ocean storage of CO2---The GOSAC study, Report PH4/37, International Energy Agency, Greenhouse Gas R&D Programme, Cheltenham, UK, 2004; 96 pp.

35. Blackford J., Widdicombe S., Lowe D., Chen B. Environmental risks and performance assessment of carbon dioxide (CO2) leakage in marine ecosystems. In: Maroto-Valer MM., editor. Developments and innovation in carbon dioxide (CO2) capture and storage technology. 2010, CRC Press.

36. Turley C., Blackford J., Harman-Mounford N., Litt E., Llewellyn C., Lowe D., Miller P., Nightingale P., Rees A., Smyth T., Tilstone G., Widdcombe S.. Carbon uptake, transport and storage by oceans and the consequences of change. In: Hester RE., Harrison RM., Carbon Capture, Sequestration and Storage. Royal Society of Chemistry, 2010.

37. Auerbach I., Caulfield JA., Adams EE., Herzog HJ. Impacts of ocean CO2 disposal on marine life: I. a toxicological assessment integrating constant-concentration laboratory assay data with variable concentration field exposure. Environmental Modeling and Assessment, 1997;2:333–343.

38. Caulfield JA., Adams EE., Auerbach DI., Herzog HJ.. Impacts of ocean CO2 disposal on marine life: II. Probablistic plume exposure model used with a time-varying dose-response analysis. Environmental Modeling and Assessment, 1997;2:345–353.

39. Israelsson PH., Chow AC., Adams EE. An updated assessment of the acute impacts of ocean carbon sequestration by direct injection. International Journal of Greenhouse Gas Control 2010;4(2): p. 262-271.

40. Israelsson PH. Studies of Lagrangian Modeling Techniques with Applications to Deep Ocean Carbon Sequestration. PhD thesis. Massachusetts Institute of Technology; 2008.

41. Sato, T. Numerical Simulation of Biological Impact Caused by Direct Injection of Carbon Dioxide in the ocean. Journal of Oceanography 2004;60, 807-816.

42. Tamburri MN., Peltzer ET., Friederich GE, Aya I., Yamane K., Brewer PG. A field study of the effects of CO2 ocean disposal on mobile deep-sea animals. Marine Chemistry 2000; 72:95–101.

43. Kita J., Watanabe Y. Impact assessment of high-CO2 environment on marine organisms. In: Proceedings of the Eighth International Conference on Greenhouse Gas Control Technologies, 2006.

44. Kikkawa T., Sato T., Kita J., Ishimatsu A. Acute toxicity of temporally varying seawater CO2 conditions on juveniles of japanese sillago (Sillago japonica). Marine Pollution Bulletin, 2006; 52:621–625.

45. Barry JP, Widdicombe S, Hall-Spencer JM. Effects of ocean acidification on marine biodiversity and ecosystem function. In: Gattuso J-P., Hansson L., eds; Ocean acidification. Oxford; 2010, p.192-209.

46. Pörtner, H.-O., Gutowska, M., Ishimatsu, A., Lucassen, M., Meizner, F. and Seibel, B.A. (2010). Effects of ocean acidification on nektonic organisms. In: Gattuso J-P., Hansson L., eds; Ocean acidification. Oxford; 2010, p.-154-175.

47. Knoll AH., Bambach RK., Lykkeboe G. Blood gas transport in the cephalopod, Sepia officinalis. Journal of Experimental Biology 1982;99:331-8.

48. Childress, JJ. Are there physiological and biochemical adaptations of metabolism in deep-sea animals? Trends in and Evolution Ecology 1995;10(1), 30-36.

49. Riebesell U., Tortell, PD. Effects of ocean acidification on pelagic organisms and ecosystems. In: Gattuso J-P., Hansson L., eds; Ocean acidification. Oxford; 2010, p.99-121.

50. Rockström, J., Steffen W., Noone K., Persson Å., Chapin FS, Lambin E., Lenton TM, Scheffer M., Folke C., Schellnhuber H., Nykvist B., De Wit CA., Hughes T., van der Leeuw S.,. Rodhe H., Sörlin S., Snyder PK., Costanza R., Svedin U., Falkenmark M., Karlberg L., Corell RW., Fabry VJ, Hansen J., Walker B., Liverman D., Richardson K., Crutzen P., Foley. M. Planetary boundaries:exploring the safe operating space for humanity. Ecology and Society 2009.;4(2): 32. [online] URL: http://www.ecologyandsociety.org/vol14/iss2/art32/

51. Seibel BA., Thuesen EV., Childress JJ., Gorodezky LA. Decline in pelagic cephalopod metabolism with habitat depth reflects differences in locomotory efficiency. Biological Bulletin 1997;192(2) 262-278.

52. Vetter EW., Smith CR. Insights into the ecological effects of deep ocean CO2 enrichment: the impacts of natural CO2 venting at Loihi seamount on deep sea scavengers. Journal of Geophysical Research 2005;110(C09S13).

53. Hall-Spencer JM., Rodolfo-Metalpa R., Martin S. Volcanic carbon dioxide vents show ecosystem effects of ocean acidification. Nature 2008; 454:96-9.

54. Herzog H., Caldeira K., Reilly J. An issue of permanence: assessing the effectiveness of ocean carbon sequestration. Climatic Change 2003;59(3), 293-310.

55. Goldberg D., Slagle AL. A global assessment of deep-sea basalt sites for carbon sequestration. Energy Proceedia 2009;1:3675-3682.

56. Akai M., Nishio N., Iijima M., Ozaki M., Minamiura J., Tanaka T.. Performance and Economic Evaluation of CO2 Capture and Sequestration Technologies. Proceedings of the Seventh International Conference on Greenhouse Gas Control Technologies, 2004.

57. International Energy Agency (IEA) and Organisation for Economic Co-operation and Development (OECD). Legal Aspects of Storing CO2, 2007.

58. OSPAR Commission for the Protection of the Marine Environment of the North-East Atlantic, Press release. Ratification of OSPAR carbon capture and storage measures. October 28, 2011. http://www.ospar.org/html_documents/ospar/news/ospar_pr_11_ratification_of_ccs_measure_en.pdf. Accessed August 6, 2013.

59. 4th Kyoto, Japan. Oslo and Paris Commission (OSPAR). OSPAR Decision 2007/1 to Prohibit the Storage of Carbon Dioxide Streams in the Water Column or on the Sea-bed. 25-29 June, 2007.

60. International Energy Agency (IEA), 2008. CO2 capture and storage: a key abatement option. 266 pp. http://www.iea.org/publications/freepublications/publication/name,35329,en.html. Accessed October 16, 2013.

61. International Energy Agency (IEA). Carbon Capture and Storage and the London Protocol. Options for Enabling Transboundary CO2 Transfer. 2011.

62. De Figueiredo MA., Reiner DM., Herzog HJ. Ocean carbon sequestration: A case study in public and institutional perceptions. Proceedings of the Sixth International Conference on Greenhouse Gas Control Technologies, 2002.

63. Reiner DM. A Looming Rhetorical Gap: A Survey of Public Communications Activities for Carbon Dioxide Capture and Storage Technologies. 2008; http://www.eprg.group.cam.ac.uk/wp-content/uploads/2008/11/eprg08012.pdf. Accessed Accessed August 6, 2013.

64. Palmgren CR., Morgan MG., de Bruin WB., Keith DW. Initial Public Perceptions of Deep Geological and Oceanic Disposal of Carbon Dioxide. In: Wilson EJ, Gerard. D., eds. Carbon Capture and Sequestration, Blackwell Publishing, Ames, Iowa; 2007.

65. Chen B., Nishio M, Song Y., Akai M. The fate of CO2 bubble leaked from seabed. In Gale J, Herzog H., Braitch J., eds. Greenhouse Gas Control Technologies 9, Proceedings of the Ninth International Conference on Greenhouse Gas Control Technologies (GHGT-9), Energy Proceedia 2009:1:4969-4976.

Index

L

M

O

P

R

S

T